STUDENT'S SOLUTIONS MANUAL

DIFFERENTIAL EQUATIONS
COMPUTING AND MODELING

DIFFERENTIAL EQUATIONS AND BOUNDARY VALUE PROBLEMS
COMPUTING AND MODELING

FIFTH EDITION

C. Henry Edwards

David E. Penney
The University of Georgia

David Calvis
Baldwin-Wallace University

PEARSON

Boston   Columbus   Hoboken   Indianapolis   New York   San Francisco

Amsterdam   Cape Town   Dubai   London   Madrid   Milan   Munich   Paris   Montreal   Toronto

Delhi   Mexico City   São Paulo   Sydney   Hong Kong   Seoul   Singapore   Taipei   Tokyo

ISBN-13: 978-0-321-79700-1
ISBN-10: 0-321-79700-0

1 2 3 4 5 6 VORW 17 16 15 14

www.pearsonhighered.com

PEARSON

Contents

Preface

This is a solutions manual to accompany the textbooks **DIFFERENTIAL EQUATIONS: Computing and Modeling** (5th edition, 2014) and **DIFFERENTIAL EQUATIONS AND BOUNDARY VALUE PROBLEMS: Computing and Modeling** (5th edition, 2014) by C. Henry Edwards, David E. Penney, and David T. Calvis. We include solutions to most of the odd problems in the text.

Our goal is to support teaching of the subject of elementary differential equations in every way that we can. We therefore invite comments and suggested improvements for future printings of this manual, as well as advice regarding features that might be added to increase its usefulness in subsequent editions. Additional supplementary material can be found at our textbook web site listed below.

Henry Edwards
David Calvis

`h.edwards@mindspring.com`
`dcalvis@bw.edu`

`http://www.pearsonhighered.com/mathstatsresources/`

CHAPTER 1

FIRST-ORDER DIFFERENTIAL EQUATIONS

SECTION 1.1

DIFFERENTIAL EQUATIONS AND MATHEMATICAL MODELS

The main purpose of Section 1.1 is simply to introduce the basic notation and terminology of differential equations, and to show the student what is meant by a solution of a differential equation. Also, the use of differential equations in the mathematical modeling of real-world phenomena is outlined.

Problems 1-12 are routine verifications by direct substitution of the suggested solutions into the given differential equations. We include here just some typical examples of such verifications.

3. If $y_1 = \cos 2x$ and $y_2 = \sin 2x$, then $y_1' = -2\sin 2x$ $y_2' = 2\cos 2x$, so
$y_1'' = -4\cos 2x = -4y_1$ and $y_2'' = -4\sin 2x = -4y_2$. Thus $y_1'' + 4y_1 = 0$ and $y_2'' + 4y_2 = 0$.

5. If $y = e^x - e^{-x}$, then $y' = e^x + e^{-x}$, so $y' - y = \left(e^x + e^{-x}\right) - \left(e^x - e^{-x}\right) = 2e^{-x}$. Thus
$y' = y + 2e^{-x}$.

11. If $y = y_1 = x^{-2}$, then $y' = -2x^{-3}$ and $y'' = 6x^{-4}$, so
$$x^2 y'' + 5x y' + 4y = x^2\left(6x^{-4}\right) + 5x\left(-2x^{-3}\right) + 4\left(x^{-2}\right) = 0.$$
If $y = y_2 = x^{-2}\ln x$, then $y' = x^{-3} - 2x^{-3}\ln x$ and $y'' = -5x^{-4} + 6x^{-4}\ln x$, so
$$x^2 y'' + 5x y' + 4y = x^2\left(-5x^{-4} + 6x^{-4}\ln x\right) + 5x\left(x^{-3} - 2x^{-3}\ln x\right) + 4\left(x^{-2}\ln x\right)$$
$$= \left(-5x^{-2} + 5x^{-2}\right) + \left(6x^{-2} - 10x^{-2} + 4x^{-2}\right)\ln x = 0.$$

13. Substitution of $y = e^{rx}$ into $3y' = 2y$ gives the equation $3r e^{rx} = 2e^{rx}$, which simplifies to $3r = 2$. Thus $r = 2/3$.

15. Substitution of $y = e^{rx}$ into $y'' + y' - 2y = 0$ gives the equation $r^2 e^{rx} + r e^{rx} - 2e^{rx} = 0$, which simplifies to $r^2 + r - 2 = (r+2)(r-1) = 0$. Thus $r = -2$ or $r = 1$.

The verifications of the suggested solutions in Problems 17-26 are similar to those in Problems 1-12. We illustrate the determination of the value of C only in some typical cases. However, we illustrate typical solution curves for each of these problems.

17. $C = 2$

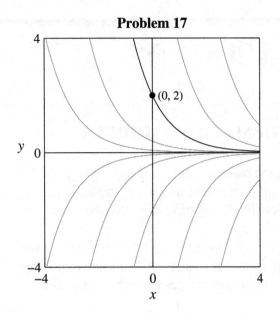

Problem 17

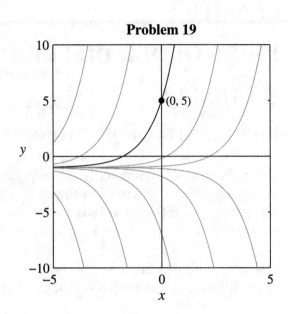

Problem 19

19. If $y(x) = Ce^x - 1$, then $y(0) = 5$ gives $C - 1 = 5$, so $C = 6$.

21. $C = 7$.

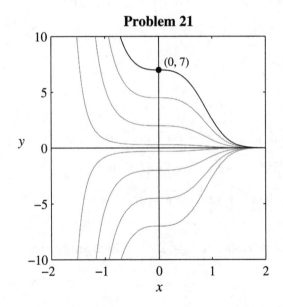

Problem 21

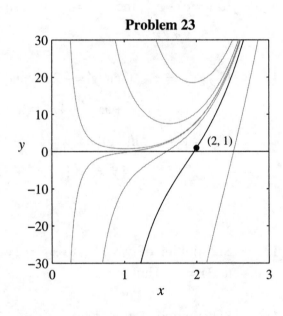

Problem 23

23. If $y(x) = \frac{1}{4}x^5 + Cx^{-2}$, then $y(2) = 1$ gives $\frac{1}{4} \cdot 32 + C \cdot \frac{1}{8} = 1$, or $C = -56$.

25. If $y = \tan(x^3 + C)$, then $y(0) = 1$ gives the equation $\tan C = 1$. Hence one value of C is $C = \pi/4$, as is this value plus any integral multiple of π.

Problem 25

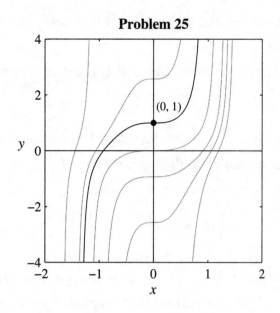

27. $y' = x + y$

29.   If $m = y'$ is the slope of the tangent line and m' is the slope of the normal line at (x, y), then the relation $m\, m' = -1$ yields $m' = -1/y' = (y-1)/(x-0)$. Solving for y' then gives the differential equation $(1-y)\, y' = x$.

31.   The slope of the line through (x, y) and $(-y, x)$ is $y' = (x-y)/(-y-x)$, so the differential equation is $(x+y)y' = y-x$.

In Problems 32-36 we get the desired differential equation when we replace the "time rate of change" of the dependent variable with its derivative with respect to time t, the word "is" with the $=$ sign, the phrase "proportional to" with k, and finally translate the remainder of the given sentence into symbols.

33. $dv/dt = kv^2$ **35.** $dN/dt = k(P-N)$

37.   The second derivative of any linear function is zero, so we spot the two solutions $y(x) \equiv 1$ and $y(x) = x$ of the differential equation $y'' = 0$.

39.   We reason that if $y = kx^2$, then each term in the differential equation is a multiple of x^2. The choice $k = 1$ balances the equation and provides the solution $y(x) = x^2$.

41.   We reason that if $y = ke^x$, then each term in the differential equation is a multiple of e^x. The choice $k = \frac{1}{2}$ balances the equation and provides the solution $y(x) = \frac{1}{2}e^x$.

43.   **(a)** We need only substitute $x(t) = 1/(C - kt)$ in both sides of the differential equation $x' = kx^2$ for a routine verification.

(b) The zero-valued function $x(t) \equiv 0$ obviously satisfies the initial value problem $x' = kx^2$, $x(0) = 0$.

45.   Substitution of $P' = 1$ and $P = 10$ into the differential equation $P' = kP^2$ gives $k = \frac{1}{100}$, so Problem 43(a) yields a solution of the form $P(t) = 1/(C - \frac{1}{100}t)$. The initial condition $P(0) = 2$ now yields $C = \frac{1}{2}$, so we get the solution

$$P(t) = \frac{1}{\dfrac{1}{2} - \dfrac{t}{100}} = \frac{100}{50 - t}.$$

We now find readily that $P = 100$ when $t = 49$ and that $P = 1000$ when $t = 49.9$. It appears that P grows without bound (and thus "explodes") as t approaches 50.

47.   **(a)** $y(10) = 10$ yields $10 = 1/(C - 10)$, so $C = 101/10$.

(b) There is no such value of C, but the constant function $y(x) \equiv 0$ satisfies the conditions $y' = y^2$ and $y(0) = 0$.

(c) It is obvious visually (in Fig. 1.1.8 of the text) that one and only one solution curve passes through each point (a, b) of the xy-plane, so it follows that there exists a unique solution to the initial value problem $y' = y^2$, $y(a) = b$.

SECTION 1.2

INTEGRALS AS GENERAL AND PARTICULAR SOLUTIONS

This section introduces **general solutions** and **particular solutions** in the very simplest situation — a differential equation of the form $y' = f(x)$ — where only direct integration and evaluation of the constant of integration are involved. Students should review carefully the elementary concepts of velocity and acceleration, as well as the fps and mks unit systems.

1.   Integration of $y' = 2x + 1$ yields $y(x) = \int (2x + 1) \, dx = x^2 + x + C$. Then substitution of $x = 0$, $y = 3$ gives $3 = 0 + 0 + C = C$, so $y(x) = x^2 + x + 3$.

3.   Integration of $y' = \sqrt{x}$ yields $y(x) = \int \sqrt{x} \, dx = \frac{2}{3}x^{3/2} + C$. Then substitution of $x = 4$, $y = 0$ gives $0 = \frac{16}{3} + C$, so $y(x) = \frac{2}{3}\left(x^{3/2} - 8\right)$.

5.   Integration of $y' = (x+2)^{-1/2}$ yields $y(x) = \int (x+2)^{-1/2} dx = 2\sqrt{x+2} + C$. Then substitution of $x = 2$, $y = -1$ gives $-1 = 2 \cdot 2 + C$, so $y(x) = 2\sqrt{x+2} - 5$.

7.   Integration of $y' = \dfrac{10}{x^2 + 1}$ yields $y(x) = \int \dfrac{10}{x^2 + 1} dx = 10 \tan^{-1} x + C$. Then substitution of $x = 0$, $y = 0$ gives $0 = 10 \cdot 0 + C$, so $y(x) = 10 \tan^{-1} x$.

9.   Integration of $y' = \dfrac{1}{\sqrt{1 - x^2}}$ yields $y(x) = \int \dfrac{1}{\sqrt{1 - x^2}} dx = \sin^{-1} x + C$. Then substitution of $x = 0$, $y = 0$ gives $0 = 0 + C$, so $y(x) = \sin^{-1} x$.

11.   If $a(t) = 50$, then $v(t) = \int 50 \, dt = 50t + v_0 = 50t + 10$. Hence
$$x(t) = \int (50t + 10) \, dt = 25t^2 + 10t + x_0 = 25t^2 + 10t + 20.$$

13.   If $a(t) = 3t$, then $v(t) = \int 3t \, dt = \tfrac{3}{2}t^2 + v_0 = \tfrac{3}{2}t^2 + 5$. Hence
$$x(t) = \int \left(\tfrac{3}{2}t^2 + 5\right) dt = \tfrac{1}{2}t^3 + 5t + x_0 = \tfrac{1}{2}t^3 + 5t.$$

15.   If $a(t) = 4(t+3)^2$, then $v(t) = \int 4(t+3)^2 \, dt = \tfrac{4}{3}(t+3)^3 + C = \tfrac{4}{3}(t+3)^3 - 37$ (taking $C = -37$ so that $v(0) = -1$). Hence
$$x(t) = \int \tfrac{4}{3}(t+3)^3 - 37 \, dt = \tfrac{1}{3}(t+3)^4 - 37t + C = \tfrac{1}{3}(t+3)^4 - 37t - 26.$$

17.   If $a(t) = (t+1)^{-3}$, then $v(t) = \int (t+1)^{-3} \, dt = -\tfrac{1}{2}(t+1)^{-2} + C = -\tfrac{1}{2}(t+1)^{-2} + \tfrac{1}{2}$ (taking $C = \tfrac{1}{2}$ so that $v(0) = 0$). Hence
$$x(t) = \int -\tfrac{1}{2}(t+1)^{-2} + \tfrac{1}{2} \, dt = \tfrac{1}{2}(t+1)^{-1} + \tfrac{1}{2}t + C = \tfrac{1}{2}\left[(t+1)^{-1} + t - 1\right]$$
(taking $C = -\tfrac{1}{2}$ so that $x(0) = 0$).

Students should understand that Problems 19-22, though different at first glance, are solved in the same way as the preceding ones, that is, by means of the fundamental theorem of calculus in the form $x(t) = x(t_0) + \int_{t_0}^{t} v(s) \, ds$ cited in the text. Actually in these problems $x(t) = \int_{0}^{t} v(s) \, ds$, since t_0 and $x(t_0)$ are each given to be zero.

19. The graph of $v(t)$ shows that $v(t) = \begin{cases} 5 & \text{if } 0 \le t \le 5 \\ 10 - t & \text{if } 5 \le t \le 10 \end{cases}$, so that

$x(t) = \begin{cases} 5t + C_1 & \text{if } 0 \le t \le 5 \\ 10t - \frac{1}{2}t^2 + C_2 & \text{if } 5 \le t \le 10 \end{cases}$. Now $C_1 = 0$ because $x(0) = 0$, and continuity of

$x(t)$ requires that $x(t) = 5t$ and $x(t) = 10t - \frac{1}{2}t^2 + C_2$ agree when $t = 5$. This implies

that $C_2 = -\frac{25}{2}$, leading to the graph of $x(t)$ shown.

Alternate solution for Problem 19 (and similar for 20-22): The graph of $v(t)$ shows

that $v(t) = \begin{cases} 5 & \text{if } 0 \le t \le 5 \\ 10 - t & \text{if } 5 \le t \le 10 \end{cases}$. Thus for $0 \le t \le 5$, $x(t) = \int_0^t v(s)\,ds$ is given by

$\int_0^t 5\,ds = 5t$, whereas for $5 \le t \le 10$ we have

$$x(t) = \int_0^t v(s)\,ds = \int_0^5 5\,ds + \int_5^t 10 - s\,ds$$

$$= 25 + \left(10s - \frac{s^2}{2} \Big|_{s=5}^{s=t} \right) = 25 + 10t - \frac{t^2}{2} - \frac{75}{2} = 10t - \frac{t^2}{2} - \frac{25}{2}.$$

The graph of $x(t)$ is shown.

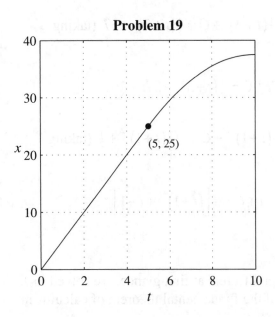

Problem 19

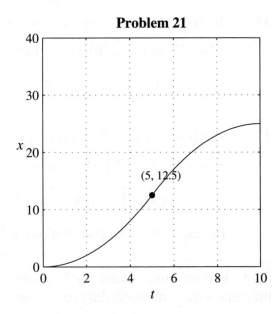

Problem 21

21. The graph of $v(t)$ shows that $v(t) = \begin{cases} t & \text{if } 0 \le t \le 5 \\ 10 - t & \text{if } 5 \le t \le 10 \end{cases}$, so that

$x(t) = \begin{cases} \frac{1}{2}t^2 + C_1 & \text{if } 0 \le t \le 5 \\ 10t - \frac{1}{2}t^2 + C_2 & \text{if } 5 \le t \le 10 \end{cases}$. Now $C_1 = 0$ because $x(0) = 0$, and continuity of

$x(t)$ requires that $x(t) = \frac{1}{2}t^2$ and $x(t) = 10t - \frac{1}{2}t^2 + C_2$ agree when $t = 5$. This implies that $C_2 = -25$, leading to the graph of $x(t)$ shown.

23. $v(t) = -9.8t + 49$, so the ball reaches its maximum height $(v = 0)$ after $t = 5$ seconds. Its maximum height then is $y(5) = -4.9(5)^2 + 49(5) = 122.5$ meters.

25. $a = -10$ m/s^2 and $v_0 = 100$ km/h ≈ 27.78 m/s, so $v = -10t + 27.78$, and hence $x(t) = -5t^2 + 27.78t$. The car stops when $v = 0$, that is $t \approx 2.78$ s, and thus the distance traveled before stopping is $x(2.78) \approx 38.59$ meters.

27. $a = -9.8$ m/s^2, so $v = -9.8t - 10$ and $y = -4.9t^2 - 10t + y_0$. The ball hits the ground when $y = 0$ and $v = -9.8t - 10 = -60$ m/s, so $t \approx 5.10$ s. Hence the height of the building is
$$y_0 = 4.9(5.10)^2 + 10(5.10) \approx 178.57 \text{ m}.$$

29.   Integration of $dv/dt = 0.12t^2 + 0.6t$ with $v(0) = 0$ gives $v(t) = 0.04t^3 + 0.3t^2$. Hence $v(10) = 70$ ft/s. Then integration of $dx/dt = 0.04t^3 + 0.3t^2$ with $x(0) = 0$ gives $x(t) = 0.01t^4 + 0.1t^3$, so $x(10) = 200$ ft. Thus after 10 seconds the car has gone 200 ft and is traveling at 70 ft/s.

31.   If $a = -20$ m/s^2 and $x_0 = 0$, then the car's velocity and position at time t are given by $v = -20t + v_0$ and $x = -10t^2 + v_0 t$. It stops when $v = 0$ (so $v_0 = 20t$), and hence when $x = 75 = -10t^2 + (20t)t = 10t^2$. Thus $t = \sqrt{7.5}$ s, so
$$v_0 = 20\sqrt{7.5} \approx 54.77 \text{ m/s} \approx 197 \text{ km/hr}.$$

33.   If $v_0 = 0$ and $y_0 = 20$, then $v = -at$ and $y = -\frac{1}{2}at^2 + 20$. Substitution of $t = 2$, $y = 0$ yields $a = 10$ ft/s^2. If $v_0 = 0$ and $y_0 = 200$, then $v = -10t$ and $y = -5t^2 + 200$. Hence $y = 0$ when $t = \sqrt{40} = 2\sqrt{10}$ s and $v = -20\sqrt{10} \approx -63.25$ ft/s.

35.   If $v_0 = 0$ and $y_0 = h$, then the stone's velocity and height are given by $v = -gt$ and $y = -0.5gt^2 + h$, respectively. Hence $y = 0$ when $t = \sqrt{2h/g}$, so $v = -g\sqrt{2h/g} = -\sqrt{2gh}$.

37.   We use units of miles and hours. If $x_0 = v_0 = 0$, then the car's velocity and position after t hours are given by $v = at$ and $x = \frac{1}{2}at^2$, respectively. Since $v = 60$ when $t = 5/6$, the

velocity equation yields . Hence the distance traveled by 12:50 pm is
$x = \frac{1}{2} \cdot 72 \cdot (5/6)^2 = 25$ miles .

39.   Integration of $y' = (9/v_S)(1 - 4x^2)$ yields $y = (3/v_S)(3x - 4x^3) + C$, and the initial condition $y(-1/2) = 0$ gives $C = 3/v_S$. Hence the swimmer's trajectory is
$y(x) = (3/v_S)(3x - 4x^3 + 1)$. Substitution of $y(1/2) = 1$ now gives $v_S = 6\,\text{mph}$.

41.   The bomb equations are $a = -32$, $v = -32t$, and $s_B = s = -16t^2 + 800$ with $t = 0$ at the instant the bomb is dropped. The projectile is fired at time $t = 2$, so its corresponding equations are $a = -32$, $v = -32(t - 2) + v_0$, and $s_P = s = -16(t - 2)^2 + v_0(t - 2)$ for $t \geq 2$ (the arbitrary constant vanishing because $s_P(2) = 0$). Now the condition
$s_B(t) = -16t^2 + 800 = 400$ gives $t = 5$, and then the further requirement that $s_P(5) = 400$ yields $v_0 = 544/3 \approx 181.33$ ft/s for the projectile's needed initial velocity.

43.   The velocity and position functions for the spacecraft are $v_S(t) = 0.0098t$ and $x_S(t) = 0.0049t^2$, and the corresponding functions for the projectile are $v_P(t) = \frac{1}{10}c = 3 \times 10^7$ and $x_P(t) = 3 \times 10^7 t$. The condition that $x_S = x_P$ when the spacecraft overtakes the projectile gives $0.0049t^2 = 3 \times 10^7 t$, whence

$$t = \frac{3 \times 10^7}{0.0049} \approx 6.12245 \times 10^9 \, \text{s} \approx \frac{6.12245 \times 10^9}{(3600)(24)(365.25)} \approx 194 \, \text{years} .$$

Since the projectile is traveling at $\frac{1}{10}$ the speed of light, it has then traveled a distance of about 19.4 light years, which is about 1.8367×10^{17} meters.

SECTION 1.3

SLOPE FIELDS AND SOLUTION CURVES

The instructor may choose to delay covering Section 1.3 until later in Chapter 1. However, before proceeding to Chapter 2, it is important that students come to grips at some point with the question of the existence of a unique solution of a differential equation — and realize that it makes no sense to look for the solution without knowing in advance that it exists. It may help some students to simplify the statement of the existence-uniqueness theorem as follows:

Suppose that the function $f(x, y)$ and the partial derivative $\partial f/\partial y$ are both continuous in some neighborhood of the point (a, b). Then the initial value problem

$$\frac{dy}{dx} = f(x, y), \ y(a) = b$$

has a unique solution in some neighborhood of the point *a*.

Slope fields and geometrical solution curves are introduced in this section as a concrete aid in visualizing solutions and existence-uniqueness questions. Instead, we provide some details of the construction of the figure for the Problem 1 answer, and then include without further comment the similarly constructed figures for Problems 2 through 9.

1.   The following sequence of *Mathematica* 7 commands generates the slope field and the solution curves through the given points. Begin with the differential equation $dy/dx = f(x, y)$, where

```
f[x_, y_] := -y - Sin[x]
```

Then set up the viewing window

```
a = -3; b = 3; c = -3; d = 3;
```

The slope field is then constructed by the command

```
dfield = VectorPlot[{1, f[x, y]}, {x, a, b}, {y, c, d},
    PlotRange -> {{a, b}, {c, d}}, Axes -> True, Frame -> True,
    FrameLabel -> {TraditionalForm[x], TraditionalForm[y]},
    AspectRatio -> 1, VectorStyle -> {Gray, "Segment"},
    VectorScale -> {0.02, Small, None},
    FrameStyle -> (FontSize -> 12), VectorPoints -> 21,
    RotateLabel -> False]
```

The original curve shown in Fig. 1.3.15 of the text (and its initial point not shown there) are plotted by the commands

```
x0 = -1.9; y0 = 0;
point0 = Graphics[{PointSize[0.025], Point[{x0, y0}]}];
soln = NDSolve[{y'[x] == f[x, y[x]], y[x0] == y0}, y[x],
    {x, a, b}];
curve0 = Plot[soln[[1, 1, 2]], {x, a, b}, PlotStyle ->
    {Thickness[0.0065], Blue}];
Show[curve0, point0]
```

(The *Mathematica* **NDSolve** command carries out an approximate numerical solution of the given differential equation. Numerical solution techniques are discussed in Sections 2.4–2.6 of the textbook.)

The coordinates of the 12 points are marked in Fig. 1.3.15 in the textbook. For instance the 7th point is $(-2.5, 1)$. It and the corresponding solution curve are plotted by the commands

```
x0 = -2.5; y0 = 1;
point7 = Graphics[{PointSize[0.025], Point[{x0, y0}]}];
soln = NDSolve[{y'[x] == f[x, y[x]], y[x0] == y0}, y[x],
    {x, a, b}];
curve7 = Plot[soln[[1, 1, 2]], {x, a, b},
    PlotStyle -> {Thickness[0.0065], Blue}];
```

```
Show[curve7, point7]
```

The following command superimposes the two solution curves and starting points found so far upon the slope field:

```
Show[dfield, point0, curve0, point7, curve7]
```

We could continue in this way to build up the entire graphic called for in the problem. Here is an alternative looping approach, variations of which were used to generate the graphics below for Problems 1-10:

```
points = {{-2.5,2}, {-1.5,2}, {-0.5,2}, {0.5,2}, {1.5,2},
    {2.5,2}, {-2,-2}, {-1,-2}, {0,-2}, {1,-2}, {2,-2}, {-2.5,1}};
curves = {}; (* start with null lists *)
dots = {};
Do [
   x0 = points[[i, 1]];
   y0 = points[[i, 2]];
   newdot = Graphics[{PointSize[0.025],Point[{x0, y0}]}];
   dots = AppendTo[dots, newdot];
   soln = NDSolve[{y'[x] == f[x, y[x]],y[x0] == y0}, y[x],
      {x, a, b}];
   newcurve = Plot[soln[[1, 1, 2]], {x, a, b},
      PlotStyle -> {Thickness[0.0065], Black}];
   AppendTo[curves, newcurve],
   {i, 1, Length[points]}];
Show[dfield, curves, dots, PlotLabel -> Style["Problem 1", Bold,
   11]]
```

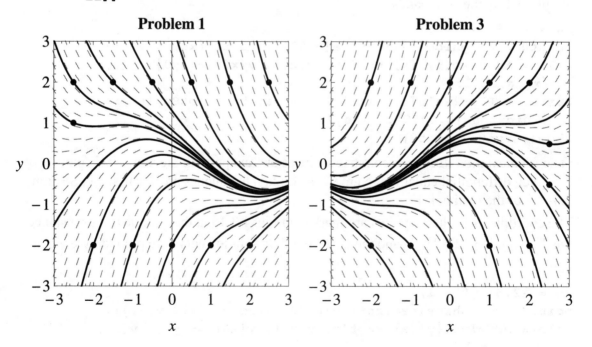

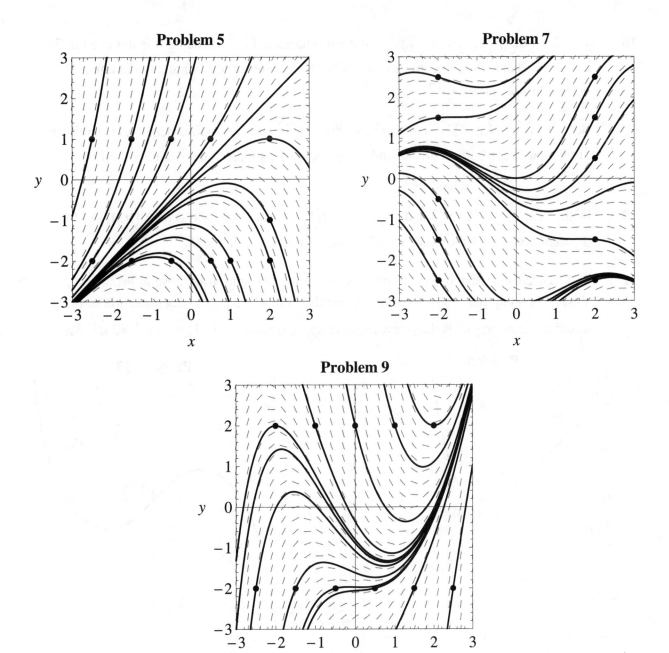

11. Because both $f(x,y) = 2x^2 y^2$ and $D_y f(x,y) = 4x^2 y$ are continuous everywhere, the existence-uniqueness theorem of Section 1.3 in the textbook guarantees the existence of a unique solution in some neighborhood of $x = 1$.

13. Both $f(x,y) = y^{1/3}$ and $\partial f / \partial y = \frac{1}{3} y^{-2/3}$ are continuous near $(0,1)$, so the theorem guarantees the existence of a unique solution in some neighborhood of $x = 0$.

15. The function $f(x, y) = (x-y)^{1/2}$ is not continuous at $(2,2)$ because it is not even defined if $y > x$. Hence the theorem guarantees neither existence nor uniqueness in any neighborhood of the point $x = 2$.

17. Both $f(x, y) = (x-1)/y$ and $\partial f/\partial y = -(x-1)/y^2$ are continuous near $(0,1)$, so the theorem guarantees both existence and uniqueness of a solution in some neighborhood of $x = 0$.

19. Both $f(x, y) = \ln(1 + y^2)$ and $\partial f/\partial y = 2y/(1 + y^2)$ are continuous near $(0,0)$, so the theorem guarantees the existence of a unique solution near $x = 0$.

21. The figure shown can be constructed using commands similar to those in Problem 1, above. Tracing this solution curve, we see that $y(-4) \approx 3$. (An exact solution of the differential equation yields the more accurate approximation $y(-4) = 3 + e^{-4} \approx 3.0183$.)

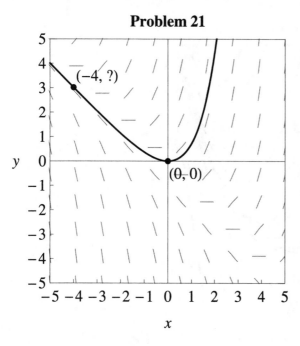

Problem 21

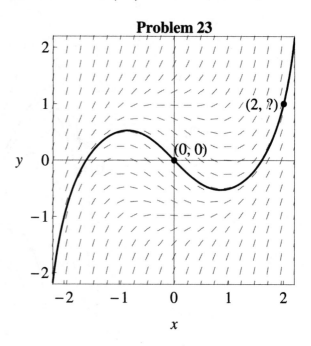

Problem 23

23. Tracing the curve in the figure shown, we see that $y(2) \approx 1$. A more accurate approximation is $y(2) \approx 1.0044$.

25. The figure indicates a limiting velocity of 20 ft/sec — about the same as jumping off a $6\frac{1}{4}$-foot wall, and hence quite survivable. Tracing the curve suggests that $v(t) = 19$ ft/sec when t is a bit less than 2 seconds. An exact solution gives $t \approx 1.8723$ then.

Problem 25

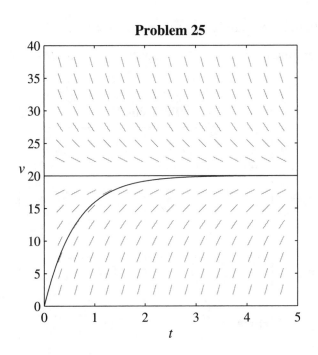

Problem 27a

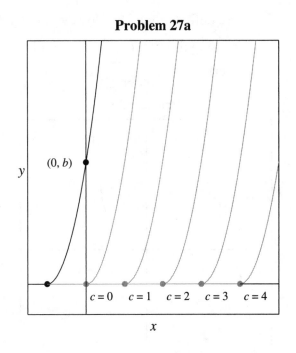

27.   **a)** It is clear that $y(x)$ satisfies the differential equation at each x with $x < c$ or $x > c$, and by examining left- and right-hand derivatives we see that the same is true at $x = c$. Thus $y(x)$ not only satisfies the differential equation for all x, it also satisfies the given initial value problem whenever $c \geq 0$. The infinitely many solutions of the initial value problem are illustrated in the figure. Note that $f(x, y) = 2\sqrt{y}$ is not continuous in any neighborhood of the origin, and so Theorem 1 guarantees neither existence nor uniqueness of solution to the given initial value problem. As it happens, existence occurs, but not uniqueness.

b) If $b < 0$, then the initial value problem $y' = 2\sqrt{y}$, $y(0) = b$ has <u>no</u> solution, because the square root of a negative number would be involved. If $b > 0$, then we get a unique solution curve through $(0, b)$ defined for all x by following a parabola (as in the figure, in black) — down (and leftward) to the x-axis and then following the x-axis to the left. Finally if $b = 0$, then starting at $(0, 0)$ we can follow the positive x-axis to the point $(c, 0)$ and then branch off on the parabola $y = (x - c)^2$, as shown in gray. Thus there are infinitely many solutions in this case.

29.   As with Problem 27, it is clear that $y(x)$ satisfies the differential equation at each x with $x < c$ or $x > c$, and by examining left- and right-hand derivatives we see that the same is true at $x = c$. Looking at the figure on the left below, we see that if, for instance, $b > 0$, then we can start at the point (a, b) and follow a branch of a cubic down to the x-axis, then follow the x-axis an arbitrary distance before branching down on another cubic. This gives infinitely many solutions of the initial value problem $y' = 3y^{2/3}$, $y(a) = b$ that

are defined for all x. However, if $b \neq 0$, then there is only a single cubic $y = (x-c)^3$ passing through (a,b), so the solution is unique near $x = a$ (as Theorem 1 would predict).

Problem 29

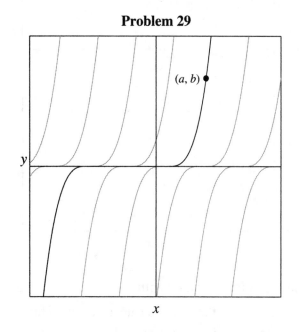

Problem 31

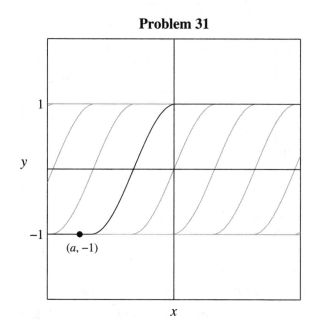

31. The function $y(x) = \begin{cases} -1 & \text{if } x < c - \pi/2 \\ \sin(x-c) & \text{if } c - \pi/2 < x < c + \pi/2 \\ +1 & \text{if } x < c + \pi/2 \end{cases}$ satisfies the given differential

equation on the interval $c - \dfrac{\pi}{2} < x < c + \dfrac{\pi}{2}$, since $y'(x) = \cos(x-c) > 0$ there and thus

$$\sqrt{1 - y^2} = \sqrt{1 - \sin^2(x-c)} = \sqrt{\cos^2(x-c)} = \cos(x-c) = y'.$$

Moreover, the same is true for $x < \dfrac{\pi}{2}$ and $x > c + \dfrac{\pi}{2}$ (since $y^2 \equiv 1$ and $y' \equiv 0$ there), and

at $x = \dfrac{\pi}{2}, c + \dfrac{\pi}{2}$ by examining one-sided derivatives. Thus $y(x)$ satisfies the given differential equation for all x.

If $|b| > 1$, then the initial value problem $y' = \sqrt{1 - y^2}$, $y(a) = b$ has no solution because the square root of a negative number would be involved. If $|b| < 1$, then there is only one curve of the form $y = \sin(x-c)$ through the point (a,b); this gives a unique solution. But if $b = \pm 1$, then we can combine a left ray of the line $y = -1$, a sine curve from the line $y = -1$ to the line $y = +1$, and then a right ray of the line $y = +1$. Looking at the figure, we see that this gives infinitely many solutions (defined for all x) through any point of the form $(a, \pm 1)$.

33. Looking at the figure provided in the answers section of the textbook, it suffices to observe that, among the pictured curves $y = x/(cx-1)$ for all possible values of c,

- there is a unique one of these curves through any point not on either coordinate axis;
- there is no such curve through any point on the y-axis other than the origin; and
- there are infinitely many such curves through the origin $(0,0)$.

But in addition we have the constant-valued solution $y(x) \equiv 0$ that "covers" the x-axis.

It follows that the given differential equation has near (a,b)

- a unique solution if $a \neq 0$;
- no solution if $a = 0$ but $b \neq 0$;
- infinitely many different solutions if $a = b = 0$.

Once again these findings are consistent with Theorem 1.

35. **(a)** With a computer algebra system we find that the solution of the initial value problem $y' = x - y + 1$, $y(-3) = -0.2$ is $y(x) = x + 2.8e^{-x-3}$, whence $y(2) \approx 2.0189$. With the same differential equation but with initial condition $y(-3) = +0.2$ the solution is $y(x) = x + 3.2e^{-x-3}$, whence $y(2) \approx 2.0216$.

(b) Similarly, the solution of the initial value problem $y' = x - y + 1$, $y(-3) = -0.5$ is $y(x) = x + 2.5e^{-x-3}$, whence $y(2) \approx 2.0168$. With the same differential equation but with initial condition $y(-3) = +0.5$ the solution is $y(x) = x + 3.5e^{-x-3}$, whence $y(2) \approx 2.0236$. Thus the initial values $y(-3) = \pm 0.5$ that are not close both yield $y(2) \approx 2.02$.

SECTION 1.4

SEPARABLE EQUATIONS AND APPLICATIONS

Of course it should be emphasized to students that the possibility of separating the variables is the first one you look for. The general concept of natural growth and decay is important for all differential equations students, but the particular applications in this section are optional. Torricelli's law in the form of Equation (24) in the text leads to some nice concrete examples and problems.

Also, in the solutions below, we make free use of the fact that if C is an arbitrary constant, then so is $5 - 3C$, for example, which we can (and usually do) replace simply with C itself. In the same way we typically replace e^C by C, with the understanding that C is then an arbitrary non-zero constant.

1.   For $y \neq 0$ separating variables gives $\int \dfrac{dy}{y} = -\int 2x \, dx$, so that $\ln|y| = -x^2 + C$, or

$y(x) = \pm e^{-x^2 + C} = Ce^{-x^2}$, where C is an arbitrary nonzero constant. (The equation also has the singular solution $y \equiv 0$.)

3.   For $y \neq 0$ separating variables gives $\int \dfrac{dy}{y} = \int \sin x \, dx$, so that $\ln|y| = -\cos x + C$, or

$y(x) = \pm e^{-\cos x + C} = Ce^{-\cos x}$, where C is an arbitrary nonzero constant. (The equation also has the singular solution $y \equiv 0$.)

5.   For $-1 < y < 1$ and $x > 0$ separating variables gives $\int \dfrac{dy}{\sqrt{1 - y^2}} = \int \dfrac{1}{2\sqrt{x}} \, dx$, so that

$\sin^{-1} y = \sqrt{x} + C$, or $y(x) = \sin\left(\sqrt{x} + C\right)$. (The equation also has the singular solutions $y \equiv 1$ and $y \equiv -1$.)

7.   For $y \neq 0$ separating variables gives $\int \dfrac{dy}{y^{1/3}} = \int 4x^{1/3} \, dx$, so that $\frac{3}{2} y^{2/3} = 3x^{4/3} + C$, or

$y(x) = \left(2x^{4/3} + C\right)^{3/2}$. (The equation also has the singular solution $y \equiv 0$.)

9.   For $y \neq 0$ separating variables and decomposing into partial fractions give

$\int \dfrac{dy}{y} = \int \dfrac{2}{1 - x^2} \, dx = \int \dfrac{1}{1 + x} + \dfrac{1}{1 - x} \, dx$, so that $\ln|y| = \ln|1 + x| - \ln|1 - x| + C$, or

$|y| = C\left|\dfrac{1 + x}{1 - x}\right|$, where C is an arbitrary positive constant, or $y(x) = C \dfrac{1 + x}{1 - x}$, where C is an arbitrary nonzero constant. (The equation also has the singular solution $y \equiv 0$.)

11.   For $y > 0$ separating variables gives $\int \dfrac{dy}{y^3} = \int x \, dx$, so that $-\dfrac{1}{2y^2} = \dfrac{x^2}{2} + C$, or

$y(x) = \left(C - x^2\right)^{-1/2}$, where C is an arbitrary constant. Likewise $y(x) = -\left(C - x^2\right)^{-1/2}$ for $y < 0$. (The equation also has the singular solution $y \equiv 0$.)

13.   Separating variables gives $\int \dfrac{y^3}{y^4 + 1} \, dy = \int \cos x \, dx$, so that $\dfrac{1}{4}\ln\left(y^4 + 1\right) = \sin x + C$, where C is an arbitrary constant.

15. For $x \neq 0$ and $y \neq 0, \frac{\sqrt{2}}{2}$ separating variables gives $\int \frac{2}{y^2} - \frac{1}{y^4}\, dy = \int \frac{1}{x} - \frac{1}{x^2}\, dx$, so that

$-\frac{2}{y} + \frac{1}{3y^3} = \ln|x| + \frac{1}{x} + C$, where C is an arbitrary constant.

17. Factoring gives $y' = 1 + x + y + xy = (1+x)(1+y)$, and then for $y \neq -1$ separating varia-

bles gives $\int \frac{1}{1+y}\, dy = \int 1 + x\, dx$, so that $\ln|1+y| = x + \frac{1}{2}x^2 + C$, where C is an arbitrary

constant. (The equation also has the singular solution $y \equiv -1$.)

19. For $y \neq 0$ separating variables gives $\int \frac{1}{y}\, dy = \int e^x dx$, so that $\ln|y| = e^x + C$, or

$|y| = C\exp(e^x)$, where C is an arbitrary positive constant, or finally $y = C\exp(e^x)$,

where C is an arbitrary nonzero constant. The initial condition $y(0) = 2e$ implies that

$C \cdot \exp(e^0) = 2e$, or $C = 2$, leading to the particular solution $y(x) = 2\exp(e^x)$.

21. For $|x| > 4$ separating variables gives $\int 2y\, dy = \int \frac{x}{\sqrt{x^2 - 16}}\, dx$, so that

$y^2 = \sqrt{x^2 - 16} + C$. The initial condition $y(5) = 2$ implies that $C = 1$, leading to the

particular solution $y(x) = \sqrt{1 + \sqrt{x^2 - 16}}$.

23. Rewriting the differential equation as $\frac{dy}{dx} = 2y - 1$, we see that for $y \neq \frac{1}{2}$ separating vari-

ables gives $\int \frac{1}{2y-1}\, dy = \int dx$, so that $\frac{1}{2}\ln|2y-1| = x + C$, or $|2y-1| = Ce^{2x}$, where C is

an arbitrary positive constant, or finally $2y - 1 = Ce^{2x}$, which is to say $y = \frac{1}{2}(Ce^{2x} + 1)$,

where C is an arbitrary nonzero constant. The initial condition $y(1) = 1$ implies that

$C = \frac{1}{e^2}$, leading to the particular solution $y(x) = \frac{1}{2}\left(\frac{1}{e^2}e^{2x} + 1\right) = \frac{1}{2}\left(e^{2x-2} + 1\right)$.

25. Rewriting the differential equation as $x\frac{dy}{dx} = 2x^2 y + y$, we see that for $x, y \neq 0$ separating

variables gives $\int \frac{1}{y}\, dy = \int 2x + \frac{1}{x}\, dx$, so that $\ln|y| = x^2 + \ln|x| + C$, or $|y| = C|x|e^{x^2}$, where

C is an arbitrary positive constant, or $y = Cxe^{x^2}$, where C is an arbitrary nonzero constant. The initial condition $y(1) = 1$ implies that $C = \dfrac{1}{e}$, leading to the particular solution $y(x) = xe^{x^2-1}$.

27. Separating variables gives $\displaystyle\int e^y\,dy = \int 6e^{2x}\,dx$, so that $e^y = 3e^{2x} + C$, or $y = \ln\left(3e^{2x} + C\right)$. The initial condition $y(0) = 0$ implies that $C = -2$, leading to the particular solution $y(x) = \ln\left(3e^{2x} - 2\right)$.

29. **(a)** For $y \neq 0$ separation of variables gives the general solution $\displaystyle\int \frac{1}{y^2}\,dy = \int dx$, so that

$$-\frac{1}{y} = x + C, \text{ or } y(x) = \frac{1}{C - x}.$$

(b) Inspection yields the singular solution $y(x) \equiv 0$ that corresponds to *no* value of the constant C.

(c) The figure illustrates that there is a unique solution curve through every point in the xy-plane.

(c) Finally, if $b > 0$, then near (a, b) there are exactly *two* solution curves through this point, corresponding to the two indicated parabolas through (a, b), one ascending, and one descending, with increasing x. (Again, see Problem 31.)

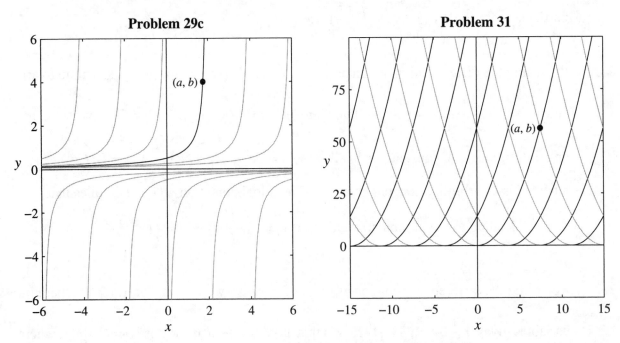

Problem 29c

Problem 31

31.   As noted in Problem 30, the solutions of the differential equation $(dy/dx)^2 = 4y$ consist of the solutions of $dy/dx = 2\sqrt{y}$ together with those of $dy/dx = -2\sqrt{y}$, and again we must have $y \geq 0$. Imposing the initial condition $y(a) = b$, where $b > 0$, upon the general solution $y(x) = (x - C)^2$ found in Problem 30 gives $b = (a - C)^2$, which leads to the two values $C = a \pm \sqrt{b}$, and thus to the two particular solutions $y(x) = \left(x - a \pm \sqrt{b}\right)^2$. For these two particular solutions we have $y'(a) = \pm 2\sqrt{b}$, where $(+)$ corresponds to $dy/dx = 2\sqrt{y}$ and $(-)$ corresponds to $dy/dx = -2\sqrt{y}$. It follows that whereas the solutions of $(dy/dx)^2 = 4y$ through (a, b) contain two parabolic segments, one ascending and one descending from left to right, the solutions of $dy/dx = 2\sqrt{y}$ through (a, b) (the black curves in the figure) contain only ascending parabolic segments, whereas for $dy/dx = -2\sqrt{y}$ the (gray) parabolic segments are strictly descending. Thus the answer to the question is "no", because the descending parabolic segments represent solutions of $(dy/dx)^2 = 4y$ but not of $dy/dx = 2\sqrt{y}$. From all this we arrive at the following answers to parts (a)-(c):

(a) No solution curve if $b < 0$;

(b) A unique solution curve if $b > 0$;

(c) Infinitely many solution curves if $b = 0$, because in this case (as noted in the solution for Problem 30) we can pick any $c > a$ and define the solution

$$y(x) = \begin{cases} 0 & \text{if } x < c \\ (x - c)^2 & \text{if } x \geq c \end{cases}.$$

33.   The population growth rate is $k = \ln(30000/25000)/10 \approx 0.01823$, so the population of the city t years after 1960 is given by $P(t) = 25000e^{0.01823t}$. The expected year 2000 population is then $P(40) = 25000e^{0.01823 \times 40} \approx 51840$.

35.   As in the textbook discussion of radioactive decay, the number of ^{14}C atoms after t years is given by $N(t) = N_0 e^{-0.0001216t}$. Hence we need only solve the equation

$\frac{1}{6} N_0 = N_0 e^{-0.0001216t}$ for the age t of the skull, finding $t = \dfrac{\ln 6}{0.0001216} \approx 14735$ years.

37.   The amount in the account after t years is given by $A(t) = 5000e^{0.08t}$. Hence the amount in the account after 18 years is given by $A(18) = 5000e^{0.08 \times 18} \approx 21{,}103.48$ dollars.

39.   To find the decay rate of this drug in the dog's blood stream, we solve the equation $\frac{1}{2} = e^{-5k}$ (half-life 5 hours) for k, finding $k = (\ln 2)/5 \approx 0.13863$. Thus the amount in the dog's bloodstream after t hours is given by $A(t) = A_0 e^{-0.13863t}$. We therefore solve the equation $A(1) = A_0 e^{-0.13863} = 50 \times 45 = 2250$ for A_0, finding $A_0 \approx 2585$ mg, the amount to anesthetize the dog properly.

41.   Taking $t = 0$ when the body was formed and $t = T$ now, we see that the amount $Q(t)$ of ^{238}U in the body at time t (in years) is given by $Q(t) = Q_0 e^{-kt}$, where $k = (\ln 2)/(4.51 \times 10^9)$. The given information implies that $\dfrac{Q(T)}{Q_0 - Q(T)} = 0.9$. Upon substituting $Q(t) = Q_0 e^{-kt}$ we solve readily for $e^{kT} = 19/9$, so that $T = (1/k)\ln(19/9) \approx 4.86 \times 10^9$. Thus the body was formed approximately 4.86 billion years ago.

43.   Because $A = 0$ in Newton's law of cooling, the differential equation reduces to $T' = -kT$, and the given initial temperature then leads to $T(t) = 25 e^{-kt}$. The fact that $T(20) = 15$ yields $k = (1/20)\ln(5/3)$, and finally we solve the equation $5 = 25 e^{-kt}$ for t to find $t = \ln 5/k \approx 63$ min.

45.   **(a)** The light intensity at a depth of x meters is given by $I(x) = I_0 e^{-1.4x}$. We solve the equation $I(x) = I_0 e^{-1.4x} = \frac{1}{2} I_0$ for x, finding $x = (\ln 2)/1.4 \approx 0.495$ meters.

   (b) At depth 10 meters the intensity is $I(10) = I_0 e^{-1.4 \times 10} \approx (8.32 \times 10^{-7}) I_0$, that is, 0.832 of one millionth of the light intensity I_0 at the surface.

   (c) We solve the equation $I(x) = I_0 e^{-1.4x} = 0.01 I_0$ for x, finding $x = (\ln 100)/1.4 \approx 3.29$ meters.

47.   If $N(t)$ denotes the number of people (in thousands) who have heard the rumor after t days, then the initial value problem is $N' = k(100 - N)$, $N(0) = 0$. Separating variables leads to $\ln(100 - N) = -kt + C$, and the initial condition $N(0) = 0$ gives $C = \ln 100$. Then $100 - N = 100 e^{-kt}$, so $N(t) = 100(1 - e^{-kt})$. Substituting $N(7) = 10$ and solving for k gives $k = \ln(100/90)/7 \approx 0.01505$. Finally, 50,000 people have heard the rumor after $t = (\ln 2)/k \approx 46.05$ days, by solving the equation $100(1 - e^{-kt}) = 50$ for t.

49. Newton's law of cooling gives $\dfrac{dT}{dt} = k(70 - T)$, and separating variables and integrating

lead to $\ln(T - 70) = -kt + C$. The initial condition $T(0) = 210$ gives $C = \ln 140$, and

then $T(30) = 140$ gives $\ln 70 = -30k + \ln 140$, or $k = (\ln 2)/30$, so that

$T(t) = e^{-kt+C} + 70 = 140e^{-kt} + 70$. Finally, setting $T(t) = 100$ gives $140e^{-kt} + 70 = 100$, or

$t = \left[\ln(14/3)\right]/k \approx 66.67$ minutes, or 66 minutes and 40 seconds.

51. **(a)** The initial condition gives $A(t) = 15e^{-kt}$, and then $A(5) = 10$ implies that $15e^{-kt} = 10$,

or $e^{kt} = \dfrac{3}{2}$, or $k = \dfrac{1}{5}\ln\dfrac{3}{2}$. Thus

$$A(t) = 15\exp\left(-\frac{t}{5}\ln\frac{3}{2}\right) = 15\left(\frac{3}{2}\right)^{-t/5} = 15\left(\frac{2}{3}\right)^{t/5}.$$

(b) After 8 months we have $A(8) = 15\left(\dfrac{2}{3}\right)^{8/5} \approx 7.84$ su .

(c) $A(t) = 1$ when $A(t) = 15\left(\dfrac{2}{3}\right)^{t/5} = 1$, that is $t = 5\dfrac{\ln(\frac{1}{15})}{\ln(\frac{2}{3})} \approx 33.3944$. Thus it will be safe

to return after about 33.4 months.

53. As in Problem 52, if $L(t)$ denotes the number of Native American language families at

time t (in years), then $L(t) = e^{kt}$ for some constant k, and the condition that

$L(6000) = e^{6000k} = 1.5$ gives $k = \dfrac{1}{6000}\ln\dfrac{3}{2}$. If "now" corresponds to time $t = T$, then we

are given that $L(T) = e^{kT} = 150$, so $T = \dfrac{1}{k}\ln 150 = 6000\dfrac{\ln 150}{\ln(3/2)} \approx 74146.48$. This result

suggests that the ancestors of today's Native Americans first arrived in the western hemi-
sphere about 74 thousand years ago.

55. With $A = \pi \cdot 3^2$ and $a = \pi(1/12)^2$, and taking $g = 32$ ft/sec^2, Equation (30) reduces to

$162y' = -\sqrt{y}$, which we solve to find $324\sqrt{y} = -t + C$. The initial condition $y(0) = 9$

leads to $C = 972$, and so $y = 0$ when $t = 972$ sec , that is 16 min 12 sec.

57. The solution of $y' = -k\sqrt{y}$ is given by $2\sqrt{y} = -kt + C$. The initial condition $y(0) = h$

(the height of the cylinder) yields $C = 2\sqrt{h}$. Then substituting $t = T$ and $y = 0$ gives

$k = 2\sqrt{h}/T$. It follows that $y = h\left(1 - \dfrac{t}{T}\right)^2$. If r denotes the radius of the cylinder, then

$$V(y) = \pi r^2 y = \pi r^2 h \left(1 - \frac{t}{T}\right)^2 = V_0 \left(1 - \frac{t}{T}\right)^2 .$$

59. (a) Since $x^2 = by$, the cross-sectional area is $A(y) = \pi x^2 = \pi b y$. Hence equation (30)

becomes $y^{1/2} y' = -k = -(a/\pi b)\sqrt{2g}$, with general solution $\frac{2}{3} y^{3/2} = -kt + C$. The initial

condition $y(0) = 4$ gives $C = 16/3$, and then $y(1) = 1$ yields $k = 14/3$. It follows that

the depth at time t is $y(t) = (8 - 7t)^{2/3}$.

(b) The tank is empty after $t = 8/7 \text{ hr}$, that is, at 1:08:34 p.m.

(c) We see above that $k = \frac{a}{\pi b}\sqrt{2g} = \frac{14}{3}$. Substitution of $a = \pi r^2$ and $b = 1$ and

$g = 32 \cdot 3600^2 \text{ ft/hr}^2$ yields $r = \frac{1}{60}\sqrt{\frac{7}{12}} \text{ ft} \approx 0.15 \text{ in}$ for the radius of the bottom hole.

61. $A(y) = \pi(8y - y^2)$ as in Example 8 in the text, but now $a = \frac{\pi}{144}$ in Equation (30), so

that the initial value problem is $18(8y - y^2) y' = -\sqrt{y}$, $y(0) = 8$. Separating variables

gives $\int 18(y^{3/2} - 8y^{1/2}) dy = \int dt$, or $18\left(\frac{2}{5} y^{5/2} - \frac{16}{3} y^{3/2}\right) = t + C$, and the initial condition

gives $C = 18\left(\frac{2}{5} 8^{5/2} - \frac{16}{3} 8^{3/2}\right)$. We seek the value of t when $y = 0$, which is given by

$-C \approx 869 \text{ sec} = 14 \text{ min } 29 \text{ sec}$.

63. (a) As in Example 6, the initial value problem is $\pi(8y - y^2)\frac{dy}{dt} = -\pi k \sqrt{y}$, $y(0) = 4$,

where $k = 0.6r^2\sqrt{2g} = 4.8r^2$. Separating variables and applying the initial condition just

as in the Example 6 solution, we find that $\frac{16}{3} y^{3/2} - \frac{2}{5} y^{5/2} = -kt + \frac{448}{15}$. When we substi-

tute $y = 2$ (ft) and $t = 1800 \text{ sec}$ (that is, 30 min) we find that $k \approx 0.009469$. Finally,

$y = 0$ when $t = \frac{448}{15k} \approx 3154 \text{ sec} = 53 \text{ min } 34 \text{ sec}$. Thus the tank is empty at 1:53:34 p.m.

(b) The radius of the bottom hole is $r = \sqrt{\frac{k}{4.8}} \approx 0.04442 \text{ ft} \approx 0.53 \text{ in}$, thus about half an

inch.

65.   The temperature $T(t)$ of the body satisfies the differential equation $\dfrac{dT}{dt} = k(70 - T)$.

Separating variables gives $\displaystyle\int \frac{1}{70 - T} dT = \int k\,dt$, or (since $T(t) > 70$ for all t)

$\ln(T - 70) = -kt + C$. If we take $t = 0$ at the (unknown) time of death, then applying the initial condition $T(0) = 98.6$ gives $C = \ln 28.6$, and so $T(t) = 70 + 28.6e^{-kt}$. Now suppose that 12 noon corresponds to $t = a$. This gives the two equations

$$T(a) = 70 + 28.6e^{-ka} = 80$$
$$T(a+1) = 70 + 28.6e^{-k(a+1)} = 75 \, ,$$

which simplify to

$$28.6e^{-ka} = 10$$
$$28.6e^{-ka}e^{-k} = 5 \, .$$

These latter equations imply that $e^{-k} = 5/10 = 1/2$, so that $k = \ln 2$. Finally, we can substitute this value of k into the first of the previous two equations to find that

$a = \dfrac{\ln 2.86}{\ln 2} \approx 1.516$ hr ≈ 1 hr 31 min, so the death occurred at 10:29 a.m.

67.   We still have $t = t_0 e^{kx}$, but now the given information yields the conditions

$$t_0 + 1 = t_0 e^{4k}$$
$$t_0 + 2 = t_0 e^{7k}$$

at 8 a.m. and 9 a.m., respectively. Elimination of t_0 gives the equation $2e^{4k} - e^{7k} - 1 = 0$, which cannot be easily factored, unlike the corresponding equation in Problem 66. Letting $u = e^k$ gives $2u^4 - u^7 - 1 = 0$, and solving this equation using MATLAB or other technology leads to three real and four complex roots. Of the three real roots, only $u \approx 1.086286$ satisfies $u > 1$, and thus represents the desired solution. This means that $k \approx \ln 1.086286 \approx 0.08276$. Using this value, we finally solve either of the preceding pair of equations for $t_0 \approx 2.5483$ hr ≈ 2 hr 33 min. Thus it began to snow at 4:27 a.m.

69.   Substitution of $v = dy/dx$ in the differential equation for $y = y(x)$ gives $a\dfrac{dv}{dx} = \sqrt{1 + v^2}$,

and separation of variables then yields $\displaystyle\int \frac{1}{\sqrt{1 + v^2}}\,dv = \int \frac{1}{a}\,dx$, or $\sinh^{-1} v = \dfrac{x}{a} + C_1$, or

$\dfrac{dy}{dx} = \sinh\left(\dfrac{x}{a} + C_1\right)$. The fact that $y'(0) = 0$ implies that $C_1 = 0$, so it follows that

$\dfrac{dy}{dx} = \sinh\left(\dfrac{x}{a}\right)$, or $y(x) = a\cosh\left(\dfrac{x}{a}\right) + C$. Of course the (vertical) position of the x-axis can be adjusted so that $C = 0$, and the units in which T and ρ are measured may be adjusted so that $a = 1$. In essence, then, the shape of the hanging cable is the hyperbolic cosine graph $y = \cosh x$.

SECTION 1.5

LINEAR FIRST-ORDER EQUATIONS

1. An integrating factor is given by $\rho = \exp\left(\int 1\,dx\right) = e^x$, and multiplying the differential equation by ρ gives $e^x y' + e^x y = 2e^x$, or $D_x\left(e^x \cdot y\right) = 2e^x$. Integrating then leads to $e^x \cdot y = \int 2e^x\,dx = 2e^x + C$, and thus to the general solution $y = 2 + Ce^{-x}$. Finally, the initial condition $y(0) = 0$ implies that $C = -2$, so the corresponding particular solution is $y(x) = 2 - 2e^{-x}$.

3. An integrating factor is given by $\rho = \exp\left(\int 3\,dx\right) = e^{3x}$, and multiplying the differential equation by ρ gives $D_x\left(y \cdot e^{3x}\right) = 2x$. Integrating then leads to $y \cdot e^{3x} = x^2 + C$, and thus to the general solution $y(x) = \left(x^2 + C\right)e^{-3x}$.

5. We first rewrite the differential equation for $x > 0$ as $y' + \dfrac{2}{x}y = 3$. An integrating factor is given by $\rho = \exp\left(\int \dfrac{2}{x}\,dx\right) = e^{2\ln x} = x^2$, and multiplying the equation by ρ gives $x^2 \cdot y' + 2xy = 3$, or $D_x\left(y \cdot x^2\right) = 3x^2$. Integrating then leads to $y \cdot x^2 = x^3 + C$, and thus to the general solution $y(x) = x + \dfrac{C}{x^2}$. Finally, the initial condition $y(1) = 5$ implies that $C = 4$, so the corresponding particular solution is $y(x) = x + \dfrac{4}{x^2}$.

7. We first rewrite the differential equation for $x > 0$ as $y' + \dfrac{1}{2x}y = \dfrac{5}{\sqrt{x}}$. An integrating factor is given by $\rho = \exp\left(\int \dfrac{1}{2x}\,dx\right) = e^{(\ln x)/2} = \sqrt{x}$, and multiplying the equation by ρ gives $\sqrt{x} \cdot y' + \dfrac{1}{2\sqrt{x}}y = 5$, or $D_x\left(y \cdot \sqrt{x}\right) = 5$. Integrating then leads to $y \cdot \sqrt{x} = 5x + C$, and thus to the general solution $y(x) = 5\sqrt{x} + \dfrac{C}{\sqrt{x}}$.

9.   We first rewrite the differential equation for $x > 0$ as $y' - \dfrac{1}{x}y = 1$. An integrating factor

is given by $\rho = \exp\left(\displaystyle\int -\dfrac{1}{x}dx\right) = \dfrac{1}{x}$, and multiplying the equation by ρ gives

$\dfrac{1}{x}y' - \dfrac{1}{x^2}y = \dfrac{1}{x}$, or $D_x\left(y \cdot \dfrac{1}{x}\right) = \dfrac{1}{x}$. Integrating then leads to $y \cdot \dfrac{1}{x} = \ln x + C$, and thus to

the general solution $y(x) = x \ln x + Cx$. Finally, the initial condition $y(1) = 7$ implies

that $C = 7$, so the corresponding particular solution is $y(x) = x \ln x + 7x$.

11.   We first collect terms and rewrite the differential equation for $x > 0$ as $y' + \left(\dfrac{1}{x} - 3\right)y = 0$

. An integrating factor is given by

$$\rho = \exp\left[\int\left(\dfrac{1}{x} - 3\right)dx\right] = e^{\ln x - 3x} = xe^{-3x},$$

and multiplying by ρ gives $xe^{-3x} \cdot y' + \left(e^{-3x} - 3xe^{-3x}\right)y = 0$, or $D_x\left(y \cdot xe^{-3x}\right) = 0$. Inte-

grating then leads to $y \cdot xe^{-3x} = C$, and thus to the genral solution $y(x) = Cx^{-1}e^{3x}$. Final-

ly, the initial condition $y(1) = 0$ implies that $C = 0$, so the corresponding particular solu-

tion is $y(x) \equiv 0$, that is, the solution is the zero function.

13.   An integrating factor is given by $\rho = \exp\left(\displaystyle\int 1dx\right) = e^x$, and multiplying by ρ gives

$e^x \cdot y' + e^x y = e^{2x}$, or $D_x\left(y \cdot e^x\right) = e^{2x}$. Integrating then leads to $y \cdot e^x = \dfrac{1}{2}e^{2x} + C$, and

thus to the general solution $y(x) = \dfrac{1}{2}e^x + Ce^{-x}$. Finally, the initial condition $y(0) = 1$

implies that $C = \dfrac{1}{2}$, so the corresponding particular solution is $y(x) = \dfrac{1}{2}e^x + \dfrac{1}{2}e^{-x}$, that

is, $y = \cosh x$.

15.   An integrating factor is given by $\rho = \exp\left(\displaystyle\int 2x\,dx\right) = e^{x^2}$, and multiplying by ρ gives

$e^{x^2} \cdot y' + 2xe^{x^2}y = xe^{x^2}$, or $D_x\left(y \cdot e^{x^2}\right) = xe^{x^2}$. Integrating then leads to $y \cdot e^{x^2} = \dfrac{1}{2}e^{x^2} + C$

, and thus to the general solution $y(x) = \dfrac{1}{2} + Ce^{-x^2}$. Finally, the initial condition

$y(0) = -2$ implies that $C = -\dfrac{5}{2}$, so the corresponding particular solution is

$y(x) = \dfrac{1}{2} - \dfrac{5}{2}e^{-x^2}$.

17. We first rewrite the differential equation for $x > -1$ as $y' + \dfrac{1}{1+x} y = \dfrac{\cos x}{1+x}$. An integrat-

ing factor is given by $\rho = \exp\left(\displaystyle\int \dfrac{1}{1+x} dx\right) = 1 + x$, and multiplying by ρ gives

$(1+x) y' + y = \cos x$ (which happens to be the original differential equation), or

$D_x\big[y \cdot (1+x)\big] = \cos x$. Integrating then leads to $y \cdot (1+x) = \sin x + C$, and thus to the

general solution $y(x) = \dfrac{\sin x + C}{1+x}$. Finally, the initial condition $y(0) = 1$ implies that

$C = 1$, so the corresponding particular solution is $y(x) = \dfrac{1 + \sin x}{1+x}$.

19. For $x > 0$ an integrating factor is given by $\rho = \exp\left(\displaystyle\int \cot x \, dx\right) = e^{\ln(\sin x)} = \sin x$, and mul-

tiplying by ρ gives $(\sin x) \cdot y' + (\cos x) y = \sin x \cos x$, or $D_x(y \cdot \sin x) = \sin x \cos x$. In-

tegrating then leads to $y \cdot \sin x = \dfrac{1}{2} \sin^2 x + C$, and thus to the general solution

$y(x) = \dfrac{1}{2} \sin x + C \csc x$.

21. We first rewrite the differential equation for $x > 0$ as $y' - \dfrac{3}{x} y = x^3 \cos x$. An integrating

factor is given by $\rho = \exp\left(\displaystyle\int -\dfrac{3}{x} dx\right) = e^{-3\ln x} = x^{-3}$, and multiplying by ρ gives

$x^{-3} \cdot y' - 3x^{-4} y = \cos x$, or $D_x(y \cdot x^{-3}) = \cos x$. Integrating then leads to

$y \cdot x^{-3} = \sin x + C$, and thus to the general solution $y(x) = x^3 \sin x + C x^3$. Finally, the

initial condition $y(2\pi) = 0$, so the corresponding particular solution is $y(x) = x^3 \sin x$.

23. We first rewrite the differential equation for $x > 0$ as $y' + \left(2 - \dfrac{3}{x}\right) y = 4x^3$. An integrat-

ing factor is given by $\rho = \exp\left(\displaystyle\int 2 - \dfrac{3}{x} dx\right) = \exp(2x - 3\ln x) = x^{-3} e^{2x}$, and multiplying

by ρ gives $x^{-3} e^{2x} \cdot y' + (2x^{-3} - 3x^{-4}) e^{2x} y = 4e^{2x}$, or $D_x(y \cdot x^{-3} e^{2x}) = 4e^{2x}$. Integrating

then leads to $y \cdot x^{-3} e^{2x} = 2e^{2x} + C$, and thus to the general solution $y(x) = 2x^3 + C x^3 e^{-2x}$.

25.   We first rewrite the differential equation as $y' + \dfrac{3x^3}{x^2+1} y = \dfrac{6x}{x^2+1} e^{-\frac{3}{2}x^2}$. An integrating

factor is given by $\rho = \exp\left(\displaystyle\int \dfrac{3x^3}{x^2+1}\,dx\right)$. Long division of polynomials shows that

$\dfrac{3x^3}{x^2+1} = 3x - \dfrac{3x}{x^2+1}$, and so

$$\rho = \exp\left(\int 3x - \dfrac{3x}{x^2+1}\,dx\right) = \exp\left[\dfrac{3}{2}x^2 - \dfrac{3}{2}\ln\left(x^2+1\right)\right] = \left(x^2+1\right)^{-3/2} e^{\frac{3}{2}x^2}.$$

Multiplying by ρ gives

$$\left(x^2+1\right)^{-3/2} e^{\frac{3}{2}x^2} \cdot y' + 3x^3 \left(x^2+1\right)^{-5/2} e^{\frac{3}{2}x^2} y = 6x\left(x^2+1\right)^{-5/2},$$

or (as can be verified using the product rule twice, together with some algebra)

$D_x\left[y \cdot \left(x^2+1\right)^{-3/2} e^{\frac{3}{2}x^2}\right] = 6x\left(x^2+1\right)^{-5/2}$. Integrating then leads to

$$y \cdot \left(x^2+1\right)^{-3/2} e^{\frac{3}{2}x^2} = \int 6x\left(x^2+1\right)^{-5/2}\,dx = -2\left(x^2+1\right)^{-3/2} + C,$$

and thus to the general solution $y = \left[-2 + C\left(x^2+1\right)^{3/2}\right] e^{-\frac{3}{2}x^2}$. Finally, the initial condi-

tion $y(0) = 1$ implies that $C = 3$, so the corresponding particular solution is

$$y = \left[-2 + 3\left(x^2+1\right)^{3/2}\right] e^{-\frac{3}{2}x^2}.$$

The strategy in each of Problems 26-28 is to use the inverse function theorem to conclude that at

points (x, y) where $\dfrac{dy}{dx} \neq 0$, x is locally a function of y with $\dfrac{dx}{dy} \cdot \dfrac{dy}{dx} = 1$. Thus the given differ-

ential equation is equivalent to one in which x is the dependent variable and y as the independent
variable, and this latter equation may be easier to solve than the one originally given. It may not
be feasible, however, to solve the resulting solution for the original dependent variable y.

27.   At points (x, y) with $x + ye^y \neq 0$, rewriting the differential equation as $\dfrac{dy}{dx} = \dfrac{1}{x + ye^y}$

shows that $\dfrac{dx}{dy} = x + ye^y$, or (putting x' for $\dfrac{dx}{dy}$) $x' - x = ye^y$, a linear equation for the

dependent variable x as a function of the independent variable y. An integrating factor is

given by $\rho = \exp\left(\displaystyle\int -1\,dy\right) = e^{-y}$, and multiplying by ρ gives $e^{-y} \cdot x' - e^{-y}x = y$, or

$D_y\left(x \cdot e^{-y}\right) = y$. Integrating then leads to $x \cdot e^{-y} = \dfrac{1}{2}y^2 + C$, and thus to the general (implicit) solution $x(y) = \left(\dfrac{1}{2}y^2 + C\right)e^y$.

29. We first rewrite the differential equation as $y' - 2xy = 1$. An integrating factor is given by $\rho = \exp\left(\int -2x\,dx\right) = e^{-x^2}$, and multiplying by ρ gives $e^{-x^2} \cdot y' - 2xe^{-x^2}y = e^{-x^2}$, or $D_x\left(y \cdot e^{-x^2}\right) = e^{-x^2}$. Integrating then leads to $y \cdot e^{-x^2} = \int e^{-x^2}\,dx$. Any antiderivative of e^{-x^2} differs by a constant (call it C) from the definite integral $\int_0^x e^{-t^2}\,dt$, and so we can write $y \cdot e^{-x^2} = \int_0^x e^{-t^2}\,dt + C$. The definition of $\mathrm{erf}\,(x)$ then gives $y \cdot e^{-x^2} = \dfrac{\sqrt{\pi}}{2}\mathrm{erf}\,(x) + C$, and thus the general solution $y(x) = e^{x^2}\left[\dfrac{\sqrt{\pi}}{2}\mathrm{erf}\,(x) + C\right]$.

31. **(a)** The fundamental theorem of calculus implies, for any value of C, that
$$y_c'(x) = Ce^{-\int P(x)\,dx}\left[-P(x)\right] = -P(x)\,y_c(x),$$
and thus that $y_c'(x) + P(x)\,y_c(x) = 0$. Therefore y_c is a general solution of $\dfrac{dy}{dx} + P(x)\,y = 0$.

(b) The product rule and the fundamental theorem of calculus imply that
$$y_p'(x) = e^{-\int P(x)\,dx} \cdot Q(x)e^{\int P(x)\,dx} + e^{-\int P(x)\,dx}\left[-P(x)\right] \cdot \int\left(Q(x)e^{\int P(x)\,dx}\right)dx$$
$$= Q(x) - P(x)e^{-\int P(x)\,dx}\int\left(Q(x)e^{\int P(x)\,dx}\right)dx$$
$$= Q(x) - P(x)\,y_p(x),$$
and thus that $y_p'(x) + P(x)\,y_p(x) = Q(x)$. Therefore y_p is a particular solution of $\dfrac{dy}{dx} + P(x)\,y = Q(x)$.

(c) The stated assumptions imply that
$$y'(x) + P(x)\,y = y_c'(x) + y_p'(x) + P(x)\left[y_c(x) + y_p(x)\right]$$
$$= \left[y_c'(x) + P(x)\,y_c(x)\right] + \left[y_p'(x) + P(x)\,y_p(x)\right]$$
$$= 0 + Q(x)$$
$$= Q(x),$$

proving that $y(x)$ is a general solution of $\dfrac{dy}{dx} + P(x)y = Q(x)$.

33.   Let $x(t)$ denote the amount of salt (in kg) in the tank after t seconds. We want to know when $x(t) = 10$. In the notation of Equation (18) of the text, the differential equation for $x(t)$ is

$$\frac{dx}{dt} = r_i c_i - \frac{r_o}{V} x = (5\,\mathrm{L/s})(0\,\mathrm{kg/L}) - \frac{5\,\mathrm{L/s}}{1000\,\mathrm{L}} \cdot x\,\mathrm{kg},$$

or $\dfrac{dx}{dt} = -\dfrac{x}{200}$. Separating variables gives the general solution $x(t) = Ce^{-t/200}$, and the initial condition $x(0) = 100$ implies that $C = 100$, and so $x(t) = 100e^{-t/200}$. Setting $x(t) = 10$ gives $10 = 100e^{-t/200}$, or $t = 200\ln 10 \approx 461$ sec, that is, about 7 min 41 sec.

35.   The only difference from the Example 4 solution in the textbook is that $V = 1640\,\mathrm{km}^3$ and $r = 410\,\mathrm{km}^3/\mathrm{yr}$ for Lake Ontario, so the time required is

$$t = \frac{V}{r}\ln 4 = 4\ln 4 \approx 5.5452 \text{ years.}$$

37.   Let $x(t)$ denote the amount of salt (in lb) after t seconds. Because the volume of liquid in the tank is increasing by 2 gallon each minute, the volume after t sec is $100 + 2t$ gallons. Thus in the notation of Equation (18) of the text, the differential equation for $x(t)$ is

$$\frac{dx}{dt} = r_i c_i - \frac{r_o}{V} x = (5\,\mathrm{gal/s})(1\,\mathrm{lb/gal}) - \frac{3\,\mathrm{gal/s}}{(100 + 2t)\,\mathrm{gal}} \cdot x\,\mathrm{lb},$$

or $\dfrac{dx}{dt} + \dfrac{3}{100 + 2t} x = 5$. An integrating factor is given by

$$\rho = \exp\left(\int \frac{3}{100 + 2t}\,dt \right) = (100 + 2t)^{3/2},$$ and multiplying the differential equation by ρ gives

$$(100 + 2t)^{3/2} \cdot \frac{dx}{dt} + 3(100 + 2t)^{1/2} x = 5(100 + 2t)^{3/2},$$

or $D_t\left[(100 + 2t)^{3/2} \cdot x \right] = 5(100 + 2t)^{3/2}$. Integrating then leads to

$$(100 + 2t)^{3/2} \cdot x = \int 5(100 + 2t)^{3/2}\,dt = (100 + 2t)^{5/2} + C,$$

and thus to the general solution $x(t) = 100 + 2t + C(100 + 2t)^{-3/2}$. The initial condition $x(0) = 50$ implies that $50 = 100 + C \cdot 100^{-3/2}$, or $C = -50000$, and so the desired particular solution is $x(t) = 100 + 2t - \dfrac{50000}{(100 + 2t)^{3/2}}$. Finally, because the tank starts out with

300 gallons of excess capacity and the volume of its contents increases at $2\,\text{gal/s}$, the

tank is full when $t = \dfrac{300\,\text{gal}}{2\,\text{gal/s}} = 150\,\text{s}$. At this time the tank contains

$x(150) = 400 - \dfrac{50000}{(400)^{3/2}} = 393.75\,\text{lb}$ of salt.

39. **(a)** In the notation of Equation (18) of the text, the differential equation for $x(t)$ is

$$\frac{dx}{dt} = r_i c_i - \frac{r_o}{V} x = (10\,\text{gal/min})(0) - (10\,\text{gal/min})\left(\frac{x}{100}\right),$$

or $\dfrac{dx}{dt} + \dfrac{1}{10} x = 0$. Separating variables leads to the general solution $x(t) = Ce^{-t/10}$, and

the initial condition $x(0) = 100$ implies that $C = 100$. Thus $x(t) = 100e^{-t/10}$. In the same

way, the differential equation for $y(t)$ is

$$\frac{dy}{dt} = (10\,\text{gal/min})\left(\frac{x}{100}\right) - (10\,\text{gal/min})\left(\frac{y}{100}\right),$$

because the volume of liquid in each tank remains constant at 2 gal. Substituting the re-

sult of part (a) gives $\dfrac{dy}{dt} + \dfrac{1}{10} y = 10e^{-t/10}$. An integrating factor is given by

$\rho = \exp\left(\displaystyle\int \frac{1}{10}\,dt\right) = e^{t/10}$, and multiplying the differential equation by ρ gives

$e^{t/10} \cdot \dfrac{dy}{dt} + \dfrac{1}{10} e^{t/10} y = 10$, or $D_t\left(e^{t/10} \cdot y\right) = 10$. Integrating then leads to $e^{t/10} \cdot y = 10t + C$,

and thus to the general solution $y(t) = (10t + C)e^{-t/10}$. The initial condition $y(0) = 0$

implies that $C = 0$, so that $y(t) = 10te^{-t/10}$.

(b) By Part (a), $y'(t) = 10\left(-\dfrac{t}{10} e^{-t/10} + e^{-t/10}\right) = e^{-t/10}(10 - t)$, which is zero for $t = 10$.

Furthermore, $y'(t) > 0$ for $0 < t < 10$, and $y'(t) < 0$ for $t > 10$, which implies that $y(t)$

reaches its absolute maximum at $t = 10$ min. The maximum amount of ethanol in tank 2

is therefore $y(10) = 100e^{-1} \approx 36.79$ gal.

41. **(a)** Between time t and time $t + \Delta t$, the amount $A(t)$ (in thousands of dollars) increases

by a deposit of $0.12S(t)\Delta t$ (12% per year of annual salary) as well as interest earnings of

$0.06A(t)\Delta t$ (6% per year of current balance). It follows that

$$\Delta A \approx 0.12 S(t) \Delta t + 0.06 A(t) \Delta t \,,$$

leading to the linear differential equation $\dfrac{dA}{dt} = 0.12S + 0.06A = 3.6e^{t/20} + 0.06A$, or

$$\dfrac{dA}{dt} - 0.06A = 3.6e^{t/20} \,.$$

(b) An integrating factor is given by $\rho = \exp\left(\int -0.06\,dt\right) = e^{-0.06t}$, and multiplying the

differential equation by ρ gives $e^{-0.06t} \cdot \dfrac{dA}{dt} - 0.06e^{-0.06t} A = 3.6e^{t/20} e^{-0.06t} = 3.6e^{-0.01t}$, or

$D_t\left(e^{-0.06t} \cdot A\right) = 3.6e^{-0.01t}$. Integrating then leads to $e^{-0.06t} \cdot A = -360e^{-0.01t} + C$, and thus to

the general solution $A(t) = -360e^{0.05t} + Ce^{0.06t}$. The initial condition $A(0) = 0$ implies

that $C = 360$, so that $A(t) = 360\left(e^{0.06t} - e^{0.05t}\right)$. At age 70 she will have

$A(40) \approx 1308.283$ thousand dollars, that is, $\$1,308,283$.

43.   **(a)** First we rewrite the differential equation as $y' + y = x$. An integrating factor is given

by $\rho = \exp\left(\int 1\,dx\right) = e^x$, and multiplying the differential equation by ρ gives

$e^x \cdot y' + e^x y = xe^x$, or $D_x\left(e^x \cdot y\right) = xe^x$. Integrating (by parts) then leads to

$e^x \cdot y = \int xe^x\,dx = xe^x - e^x + C$, and thus to the general solution $y(x) = x - 1 + Ce^{-x}$.

Then the fact that $\lim\limits_{x \to \infty} e^{-x} = 0$ implies that every solution curve approaches the straight

line $y = x - 1$ as $x \to \infty$.

(b) The initial condition $y(-5) = y_0$ imposed upon the general solution in part (a) im-

plies that $y_0 = -5 - 1 + Ce^5$, and thus that $C = e^{-5}(y_0 + 6)$. Hence the solution of the ini-

tial value problem $y' = x - y$, $y(-5) = y_0$ is $y(x) = x - 1 + (y_0 + 6)e^{-x-5}$. Substituting

$x = 5$, we therefore solve the equation $4 + (y_0 + 6)e^{-10} = y_1$ with

$$y_1 = 3.998, 3.999, 4, 4.001, 4.002$$

for the desired initial values

$$y_0 = -50.0529, -28.0265, -6.0000, 16.0265, 38.0529 \,,$$

respectively.

45.   The volume of the reservoir (in millions of cubic meters, denoted m-m^3) is 2. In the notation of Equation (18) of the text, the differential equation for $x(t)$ is

$$\frac{dx}{dt} = r_i c_i - \frac{r_o}{V} x = \left(0.2\,\text{m-m}^3/\text{month}\right)\left(10\,\text{L}/\text{m}^3\right) - \left(0.2\,\text{m-m}^3/\text{month}\right)\left(\frac{x}{2}\,\text{L}/\text{m}^3\right),$$

or $\dfrac{dx}{dt} + \dfrac{1}{10}x = 2$. An integrating factor is given by $\rho = e^{t/10}$, and multiplying the differ-

ential equation by ρ gives $e^{t/10} \cdot \dfrac{dx}{dt} + \dfrac{1}{10}e^{t/10}x = 2e^{t/10}$, or $D_t\left(e^{t/10} \cdot x\right) = 2e^{t/10}$. Integrating

then leads to $e^{t/10} \cdot x = 20e^{t/10} + C$, and thus to the general solution $x(t) = 20 + Ce^{-t/10}$.

The initial condition $x(0) = 0$ implies that $C = -20$, and so $x(t) = 20\left(1 - e^{-t/10}\right)$, which

shows that indeed $\lim\limits_{t \to \infty} x(t) = 20$ (million liters). This was to be expected because the

reservoir's pollutant concentration should ultimately match that of the incoming water,

namely $10\,\text{L/m}^3$. Finally, since the volume of reservoir remains constant at 2 m-m^3, a

pollutant concentration of $5\,\text{L/m}^3$ is reached when $\dfrac{x(t)}{2} = 5$, that is, when

$10 = 20\left(1 - e^{-t/10}\right)$, or $t = 10\ln 2 \approx 6.93\,\text{months}$.

SECTION 1.6

SUBSTITUTION METHODS AND EXACT EQUATIONS

It is traditional for every elementary differential equations text to include the particular types of equation that are found in this section. However, no one of them is vitally important solely in its own right. Their main purpose (at this point in the course) is to familiarize students with the technique of transforming a differential equation by substitution. The subsection on airplane flight trajectories (together with Problems 56–59) is included as an application, but is optional material and may be omitted if the instructor desires.

The differential equations in Problems 1-15 are homogeneous, and so we solve by means of the substitution $v = y/x$ indicated in Equation (8) of the text. In some cases we present solutions by other means, as well.

1. For $x \neq 0$ and $x + y \neq 0$ we rewrite the differential equation as $\dfrac{dy}{dx} = \dfrac{x - y}{x + y} = \dfrac{1 - \dfrac{y}{x}}{1 + \dfrac{y}{x}}$.

Substituting $v = \dfrac{y}{x}$ then gives $v + x\dfrac{dv}{dx} = \dfrac{1 - v}{1 + v}$, or $x\dfrac{dv}{dx} = \dfrac{1 - v}{1 + v} - v = \dfrac{1 - 2v - v^2}{1 + v}$. Sepa-

rating variables leads to $\displaystyle\int \dfrac{v + 1}{v^2 + 2v - 1}\,dv = -\int \dfrac{1}{x}\,dx$, or $\dfrac{1}{2}\ln\left|v^2 + 2v - 1\right| = -\ln|x| + C$, or

$\left|v^2 + 2v - 1\right| = Cx^{-2}$, where C is an arbitrary positive constant, or finally $v^2 + 2v - 1 = Cx^{-2}$

, where C is an arbitrary nonzero constant. Back-substituting $\dfrac{y}{x}$ for v then gives the so-

lution $\left(\dfrac{y}{x}\right)^2 + 2\dfrac{y}{x} - 1 = Cx^{-2}$, or $y^2 + 2xy - x^2 = C$.

3.   For x, y with $xy > 0$ we rewrite the differential equation as $\dfrac{dy}{dx} = \dfrac{y}{x} + 2\sqrt{\dfrac{y}{x}}$. Substitut-

ing $v = \dfrac{y}{x}$ then gives $v + x\dfrac{dv}{dx} = v + 2\sqrt{v}$, or $x\dfrac{dv}{dx} = 2\sqrt{v}$. Separating variables leads to

$\displaystyle\int \dfrac{1}{\sqrt{v}}\,dv = \int \dfrac{2}{x}\,dx$, or $2\sqrt{v} = 2\ln|x| + C$, or $v = \left(\ln|x| + C\right)^2$. Back-substituting $\dfrac{y}{x}$ for v

then gives the solution $y = x\left(\ln|x| + C\right)^2$.

5.   For $x \neq 0$ and $x + y \neq 0$ we rewrite the differential equation as $\dfrac{dy}{dx} = \dfrac{y}{x} \cdot \dfrac{x - y}{x + y} = \dfrac{y}{x} \cdot \dfrac{1 - \dfrac{y}{x}}{1 + \dfrac{y}{x}}$

. Substituting $v = \dfrac{y}{x}$ then gives $v + x\dfrac{dv}{dx} = v \cdot \dfrac{1 - v}{1 + v}$, or $x\dfrac{dv}{dx} = v \cdot \dfrac{1 - v}{1 + v} - v = \dfrac{-2v^2}{1 + v}$. Sepa-

rating variables leads to $\displaystyle\int \dfrac{1 + v}{v^2}\,dv = -2\int \dfrac{1}{x}\,dx$, or $-\dfrac{1}{v} + \ln|v| = -2\ln|x| + C$. Back-

substituting $\dfrac{y}{x}$ for v then gives $-\dfrac{x}{y} + \ln\left|\dfrac{y}{x}\right| = -2\ln|x| + C$, or $\ln\left|\dfrac{y}{x}\right| + 2\ln|x| = \dfrac{x}{y} + C$, or

$\ln|xy| = \dfrac{x}{y} + C$.

7.   For $x, y \neq 0$ we rewrite the differential equation as $\dfrac{dy}{dx} = \left(\dfrac{x}{y}\right)^2 + \dfrac{y}{x}$. Substituting $v = \dfrac{y}{x}$

then gives $v + x\dfrac{dv}{dx} = \left(\dfrac{1}{v}\right)^2 + v$, or $x\dfrac{dv}{dx} = \left(\dfrac{1}{v}\right)^2$. Separating variables leads to

$\displaystyle\int v^2\,dv = \int \dfrac{1}{x}\,dx$, or $v^3 = 3\ln|x| + C$. Back-substituting $\dfrac{y}{x}$ for v then gives

$\left(\dfrac{y}{x}\right)^3 = 3\ln|x| + C$, or $y^3 = x^3\left(3\ln|x| + C\right)$.

Alternatively, the substitution $v = y^3$, which implies that $v' = 3y^2 y'$, gives $\dfrac{1}{3}xv' = x^3 + v$,

or $v' - \dfrac{3}{x}v = 3x^2$, a linear equation in v as a function of x. An integrating factor is given by $\rho = \exp\left(-\int \dfrac{3}{x}dx\right) = x^{-3}$, and multiplying the differential equation by ρ gives

$x^{-3}\cdot v' - 3x^{-4}v = 3x^{-1}$, or $D_x\left(x^{-3}\cdot v\right) = 3x^{-1}$. Integrating then gives $x^{-3}\cdot v = 3\ln|x| + C$,

and finally back-substituting y^3 for v yields $y^3 = x^3\left(3\ln|x| + C\right)$, as determined above.

9. For $x \neq 0$ we rewrite the differential equation as $\dfrac{dy}{dx} = \dfrac{y}{x} + \left(\dfrac{y}{x}\right)^2$. Substituting $v = \dfrac{y}{x}$

then gives $v + x\dfrac{dv}{dx} = v + v^2$, or $x\dfrac{dv}{dx} = v^2$. Separating variables leads to $\displaystyle\int \dfrac{1}{v^2}dv = \int \dfrac{1}{x}dx$

, or $-\dfrac{1}{v} = \ln|x| + C$. Back-substituting $\dfrac{y}{x}$ for v then gives the solution $y = \dfrac{x}{C - \ln|x|}$.

11. For $x^2 - y^2 \neq 0$ and $x \neq 0$ we rewrite the differential equation as

$\dfrac{dy}{dx} = \dfrac{2xy}{x^2 - y^2} = \dfrac{2\dfrac{y}{x}}{1 - \left(\dfrac{y}{x}\right)^2}$. Substituting $v = \dfrac{y}{x}$ then gives $v + x\dfrac{dv}{dx} = \dfrac{2v}{1 - v^2}$, or

$x\dfrac{dv}{dx} = \dfrac{2v}{1 - v^2} - v = \dfrac{v + v^3}{1 - v^2}$. Separating variables leads to $\displaystyle\int \dfrac{1 - v^2}{v + v^3}dv = \int \dfrac{1}{x}dx$, or (after

decomposing into partial fractions) $\displaystyle\int \dfrac{1}{v} - \dfrac{2v}{v^2 + 1}dv = \int \dfrac{1}{x}dx$, or

$\ln|v| - \ln\left(v^2 + 1\right) = \ln|x| + C$, or $\dfrac{v}{v^2 + 1} = Cx$. Back-substituting $\dfrac{y}{x}$ for v then gives the

solution $\dfrac{y}{x} = Cx\left[\left(\dfrac{y}{x}\right)^2 + 1\right]$, or finally $y = C\left(x^2 + y^2\right)$.

13. For $x > 0$ we rewrite the differential equation as

$$\dfrac{dy}{dx} = \dfrac{y}{x} + \dfrac{\sqrt{x^2 + y^2}}{x} = \dfrac{y}{x} + \sqrt{1 + \left(\dfrac{y}{x}\right)^2}.$$

Substituting $v = \dfrac{y}{x}$ then gives $v + x\dfrac{dv}{dx} = v + \sqrt{1 + v^2}$, or $x\dfrac{dv}{dx} = \sqrt{1 + v^2}$. Separating vari-

ables leads to $\displaystyle\int \dfrac{1}{\sqrt{1 + v^2}}dv = \int \dfrac{1}{x}dx$, or (by means of either the substitution $v = \tan\theta$ or

an integral table) $\ln\left(v + \sqrt{v^2 + 1}\right) = \ln|x| + C$, or finally $v + \sqrt{v^2 + 1} = Cx$. Back-

substituting $\dfrac{y}{x}$ for v then gives the solution $y + \sqrt{y^2 + x^2} = Cx^2$.

15. For $x \neq 0$ and $x + y \neq 0$ we rewrite the differential equation as

$$\frac{dy}{dx} = \frac{-y(3x+y)}{x(x+y)} = -\frac{y}{x} \cdot \frac{3+\dfrac{y}{x}}{1+\dfrac{y}{x}}.$$

Substituting $v = \dfrac{y}{x}$ then gives $v + x\dfrac{dv}{dx} = -v\dfrac{3+v}{1+v} = -\dfrac{3v+v^2}{1+v}$, or

$$x\frac{dv}{dx} = -\frac{3v+v^2}{1+v} - v = \frac{-4v - 2v^2}{1+v}.$$

Separating variables leads to $\displaystyle\int \frac{1+v}{4v+2v^2}\,dv = -\int \frac{1}{x}\,dx$, or $\dfrac{1}{4}\ln\left|4v+2v^2\right| = -\ln|x| + C$, or

$x^4\left(4v+2v^2\right) = C$, or simply $x^4\left(2v+v^2\right) = C$. Back-substituting $\dfrac{y}{x}$ for v then gives the

solution $x^2\left(2xy + y^2\right) = C$.

17. The expression $4x + y$ suggests the substitution $v = 4x + y$, which implies that
$y = v - 4x$, and thus that $y' = v' - 4$. Substituting gives $v' - 4 = v^2$, or $v' = v^2 + 4$, a separable equation for v as a function of x. Separating variables gives $\displaystyle\int \frac{1}{v^2+4}\,dv = \int dx$, or

$\dfrac{1}{2}\tan^{-1}\dfrac{v}{2} = x + C$, or $v = 2\tan(2x+C)$. Finally, back-substituting $4x + y$ for v leads to
the solution $y = 2\tan(2x+C) - 4x$.

The differential equations in Problems 19-25 are Bernoulli equations, and so we solve by means
of the substitution $v = y^{1-n}$ indicated in Equation (10) of the text. (Problem 25 also admits of
another solution.)

19. We first rewrite the differential equation for $x, y > 0$ as $y' + \dfrac{2}{x}y = \dfrac{5}{x^2}y^3$, a Bernoulli

equation with $n = 3$. The substitution $v = y^{1-3} = y^{-2}$ implies that $y = v^{-1/2}$ and thus that

$y' = -\dfrac{1}{2}v^{-3/2}v'$. Substituting gives $-\dfrac{1}{2}v^{-3/2}v' + \dfrac{2}{x}v^{-1/2} = \dfrac{5}{x^2}v^{-3/2}$, or $v' - \dfrac{4}{x}v = -\dfrac{10}{x^2}$, a lin-

ear equation for v as a function of x. An integrating factor is given by

$\rho = \exp\left(\displaystyle\int -\frac{4}{x}\,dx\right) = x^{-4}$, and multiplying the differential equation by ρ gives

$\dfrac{1}{x^4}v' - \dfrac{4}{x^5}v = -\dfrac{10}{x^6}$, or $D_x\left(\dfrac{1}{x^4}\cdot v\right) = -\dfrac{10}{x^6}$. Integrating then leads to $\dfrac{1}{x^4}\cdot v = \dfrac{2}{x^5} + C$, or

$v = \dfrac{2}{x} + Cx^4 = \dfrac{2 + Cx^5}{x}$. Finally, back-substituting y^{-2} for v gives the general solution

$y^{-2} = \dfrac{2 + Cx^5}{x}$, or $y^2 = \dfrac{x}{2 + Cx^5}$.

21. We first rewrite the differential equation as $y' - y = y^3$, a Bernoulli equation with $n = 3$. The substitution $v = y^{1-3} = y^{-2}$ implies that $y = v^{-1/2}$ and thus that $y' = -\dfrac{1}{2} v^{-3/2} v'$. Substituting gives $-\dfrac{1}{2} v^{-3/2} v' - v^{-1/2} = v^{-3/2}$, or $v' + 2v = -2$, a linear equation for v as a function of x. An integrating factor is given by $\rho = \exp\left(\int 2\,dx\right) = e^{2x}$, and multiplying the differential equation by ρ gives $e^{2x} \cdot v' + 2e^{2x} v = -2e^{2x}$, or $D_x\left(e^{2x} \cdot v\right) = -2e^{2x}$. Integrating then leads to $e^{2x} \cdot v = -e^{2x} + C$, or $v = -1 + Ce^{-2x}$. Finally, back-substituting y^{-2} for v gives the general solution $y^{-2} = -1 + Ce^{-2x}$, or $y^2 = \dfrac{1}{Ce^{-2x} - 1}$.

23. We first rewrite the differential equation for $x > 0$ as $y' + \dfrac{6}{x} y = 3y^{4/3}$, a Bernoulli equation with $n = 4/3$. The substitution $v = y^{1-(4/3)} = y^{-1/3}$ implies that $y = v^{-3}$ and thus that $y' = -3v^{-4} v'$. Substituting gives $-3v^{-4} v' + \dfrac{6}{x} v^{-3} = 3v^{-4}$, or $v' - \dfrac{2}{x} v = -1$, a linear equation for v as a function of x. An integrating factor is given by $\rho = \exp\left(\int -\dfrac{2}{x} dx\right) = \dfrac{1}{x^2}$, and multiplying the differential equation by ρ gives $\dfrac{1}{x^2} v' - \dfrac{2}{x^3} v = -\dfrac{1}{x^2}$, or $D_x\left(\dfrac{1}{x^2} v\right) = -\dfrac{1}{x^2}$. Integrating then leads to $\dfrac{1}{x^2} v = \dfrac{1}{x} + C$, or $v = x + Cx^2$. Finally, back-substituting $y^{-1/3}$ for v gives the general solution $y^{-1/3} = x + Cx^2$, or $y = \left(x + Cx^2\right)^{-3}$.

25. We first rewrite the differential equation for $x, y > 0$ as $y' + \dfrac{1}{x} y = \dfrac{1}{\left(1 + x^4\right)^{1/2}} y^{-2}$, a Bernoulli equation with $n = -2$. The substitution $v = y^{1-(-2)} = y^3$ implies that $y = v^{1/3}$ and thus that $y' = \dfrac{1}{3} v^{-2/3} v'$. Substituting gives $\dfrac{1}{3} v^{-2/3} v' + \dfrac{1}{x} v^{1/3} = \dfrac{1}{\left(1 + x^4\right)^{1/2}} v^{-2/3}$, or

$v' + \dfrac{3}{x} v = \dfrac{3}{\left(1 + x^4\right)^{1/2}}$, a linear equation for v as a function of x. An integrating factor is given by $\rho = \exp\left(\int \dfrac{3}{x} dx\right) = x^3$, and multiplying the differential equation by ρ gives

$x^3 \cdot v' + 3x^2 v = \dfrac{3x^3}{\left(1+x^4\right)^{1/2}}$, or $D_x\left(x^3 \cdot v\right) = \dfrac{3x^4}{\left(1+x^4\right)^{1/2}}$. Integrating then leads to

$x^3 \cdot v = \dfrac{3}{2}\left(1+x^4\right)^{1/2} + C$, or $v = \dfrac{3\left(1+x^4\right)^{1/2} + C}{2x^3}$. Finally, back-substituting y^3 for v gives

the general solution $y^3 = \dfrac{3\left(1+x^4\right)^{1/2} + C}{2x^3}$.

Alternatively, for $x \neq 0$, the substitution $v = xy$, which implies that $v' = xy' + y$ and that

$y = \dfrac{v}{x}$, gives $\dfrac{v^2}{x^2} v'\left(1+x^4\right)^{1/2} = x$. Separating variables leads to $\displaystyle\int v^2\, dv = \int \dfrac{x^3}{\left(1+x^4\right)^{1/2}} dx$,

or $\dfrac{1}{3}v^3 = \dfrac{1}{2}\left(1+x^4\right)^{1/2} + C$, or $v^3 = \dfrac{3\left(1+x^4\right)^{1/2} + C}{2}$. Back-substituting xy for v then gives

the solution $y^3 = \dfrac{3\left(1+x^4\right)^{1/2} + C}{2x^3}$, as determined above.

As with Problems 16-18, the differential equations in Problems 27-29 rely upon substitutions that are generally suggested by the equations themselves. Two of these equations are also Bernoulli equations.

27.   The substitution $v = y^3$, which implies that $v' = 3y^2 y'$, gives $xv' - v = 3x^4$, or (for $x > 0$)

$v' - \dfrac{1}{x}v = 3x^3$, a linear equation for v as a function of x. An integrating factor is given by

$\rho = \exp\left(\int -\dfrac{1}{x} dx\right) = \dfrac{1}{x}$, and multiplying the differential equation by ρ gives

$\dfrac{1}{x} \cdot v' - \dfrac{1}{x^2}v = 3x^2$, or $D_x\left(\dfrac{1}{x} \cdot v\right) = 3x^2$. Integrating then leads to $\dfrac{1}{x} \cdot v = x^3 + C$, or

$v = x^4 + Cx$. Finally, back-substituting y^3 for v gives the general solution $y^3 = x^4 + Cx$,

or $y = \left(x^4 + Cx\right)^{1/3}$.

Alternatively, for $x, y > 0$ we can first rewrite the differential equation as

$y' - \dfrac{1}{3x}y = x^3 y^{-2}$, a Bernoulli equation with $n = -2$. This leads to the substitution

$v = y^{1-(-2)} = y^3$ used above.

29. The substitution $v = \sin^2 y$, which implies that $v' = (2\sin y \cos y)\, y'$, gives $xv' = 4x^2 + v$, or (for $x > 0$) $v' - \dfrac{1}{x}v = 4x$, a linear equation for v as a function of x. An integrating factor is given by $\rho = \exp\!\left(\displaystyle\int -\dfrac{1}{x}dx\right) = \dfrac{1}{x}$, and multiplying the differential equation by ρ gives $\dfrac{1}{x}\cdot v' - \dfrac{1}{x^2}v = 4$, or $D_x\!\left(\dfrac{1}{x}\cdot v\right) = 4$. Integrating then leads to $\dfrac{1}{x}\cdot v = 4x + C$, or $v = 4x^2 + Cx$. Finally, back-substituting $\sin^2 y$ for v gives the general solution $\sin^2 y = 4x^2 + Cx$.

Each of the differential equations in Problems 31–41 is of the form $M\,dx + N\,dy = 0$, and the exactness condition $\partial M/\partial y = \partial N/\partial x$ is routine to verify. For each problem we give the principal steps in the calculation corresponding to the method of Example 9 in this section.

31. The condition $F_x = M$ implies that $F(x,y) = \displaystyle\int 2x + 3y\,dx = x^2 + 3xy + g(y)$, and then the condition $F_y = N$ implies that $3x + g'(y) = 3x + 2y$, or $g'(y) = 2y$, or $g(y) = y^2$. Thus the solution is given by $x^2 + 3xy + y^2 = C$.

33. The condition $F_x = M$ implies that $F(x,y) = \displaystyle\int 3x^2 + 2y^2\,dx = x^3 + xy^2 + g(y)$, and then the condition $F_y = N$ implies that $4xy + g'(y) = 4xy + 6y^2$, or $g'(y) = 6y^2$, or $g(y) = 2y^3$. Thus the solution is given by $x^3 + 2xy^2 + 2y^3 = C$.

35. The condition $F_x = M$ implies that $F(x,y) = \displaystyle\int x^3 + \dfrac{y}{x}\,dx = \dfrac{1}{4}x^4 + y\ln x + g(y)$, and then the condition $F_y = N$ implies that $\ln x + g'(y) = y^2 + \ln x$, or $g'(y) = y^2$, or $g(y) = \dfrac{1}{3}y^3$. Thus the solution is given by $\dfrac{1}{4}x^4 + \dfrac{1}{3}y^3 + y\ln x = C$.

37. The condition $F_x = M$ implies that $F(x,y) = \displaystyle\int \cos x + \ln y\,dx = \sin x + x\ln y + g(y)$, and then the condition $F_y = N$ implies that $\dfrac{x}{y} + g'(y) = \dfrac{x}{y} + e^y$, or $g'(y) = e^y$, or $g(y) = e^y$. Thus the solution is given by $\sin x + x\ln y + e^y = C$.

39.   The condition $F_x = M$ implies that $F(x, y) = \int 3x^2 y^3 + y^4 \, dx = x^3 y^3 + xy^4 + g(y)$, and then the condition $F_y = N$ implies that $3x^3 y^2 + 4xy^3 + g'(y) = 3x^3 y^2 + y^4 + 4xy^3$, or $g'(y) = y^4$, or $g(y) = \frac{1}{5} y^5$. Thus the solution is given by $x^3 y^3 + xy^4 + \frac{1}{5} y^5 = C$.

41.   The condition $F_x = M$ implies that $F(x, y) = \int \frac{2x}{y} - \frac{3y^2}{x^4} \, dx = \frac{x^2}{y} + \frac{y^2}{x^3} + g(y)$, and then the condition $F_y = N$ implies that $-\frac{x^2}{y^2} + \frac{2y}{x^3} + g'(y) = -\frac{x^2}{y^2} + \frac{2y}{x^3} + \frac{1}{\sqrt{y}}$, or $g'(y) = \frac{1}{\sqrt{y}}$, or $g(y) = 2\sqrt{y}$. Thus the solution is given by $\frac{x^2}{y} + \frac{y^2}{x^3} + 2\sqrt{y} = C$.

In Problems 43-47 either the dependent variable y or the independent variable x (or both) is missing, and so we use the substitutions in equations (34) and/or (36) of the text to reduce the given differential equation to a first-order equation for $p = y'$.

43.   Since the dependent variable y is missing, we can substitute $y' = p$ and $y'' = p'$ as in Equation (34) of the text. This leads to $xp' = p$, a separable equation for p as a function of x. Separating variables gives $\int \frac{dp}{p} = \int \frac{dx}{x}$, or $\ln p = \ln x + \ln C$, or $p = Cx$, that is, $y' = Cx$. Finally, integrating gives the solution $y(x) = \frac{1}{2} Cx^2 + B$, which we rewrite as $y(x) = Ax^2 + B$.

45.   Since the independent variable x is missing, we can substitute $y' = p$ and $y'' = p\frac{dp}{dy}$ as in Equation (36) of the text. This leads to $p\frac{dp}{dy} + 4y = 0$, or $\int p \, dp = -\int 4y \, dy$, or $\frac{1}{2} p^2 = -2y^2 + C$, or $p = \sqrt{2C - 4y^2} = 2\sqrt{C - y^2}$ (replacing $\frac{C}{2}$ simply with C in the last step). Thus $\frac{dy}{dx} = 2\sqrt{C - y^2}$. Separating variables once again yields $\int \frac{dy}{2\sqrt{C - y^2}} = \int dx$, or $\int \frac{dy}{2\sqrt{k^2 - y^2}} = \int dx$, upon replacing C with k^2. Integrating gives

$x = \int \frac{dy}{2\sqrt{k^2 - y^2}} = \frac{1}{2} \sin^{-1} \frac{y}{k} + D$; solving for y leads to the solution

$$y(x) = k \sin(2x - 2D) = k(\sin 2x \cos 2D - \cos 2x \sin 2D),$$

or simply $y(x) = A\cos 2x + B\sin 2x$. (A much easier method of solution for this equation will be introduced in Chapter 3.)

47.   Since the dependent variable y is missing, we can substitute $y' = p$ and $y'' = p'$ as in Equation (34) of the text. This leads to $p' = p^2$, a separable equation for p as a function of x. Separating variables gives $\int \dfrac{dp}{p^2} = \int x\,dx$, or $-\dfrac{1}{p} = x + B$, or $p = -\dfrac{1}{x+B}$, that is, $\dfrac{dy}{dx} = -\dfrac{1}{x+B}$. Finally, integrating gives the solution $y(x) = A - \ln|x + B|$.

Alternatively, since the independent variable x is also missing, we can instead substitute $y' = p$ and $y'' = p\dfrac{dp}{dy}$ as in Equation (36) of the text. This leads to $p\dfrac{dp}{dy} = p^2$, or $\int \dfrac{dp}{p} = \int dy$, or $\ln p = y + C$, or $p = Ce^y$, that is, $\dfrac{dy}{dx} = Ce^y$. Separating variables once again leads to $\int e^{-y}\,dy = C\int dx$, or $-e^{-y} = Cx + D$, or

$$y = -\ln(Cx + D) = -\ln\left[C\left(x + \dfrac{D}{C}\right)\right] = -\ln C - \ln\left(x + \dfrac{D}{C}\right).$$

Putting $A = -\ln C$ and $B = \dfrac{D}{C}$ gives the same solution as found above.

49.   Since the independent variable x is missing, we can substitute $y' = p$ and $y'' = p\dfrac{dp}{dy}$ as in Equation (36) of the text. This leads to $py\dfrac{dp}{dy} + 4p^2 = yp$, or $y\dfrac{dp}{dy} + p = y$, a linear equation for p as a function as a function of y which we can rewrite as $D_y(y \cdot p) = y$, or $y \cdot p = \dfrac{1}{2}y^2 + C$, or $p = \dfrac{y^2 + C}{2y}$, that is, $\dfrac{dy}{dx} = \dfrac{y^2 + C}{2y}$. Separating variables leads to $\int \dfrac{2y}{y^2 + C}\,dy = \int dx$, or $x = \int \dfrac{2y\,dy}{y^2 + C} = \ln(y^2 + C) + B$. Solving for y leads to the solution $y^2 + C = e^{x+B} = Be^x$, or finally $y(x) = \pm\sqrt{A + Be^x}$.

51.   Since the independent variable x is missing, we can substitute $y' = p$ and $y'' = p\dfrac{dp}{dy}$ as in Equation (36) of the text. This leads to $pp' = 2yp^3$, or $p' = 2yp^2$, or $\int \dfrac{1}{p^2}\,dp = \int 2y\,dy$, or $-\dfrac{1}{p} = y^2 + C$, or $p = -\dfrac{1}{y^2 + C}$, that is, $\dfrac{dy}{dx} = -\dfrac{1}{y^2 + C}$. Separating variables once again leads to $\int y^2 + C\,dy = -\int dx$, or $\dfrac{1}{3}y^3 + Cy = -x + D$, or finally the solution

$y^3 + 3x + Ay + B = 0$.

53.   Since the independent variable x is missing, we can substitute $y' = p$ and $y'' = p\dfrac{dp}{dy}$ as in

Equation (36) of the text. This leads to $pp' = 2yp$, or $\int dp = \int 2y\,dy$, or $p = y^2 + A$,

that is, $\dfrac{dy}{dx} = y^2 + A$. Separating variables once again yields $\int \dfrac{1}{y^2 + A}\,dy = \int dx$, or

$A\arctan\dfrac{y}{A} = x + C$, or $\dfrac{y}{A} = \tan\left(Ax + B\right)$, or finally the solution $y(x) = A\tan\left(Ax + B\right)$.

55.   The proposed substitution $v = ax + by + c$ implies that $y = \dfrac{1}{b}(v - ax - c)$, so that

$y' = \dfrac{1}{b}(v' - a)$. Substituting into the given differential equation gives $\dfrac{1}{b}(v' - a) = F(v)$,

that is $\dfrac{dv}{dx} = bF(v) + a$, a separable equation for v as a function of x.

Problems 57-62 illustrate additional substitutions that are helpful in solving certain types of first-order differential equation.

57.   The proposed substitution $v = \ln y$ implies that $y = e^v$, and thus that $\dfrac{dy}{dx} = e^v\dfrac{dv}{dx}$. Substi-

tuting into the given equation yields $e^v\dfrac{dv}{dx} + P(x)e^v = Q(x)ve^v$. Cancellation of the fac-

tor e^v then yields the linear differential equation $\dfrac{dv}{dx} - Q(x)v = P(x)$.

59.   The substitution $y = v + k$ implies that $\dfrac{dy}{dx} = \dfrac{dv}{dx}$, leading to

$$\dfrac{dv}{dx} = \dfrac{x - (v + k) - 1}{x + (v + k) + 3} = \dfrac{x - v - (k + 1)}{x + v + (k + 3)}.$$

Likewise the substitution $x = u + h$ implies that $u = x - h$ and thus that $\dfrac{dv}{dx} = \dfrac{dv}{du}\dfrac{du}{dx} = \dfrac{dv}{du}$

(since $\dfrac{du}{dx} = 1$), giving

$$\dfrac{dv}{du} = \dfrac{(u + h) - (v + k) - 1}{(u + h) + (v + k) + 3} = \dfrac{u - v + (h - k - 1)}{u + v + (h + k + 3)}.$$

Thus h and k must be chosen to satisfy the system

$$h - k - 1 = 0$$
$$h + k + 3 = 0,$$

which means that $h = -1$ and $k = -2$. These choices for h and k lead to the homogeneous equation

$$\frac{dv}{du} = \frac{u - v}{u + v} = \frac{1 - \dfrac{v}{u}}{1 + \dfrac{v}{u}},$$

which calls for the further substitution $p = \dfrac{v}{u}$, so that $v = pu$ and thus $\dfrac{dv}{du} = p + u\dfrac{dp}{du}$.

Substituting gives $p + u\dfrac{dp}{du} = \dfrac{1 - p}{1 + p}$, or

$$u\frac{dp}{du} = \frac{1 - p}{1 + p} - \frac{p + p^2}{1 + p} = \frac{1 - 2p - p^2}{1 + p}.$$

Separating variables yields $\displaystyle\int \frac{1 + p}{1 - 2p - p^2}\, dp = \int \frac{1}{u}\, du$, or $-\dfrac{1}{2}\ln\left(1 - 2p - p^2\right) = \ln u + C$,

or $\left(p^2 + 2p - 1\right)u^2 = C$. Back-substituting $\dfrac{v}{u}$ for p leads to $\left[\left(\dfrac{v}{u}\right)^2 + 2\dfrac{v}{u} - 1\right]u^2 = C$, or

$v^2 + 2uv - u^2 = C$. Finally, back-substituting $x + 1$ for u and $y + 2$ for v gives the implicit solution

$$(y + 2)^2 + 2(x + 1)(y + 2) - (x + 1)^2 = C,$$

which reduces to $y^2 + 2xy - x^2 + 2x + 6y = C$.

61. The expression $x - y$ appearing on the right-hand side suggests that we try the substitution $v = x - y$, which implies that $y = x - v$, and thus that $\dfrac{dy}{dx} = 1 - \dfrac{dv}{dx}$. This gives the separable equation $1 - \dfrac{dv}{dx} = \sin v$, or $\dfrac{dv}{dx} = 1 - \sin v$. Separating variables leads to

$\displaystyle\int \frac{1}{1 - \sin v}\, dv = \int dx$. The left-hand integral is carried out with the help of the trigonometric identities

$$\frac{1}{1 - \sin v} = \frac{1 + \sin v}{\cos^2 v} = \sec^2 v + \sec v \tan v;$$

the solution is given by $\displaystyle\int \sec^2 v + \sec v \tan v\, dv = \int dx$, or $x = \tan v + \sec v + C$. Finally, back-substituting $x - y$ for v gives the implicit solution $x = \tan(x - y) + \sec(x - y) + C$. However, for no value of the constant C does this general solution include the "basic" solution $y(x) = x - \dfrac{\pi}{2}$. The reason is that for this solution, $v = x - y$ is the constant $\dfrac{\pi}{2}$, so that the expression $1 - \sin v$ (by which we divided above) is identically zero. Thus the solution $y(x) = x - \dfrac{\pi}{2}$ is singular for this solution procedure.

63. The substitution $y = y_1 + \dfrac{1}{v}$, which implies that $\dfrac{dy}{dx} = y_1' - \dfrac{1}{v^2}\dfrac{dv}{dx}$, gives

$$y_1' - \frac{1}{v^2}\frac{dv}{dx} = A(x)\left(y_1 + \frac{1}{v}\right)^2 + B(x)\left(y_1 + \frac{1}{v}\right) + C(x),$$

which upon expanding becomes

$$\underline{y_1'} - \frac{1}{v^2}\frac{dv}{dx} = A(x)\left(y_1^2 + 2\frac{y_1}{v} + \frac{1}{v^2}\right) + B(x)y_1 + B(x)\frac{1}{v} + C(x)$$

$$= \underline{A(x)y_1^2 + B(x)y_1 + C(x)} + A(x)\left(2\frac{y_1}{v} + \frac{1}{v^2}\right) + B(x)\frac{1}{v}.$$

The underlined terms cancel because y_1 is a solution of the given equation $\dfrac{dy}{dx} = A(x)y^2 + B(x)y + C(x)$, resulting in

$$-\frac{1}{v^2}\frac{dv}{dx} = A(x)\left(2\frac{y_1}{v} + \frac{1}{v^2}\right) + B(x)\frac{1}{v},$$

or $\dfrac{dv}{dx} = -A(x)(2vy_1 + 1) - B(x)v$, that is, $\dfrac{dv}{dx} + (B + 2Ay_1)v = -A$, a linear equation for v as a function of x.

In Problem 65 we outline the application of the method of Problem 63 to the given Riccati equation.

65. Here $A(x) = 1$, $B(x) = -2x$, and $C(x) = 1 + x^2$. Thus the substitution $y = y_1 + \dfrac{1}{v} = x + \dfrac{1}{v}$ yields the trivial linear equation $\dfrac{dv}{dx} = -1$, with immediate solution $v(x) = C - x$. Hence the general solution of our Riccati equation is given by

$$y(x) = x + \frac{1}{C - x}.$$

67. First, the line $y = Cx - \dfrac{1}{4}C^2$ has slope C and passes through the point $\left(\tfrac{1}{2}C, \tfrac{1}{4}C^2\right)$; the same is true of the parabola $y = x^2$ at the point $\left(\tfrac{1}{2}C, \tfrac{1}{4}C^2\right)$, because

$\dfrac{dy}{dx} = 2x = 2 \cdot \frac{1}{2}C = C$. Thus the line is tangent to the parabola at this point. It follows that $y = x^2$ is in fact a solution to the differential equation, since for each x, the parabola has the same values of y and y' as the known solution $y = Cx - \dfrac{1}{4}C^2$. Finally, $y = x^2$ is a singular solution with respect to the general solution $y = Cx - \dfrac{1}{4}C^2$, since for no value of C does $Cx - \dfrac{1}{4}C^2$ equal x^2 for all x.

69.   With $a = 100$ and $k = \dfrac{1}{10}$, Equation (19) in the text is $y = 50\left[\left(\dfrac{x}{100}\right)^{9/10} - \left(\dfrac{x}{100}\right)^{11/10}\right]$. We find the maximum northward displacement of plane by setting

$$y'(x) = 50\left[\dfrac{9}{10}\left(\dfrac{x}{100}\right)^{-1/10} - \dfrac{11}{10}\left(\dfrac{x}{100}\right)^{1/10}\right] = 0,$$

which yields $\left(\dfrac{x}{100}\right)^{1/10} = \left(\dfrac{9}{11}\right)^{1/2}$. Because

$$y''(x) = 50\left[\dfrac{-9}{100}\left(\dfrac{x}{100}\right)^{-11/10} - \dfrac{11}{100}\left(\dfrac{x}{100}\right)^{-9/10}\right] < 0$$

for all x, this critical point in fact represents the absolute maximum value of y. Substituting this value of x into $y(x)$ gives $y_{max} = 50\left[\left(\dfrac{9}{11}\right)^{9/2} - \left(\dfrac{9}{11}\right)^{11/2}\right] \approx 3.68\,\text{mi}$.

71.   Equations (12)-(19) apply to this situation as with the airplane in flight.

(a) With $a = 100$ and $k = \dfrac{w}{v_0} = \dfrac{2}{4} = \dfrac{1}{2}$, the solution given by Equation (19) is

$y = 50\left[\left(\dfrac{x}{100}\right)^{1/2} - \left(\dfrac{x}{100}\right)^{3/2}\right]$. The fact that $y(0) = 0$ means that this trajectory goes through the origin where the tree is located.

(b) With $k = \dfrac{4}{4} = 1$, the solution is $y = 50\left[1 - \left(\dfrac{x}{100}\right)^2\right]$, and we see that the dog hits the bank at a distance $y(0) = 50\,\text{ft}$ north of the tree.

(c) With $k = \dfrac{6}{4} = \dfrac{3}{2}$, the solution is $y = 50\left[\left(\dfrac{x}{100}\right)^{-1/2} - \left(\dfrac{x}{100}\right)^{5/2}\right]$. This trajectory is asymptotic to the positive x-axis, so we see that the dog never reaches the west bank of the river.

CHAPTER 1 Review Problems

The main objective of this set of review problems is practice in the identification of the different types of first-order differential equations discussed in this chapter. In each of Problems 1–36 we identify the type of the given equation and indicate one or more appropriate method(s) of solution.

1.   We first rewrite the differential equation for $x > 0$ as $y' - \dfrac{3}{x} y = x^2$, showing that the equation is *linear*. An integrating factor is given by $\rho = \exp\left(-\int \dfrac{3}{x} dx\right) = e^{-3\ln x} = x^{-3}$, and multiplying the equation by ρ gives $x^{-3} \cdot y' - 3x^{-4} y = x^{-1}$, or $D_x\left(x^{-3} \cdot y\right) = x^{-1}$. Integrating then leads to $x^{-3} \cdot y = \ln x + C$, and thus to the general solution $y = x^3\left(\ln x + C\right)$.

3.   Rewriting the differential equation for $x \neq 0$ as $y' = \dfrac{xy + y^2}{x^2} = \dfrac{y}{x} + \left(\dfrac{y}{x}\right)^2$ shows that the equation is *homogeneous*. Actually the equation is identical to Problem 9 in Section 1.6; the general solution found there is $y = \dfrac{x}{C - \ln|x|}$.

5.   We first rewrite the differential equation for $x, y \neq 0$ as $\dfrac{y'}{y} = \dfrac{2x - 3}{x^4}$, showing that the equation is *separable*. Separating variables yields $\displaystyle\int \dfrac{1}{y} dy = \int \dfrac{2x - 3}{x^4} dx$, or $\ln|y| = -\dfrac{1}{x^2} + \dfrac{1}{x^3} + C = \dfrac{1 - x}{x^3} + C$, leading to the general solution $y = C\exp\left(\dfrac{1 - x}{x^3}\right)$, where C is an arbitrary nonzero constant.

7.   We first rewrite the differential equation for $x > 0$ as $y' + \dfrac{2}{x} y = \dfrac{1}{x^3}$, showing that the equation is *linear*. An integrating factor is given by $\rho = \exp\left(\int \dfrac{2}{x} dx\right) = e^{2\ln x} = x^2$, and multiplying the equation by ρ gives $x^2 \cdot y' + 2xy = \dfrac{1}{x}$, or $D_x\left(x^2 \cdot y\right) = \dfrac{1}{x}$. Integrating then leads to $x^2 \cdot y = \ln x + C$, and thus to the general solution $y = \dfrac{\ln x + C}{x^2}$.

9.   We first rewrite the differential equation for $x, y > 0$ as $y' + \dfrac{2}{x}y = 6xy^{1/2}$, showing that it is a *Bernoulli* equation with $n = 1/2$. The substitution $v = y^{1/2}$ implies that $y = v^2$ and thus that $y' = 2vv'$. Substituting gives $2vv' + \dfrac{2}{x}v^2 = 6xv$, or $v' + \dfrac{1}{x}v = 3x$, a linear equation for v as a function of x. An integrating factor is given by $\rho = \exp\left(\displaystyle\int \dfrac{1}{x}dx\right) = x$, and multiplying the differential equation by ρ gives $xv' + v = 3x^2$, or $D_x(xv) = 3x^2$. Integrating then leads to $xv = x^3 + C$, or $v = x^2 + \dfrac{C}{x}$. Finally, back-substituting $y^{1/2}$ for v gives the general solution $y^{1/2} = x^2 + \dfrac{C}{x}$, or $y = \left(x^2 + \dfrac{C}{x}\right)^2$.

11.   We first rewrite the differential equation for $x, y > 0$ as $\dfrac{dy}{dx} = \dfrac{y}{x} + 3\left(\dfrac{y}{x}\right)^2$, showing that it is *homogeneous*. Substituting $v = \dfrac{y}{x}$ then gives $v + x\dfrac{dv}{dx} = v + 3v^2$, or $x\dfrac{dv}{dx} = 3v^2$. Separating variables leads to $\displaystyle\int \dfrac{1}{v^2}dv = \int \dfrac{3}{x}dx$, or $-\dfrac{1}{v} = 3\ln x + C$, or $v = \dfrac{1}{C - 3\ln x}$. Back-substituting $\dfrac{y}{x}$ for v then gives the solution $\dfrac{y}{x} = \dfrac{1}{C - 3\ln x}$, or $y = \dfrac{x}{C - 3\ln x}$.

Alternatively, writing the equation in the form $y' - \dfrac{1}{x}y = \dfrac{3}{x^2}y^2$ for $x, y > 0$ shows that it is also a *Bernoulli* equation with $n = 2$. The substitution $v = y^{-1}$ implies that $y = v^{-1}$ and thus that $y' = -v^{-2}v'$. Substituting gives $-v^{-2}v' - \dfrac{1}{x}v^{-1} = \dfrac{3}{x^2}v^{-2}$, or $v' + \dfrac{1}{x}v = -\dfrac{3}{x^2}$, a linear equation for v as a function of x. An integrating factor is given by $\rho = \exp\left(\displaystyle\int \dfrac{1}{x}dx\right) = x$, and multiplying the differential equation by ρ gives $xv' + v = -\dfrac{3}{x}$, or $D_x(xv) = -\dfrac{3}{x}$. Integrating then leads to $xv = -3\ln x + C$, or $v = \dfrac{-3\ln x + C}{x}$. Finally, back-substituting y^{-1} for v gives the same general solution as found above.

13. We first rewrite the differential equation for $y > 0$ as $\dfrac{y'}{y^2} = 5x^4 - 4x$, showing that the

equation is *separable*. Separating variables yields $\displaystyle\int \frac{1}{y^2}\,dy = \int 5x^4 - 4x\,dx$, or

$-\dfrac{1}{y} = x^5 - 2x^2 + C$, leading to the general solution $y = \dfrac{1}{C + 2x^2 - x^5}$.

15. This is a *linear* differential equation. An integrating factor is given by

$\rho = \exp\!\left(\displaystyle\int 3\,dx\right) = e^{3x}$, and multiplying the equation by ρ gives $e^{3x} \cdot y' + 3e^{3x}y = 3x^2$, or

$D_x\!\left(e^{3x} \cdot y\right) = 3x^2$. Integrating then leads to $e^{3x} \cdot y = x^3 + C$, and thus to the general solu-

tion $y = \left(x^3 + C\right)e^{-3x}$.

17. Rewriting the differential equation in differential form gives

$$\left(e^x + ye^{xy}\right)dx + \left(e^y + xe^{xy}\right)dy = 0,$$

and because $\dfrac{\partial}{\partial y}\!\left(e^x + ye^{xy}\right) = xye^{xy} = \dfrac{\partial}{\partial x}\!\left(e^y + xe^{xy}\right)$, the given equation is *exact*. We ap-

ply the method of Example 9 in Section 1.6 to find a solution in the form $F(x,y) = C$.

First, the condition $F_x = M$ implies that

$$F(x,y) = \int e^x + ye^{xy}\,dx = e^x + e^{xy} + g(y),$$

and then the condition $F_y = N$ implies that $xe^{xy} + g'(y) = e^y + xe^{xy}$, or $g'(y) = e^y$, or

$g(y) = e^y$. Thus the solution is given by $e^x + e^{xy} + e^y = C$.

19. We first rewrite the differential equation for $x, y \neq 0$ as $\dfrac{y'}{y^2} = 2x^{-3} - 3x^2$, showing that

the equation is *separable*. Separating variables yields $\displaystyle\int \frac{1}{y^2}\,dy = \int 2x^{-3} - 3x^2\,dx$, or

$-\dfrac{1}{y} = -x^{-2} - x^3 + C$, leading to the general solution $y = \dfrac{1}{x^{-2} + x^3 + C} = \dfrac{x^2}{x^5 + Cx^2 + 1}$.

21. We first rewrite the differential equation for $x > 1$ as $y' + \dfrac{1}{x+1}y = \dfrac{1}{x^2 - 1}$, showing that

the equation is *linear*. An integrating factor is given by $\rho = \exp\!\left(\displaystyle\int \frac{1}{x+1}\,dx\right) = x+1$, and

multiplying the equation by ρ gives $(x+1)y' + y = \dfrac{1}{x-1}$, or $D_x\!\left[(x+1)y\right] = \dfrac{1}{x-1}$. Integrating then leads to $(x+1)y = \ln(x-1)+C$, and thus to the general solution

$$y = \frac{1}{x+1}\Big[\ln(x-1)+C\Big].$$

23. Rewriting the differential equation in differential form gives

$$\big(e^y + y\cos x\big)dx + \big(xe^y + \sin x\big)dy = 0,$$

and because $\dfrac{\partial}{\partial y}\big(e^y + y\cos x\big) = e^y + \cos x = \dfrac{\partial}{\partial x}\big(xe^y + \sin x\big)$, the given equation is exact.

We apply the method of Example 9 in Section 1.6 to find a solution in the form $F(x,y) = C$. First, the condition $F_x = M$ implies that

$$F(x,y) = \int e^y + y\cos x\, dx = xe^y + y\sin x + g(y),$$

and then the condition $F_y = N$ implies that $xe^y + \sin x + g'(y) = xe^y + \sin x$, or $g'(y) = 0$, that is, g is constant. Thus the solution is given by $xe^y + y\sin x = C$.

25. We first rewrite the differential equation for $x > -1$ as $y' + \dfrac{2}{x+1}y = 3$, showing that the

equation is *linear*. An integrating factor is given by $\rho = \exp\!\left(\displaystyle\int \frac{2}{x+1}dx\right) = (x+1)^2$, and

multiplying the equation by ρ gives $(x+1)^2\, y' + 2(x+1)y = 3(x+1)^2$, or

$D_x\Big[(x+1)^2 \cdot y\Big] = 3(x+1)^2$. Integrating then leads to $(x+1)^2 \cdot y = (x+1)^3 + C$, and thus

to the general solution $y = x+1+\dfrac{C}{(x+1)^2}$.

27. Writing the given equation for $x > 0$ as $\dfrac{dy}{dx} + \dfrac{1}{x}y = -\dfrac{x^2}{3}y^4$ shows that it is a *Bernoulli*

equation with $n = 4$. The substitution $v = y^{-3}$ implies that $y = v^{-1/3}$ and thus that

$y' = -\dfrac{1}{3}v^{-4/3}v'$. Substituting gives $-\dfrac{1}{3}v^{-4/3}v' + \dfrac{1}{x}v^{-1/3} = -\dfrac{x^2}{3}v^{-4/3}$, or $v' - \dfrac{3}{x}v = x^2$, a linear equation for v as a function of x. An integrating factor is given by

$\rho = \exp\!\left(\displaystyle\int -\frac{3}{x}dx\right) = x^{-3}$, and multiplying the differential equation by ρ gives

$x^{-3}\cdot v' - 3x^{-4}v = x^{-1}$, or $D_x\big(x^{-3}\cdot v\big) = x^{-1}$. Integrating then leads to $x^{-3}\cdot v = \ln x + C$, or

$v = x^3\big(\ln x + C\big)$. Finally, back-substituting y^{-3} for v gives the general solution

$y = x^{-1}\big(\ln x + C\big)^{-1/3}$.

29. We first rewrite the differential equation for $x > -\dfrac{1}{2}$ as $y' + \dfrac{1}{2x+1}y = (2x+1)^{1/2}$, show-ing that the equation is linear. An integrating factor is given by

$\rho = \exp\left(\displaystyle\int \dfrac{1}{2x+1}\,dx\right) = (2x+1)^{1/2}$, and multiplying the equation by ρ gives

$(2x+1)^{1/2}\,y' + (2x+1)^{-1/2}\,y = 2x+1$, or $D_x\left[(2x+1)^{1/2}\cdot y\right] = 2x+1$. Integrating then

leads to $(2x+1)^{1/2}\cdot y = x^2 + x + C$, and thus to the general solution

$y = \left(x^2 + x + C\right)(2x+1)^{-1/2}$.

31. Rewriting the differential equation as $y' - 3x^2 y = 21x^2$ shows that it is *linear*. An inte-grating factor is given by $\rho = \exp\left(\displaystyle\int -3x^2\,dx\right) = e^{-x^3}$, and multiplying the equation by ρ

gives $e^{-x^3}\cdot y' - 3x^2 e^{-x^3} y = 21x^2 e^{-x^3}$, or $D_x\left(e^{-x^3}\cdot y\right) = 21x^2 e^{-x^3}$. Integrating then leads to

$e^{-x^3}\cdot y = -7e^{-x^3} + C$, and thus to the general solution $y = -7 + Ce^{x^3}$.

Alternatively, writing the equation for $y > -7$ as $\dfrac{dy}{y+7} = 3x^2\,dx$ shows that it is *separa-*

ble. Integrating yields the general solution $\ln(y+7) = x^3 + C$, that is, $y = Ce^{x^3} - 7$, as found above.

(Note that the restriction $y > -7$ in the second solution causes no loss of generality. The general solution as found by the first method shows that either $y < -7$ for all x or $y > -7$ for all x. Of course, the second solution could be carried out under the assumption $y < -7$ as well.)

33. Rewriting the differential equation for $x, y > 0$ in differential form gives

$$\left(3x^2 + 2y^2\right)dx + 4xy\,dy = 0,$$

and because $\dfrac{\partial}{\partial y}\left(3x^2 + 2y^2\right) = 4y = \dfrac{\partial}{\partial x}4xy$, the given equation is *exact*. We apply the

method of Example 9 in Section 1.6 to find a solution in the form $F(x, y) = C$. First, the condition $F_x = M$ implies that

$$F(x, y) = \int 3x^2 + 2y^2\,dx = x^3 + 2xy^2 + g(y),$$

and then the condition $F_y = N$ implies that $4xy + g'(y) = 4xy$, or $g'(y) = 0$, that is, g is constant. Thus the solution is given by $x^3 + 2xy^2 = C$.

Alternatively, rewriting the given equation for $x, y > 0$ as $\dfrac{dy}{dx} = -\dfrac{3}{4}\dfrac{x}{y} - \dfrac{1}{2}\dfrac{y}{x}$ shows that it

is *homogeneous*. Substituting $v = \dfrac{y}{x}$ then gives $v + x\dfrac{dv}{dx} = -\dfrac{3}{4v} - \dfrac{1}{2}v$, or

$x\dfrac{dv}{dx} = -\dfrac{3}{4v} - \dfrac{3}{2}v = -\dfrac{3+6v^2}{4v}$. Separating variables leads to $\displaystyle\int \dfrac{4v}{6v^2+3}dv = -\int\dfrac{1}{x}dx$, or

$\ln(6v^2+3) = -3\ln x + C$, or $(2v^2+1)x^3 = C$. Back-substituting $\dfrac{y}{x}$ for v then gives the

solution $\left[2\left(\dfrac{y}{x}\right)^2 + 1\right]x^3 = C$, or finally $2y^2x + x^3 = C$, as found above.

Still another solution arises from writing the differential equation for $x, y > 0$ as

$\dfrac{dy}{dx} + \dfrac{1}{2x}y = -\dfrac{3x}{4}y^{-1}$, which shows that it is *Bernoulli* with $n = -1$. The substitution

$v = y^2$ implies that $y = v^{1/2}$ and thus that $y' = \dfrac{1}{2}v^{-1/2}v'$. Substituting gives

$\dfrac{1}{2}v^{-1/2}v' + \dfrac{1}{2x}v^{1/2} = -\dfrac{3x}{4}v^{-1/2}$, or $v' + \dfrac{1}{x}v = -\dfrac{3x}{2}$, a linear equation for v as a function of x.

An integrating factor is given by $\rho = \exp\left(\displaystyle\int\dfrac{1}{x}dx\right) = x$, and multiplying the differential

equation by ρ gives $x\cdot v' + v = -\dfrac{3x^2}{2}$, or $D_x(x\cdot v) = -\dfrac{3x^2}{2}$. Integrating then leads to

$x\cdot v = -\dfrac{x^3}{2} + C$. Finally, back-substituting y^2 for v leads to the general solution

$x\cdot y^2 = -\dfrac{x^3}{2} + C$, that is, $2xy^2 + x^3 = C$, as found above.

35. Rewriting the differential equation as $\dfrac{dy}{dx} = \dfrac{2x}{x^2+1}(y+1)$ shows that it is *separable*. For

$y > -1$ separating variables gives $\displaystyle\int\dfrac{1}{y+1}dy = \int\dfrac{2x}{x^2+1}dx$, or $\ln(y+1) = \ln(x^2+1)+C$,

leading to the general solution $y = C(x^2+1)-1$.

Alternatively, writing the differential equation as $y' - \dfrac{2x}{x^2+1}y = \dfrac{2x}{x^2+1}$ shows that it is

linear. An integrating factor is given by $\rho = \exp\left(\displaystyle\int -\dfrac{2x}{x^2+1}dx\right) = \dfrac{1}{x^2+1}$, and multiply-

ing the equation by ρ gives $\dfrac{1}{x^2+1}y' - \dfrac{2x}{(x^2+1)^2}y = \dfrac{2x}{(x^2+1)^2}$, or

$D_x\left(\dfrac{1}{x^2+1}\,y\right)=\dfrac{2x}{\left(x^2+1\right)^2}$. Integrating then leads to $\dfrac{1}{x^2+1}\,y=-\dfrac{1}{x^2+1}+C$, or thus to the

general solution $y=-1+C\left(x^2+1\right)$ found above.

CHAPTER 2

MATHEMATICAL MODELS AND NUMERICAL METHODS

SECTION 2.1

POPULATION MODELS

Section 2.1 introduces the first of the two major classes of mathematical models studied in the textbook, and is a prerequisite to the discussion of equilibrium solutions and stability in Section 2.2. In Problems 1-8 we find the desired particular solution and sketch some typical solution curves, with the desired particular solution highlighted.

1.   Separating variables gives $\int \dfrac{1}{x(1-x)}\,dx = \int dt$. By the method of partial fractions

$$\int \frac{1}{x(1-x)}\,dx = \int \frac{1}{x} - \frac{1}{x-1}\,dx = \ln|x| - \ln|x-1|,$$

and so the general solution of the differential equation is $\ln|x| - \ln|x-1| = t + C$, or

$\dfrac{x}{x-1} = Ce^{t}$. The initial condition $x(0) = 2$ implies that $C = 2$, leading to the particular

solution $\dfrac{x}{x-1} = 2e^{t}$, or $x = 2(x-1)e^{t}$, or finally $x(t) = \dfrac{2e^{t}}{2e^{t}-1} = \dfrac{2}{2-e^{-t}}$.

Problem 1

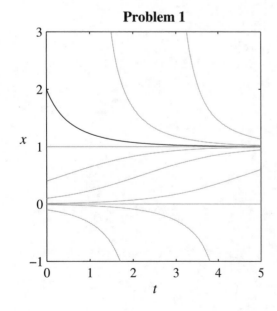

Problem 3

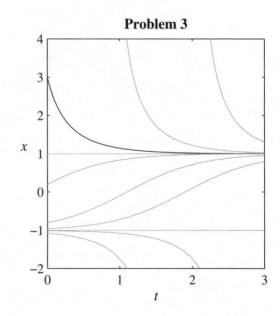

3.   Separating variables gives $\displaystyle\int \frac{1}{(1+x)(1-x)}\,dx = \int dt$. By the method of partial fractions

$$\int \frac{1}{(x+1)(x-1)}\,dx = -\frac{1}{2}\int \frac{1}{x-1} - \frac{1}{x+1}\,dx = -\frac{1}{2}\left(\ln|x-1| - \ln|x+1|\right),$$

and so the general solution of the differential equation is $\ln|x-1| - \ln|x+1| = -2t + C$, or

$\displaystyle\frac{x-1}{x+1} = Ce^{-2t}$. The initial condition $x(0) = 3$ implies that that $C = \dfrac{1}{2}$, leading to the

particular solution $\displaystyle\frac{x-1}{x+1} = \frac{1}{2}e^{-2t}$, or $2(x-1) = (x+1)e^{-2t}$, or finally

$$x(t) = \frac{2 + e^{-2t}}{2 - e^{-2t}} = \frac{2e^{2t} + 1}{2e^{2t} - 1}.$$

5.   Separating variables gives $\displaystyle\int \frac{1}{x(x-5)}\,dx = \int -3\,dt$. By the method of partial fractions,

$$\int \frac{1}{x(x-5)}\,dx = -\frac{1}{5}\int \frac{1}{x} - \frac{1}{x-5}\,dx = -\frac{1}{5}\left(\ln|x| - \ln|x-5|\right),$$

and so the general solution of the differential equation is $-\dfrac{1}{5}\left(\ln|x| - \ln|x-5|\right) = -3t + C$,

or $\displaystyle\frac{x}{x-5} = Ce^{15t}$. The initial condition $x(0) = 8$ implies that $C = \dfrac{8}{3}$, leading to the

particular solution $\displaystyle\frac{x}{x-5} = \frac{8}{3}e^{15t}$, or $3x = 8(x-5)e^{15t}$, or finally

$$x(t) = \frac{-40e^{15t}}{3 - 8e^{15t}} = \frac{40}{8 - 3e^{-15t}}.$$

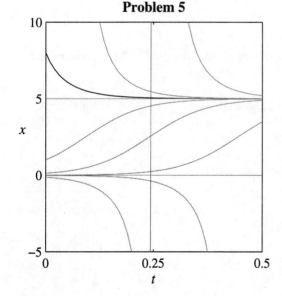

Problem 5

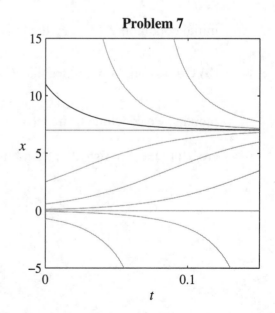

Problem 7

7. Separating variables gives $\int \dfrac{1}{x(x-7)}\,dx = \int -4\,dt$. By the method of partial fractions,

$$\int \dfrac{1}{x(x-7)}\,dx = -\dfrac{1}{7}\int \dfrac{1}{x} - \dfrac{1}{x-7}\,dx = -\dfrac{1}{7}\left(\ln|x| - \ln|x-7|\right),$$

and so the general solution of the differential equation is $\ln|x| - \ln|x-7| = 28t + C$, or

$\dfrac{x}{x-7} = Ce^{28t}$. The initial condition $x(0) = 11$ implies that $C = \dfrac{11}{4}$, leading to the

particular solution $\dfrac{x}{x-7} = \dfrac{11}{4}e^{28t}$, or $4x = 11(x-7)e^{28t}$, or finally

$$x(t) = \dfrac{-77e^{28t}}{4 - 11e^{28t}} = \dfrac{77}{11 - 4e^{-28t}}.$$

9. Substitution of $P(0) = 100$ and $P'(0) = 20$ into $P' = k\sqrt{P}$ yields $k = 2$, so the

differential equation is $P' = 2\sqrt{P}$. Separation of variables gives $\int \dfrac{1}{2\sqrt{P}}\,dP = \int dt$,

which upon integrating is $\sqrt{P} = t + C$. Then $P(0) = 100$ implies $C = 10$, so that

$P(t) = (t+10)^2$. Hence the number of rabbits after one year is $P(12) = 484$.

11. **(a)** Substituting our assumptions that $\beta = \dfrac{k_1}{\sqrt{P}}$ and $\delta = \dfrac{k_2}{\sqrt{P}}$ into the general population

equation gives $\dfrac{dP}{dt} = \dfrac{(k_1 - k_2)}{\sqrt{P}} \cdot P = k\sqrt{P}$. Separation of variables leads to

$\int \dfrac{1}{\sqrt{P}}\,dP = \int k\,dt$, which upon integrating is $2\sqrt{P} = kt + C$, or $P = \left(\dfrac{kt}{2} + C\right)^2$. The

initial condition $P(0) = P_0$ then gives $C = \sqrt{P_0}$.

(b) Our assumption implies that $C = \sqrt{P_0} = 10$, so that $P = \left(\dfrac{kt}{2} + 10\right)^2$. Measuring t in

months, we conclude from $P(6) = 169$ that $k = 1$, so that $P = \left(\dfrac{t}{2} + 10\right)^2$. Hence there

are $P(12) = 256$ fish after 12 months.

13.   **(a)** Substituting our assumptions that $\beta = k_1 P$ and $\delta = k_2 P$ into the general population

equation gives $\dfrac{dP}{dt} = \left[\left(k_1 - k_2 \right) P \right] P = kP^2$, where $k = k_1 - k_2 > 0$ by our assumption that

$\beta > \delta$. Solving as in Problem 12 leads to $P = \dfrac{1}{C - kt}$. The initial condition $P(0) = P_0$

implies that $C = \dfrac{1}{P_0}$, so that $P(t) = \dfrac{P_0}{1 - kP_0 t}$. As $t \to \dfrac{1}{kP_0}$ we find that $P(t) \to \infty$.

(b) Our assumption that $P_0 = 6$ gives $P(t) = \dfrac{6}{1 - 6kt}$. Then, with t measured in months,

we conclude from $P(10) = 9$ that $k = 180$, so that $P(t) = \dfrac{6}{1 - (t/30)} = \dfrac{180}{30 - t}$. From this

we can see that doomsday occurs after 30 months.

15.   Writing $\dfrac{dP}{dt} = bP \left(\dfrac{a}{b} - P \right)$ shows that the limiting population M is $\dfrac{a}{b}$. Then the facts that

$B_0 = aP_0$ and $D_0 = aP_0^2$ give $\dfrac{B_0 P_0}{D_0} = \dfrac{(aP_0) P_0}{bP_0^2} = \dfrac{a}{b} = M$. With Problems 16 and 17 in

mind, we note also that $a = \dfrac{B_0}{P_0}$ and $b = \dfrac{D_0}{P_0^2} = k$.

17.   The relations in Problem 15 give $k = \dfrac{D_0}{P_0^2} = \dfrac{12}{240^2} = \dfrac{1}{2400}$ and a limiting population of

$M = \dfrac{B_0 P_0}{D_0} = \dfrac{9 \cdot 240}{12} = 180$ rabbits. The solution is then

$$P(t) = \dfrac{180 \cdot 240}{240 + (180 - 240) e^{-t/15}} = \dfrac{43200}{120 - 60 e^{-t/15}},$$

again by Equation (7). Setting $P(t) = 1.05M = 189$ rabbits yields $t \approx 44.22$ months.

19.   The relations in Problem 18 give $k = \dfrac{B_0}{P_0^2} = \dfrac{10}{100^2} = \dfrac{1}{1000}$ and $M = \dfrac{D_0 P_0}{B_0} = \dfrac{9 \cdot 100}{10} = 90$.

Problem 33 below then gives the solution

$$P(t) = \dfrac{90 \cdot 100}{100 + (90 - 100) e^{9t/100}} = \dfrac{9000}{100 - 10 e^{9t/100}}.$$

Setting $P(t) = 10M = 900$ rabbits yields $t \approx 24.41$ months.

21.   Separating variables in our assumption that $\dfrac{dP}{dt} = kP(200 - P)$ gives

$$\int \frac{1}{P(200 - P)} dP = \int k \, dt.$$ By the method of partial fractions

$$\int \frac{1}{P(200 - P)} dP = \frac{1}{200} \int \frac{1}{P} + \frac{1}{200 - P} dP = \frac{1}{200}\left(\ln P - \ln|200 - P|\right),$$

and so the general solution of the differential equation is $\ln P - \ln|200 - P| = 200kt$, or

$\ln \dfrac{P}{|200 - P|} = 200kt + C$, or $\dfrac{P}{200 - P} = Ce^{200kt}$. The initial condition $P(0) = 100$ (taking

$t = 0$ in 1960) implies that $C = 1$. Further, $P'(0) = 1$, when substituted into the original

differential equation along with $P(0) = 100$, implies that $1 = k \cdot 100(200 - 100)$, or

$k = \dfrac{1}{10000}$. Substituting these values into the general solution gives $\dfrac{P}{200 - P} = e^{t/50}$, or

$P = e^{t/50}(200 - P)$, or $P(t) = \dfrac{200}{1 + e^{-t/50}}$. Finally, in the year 2020 the country's

population will be $P(60) = \dfrac{200}{1 + e^{-6/5}} \approx 153.7$ million.

23.   **(a)** The given differential equation implies that

$$x' = 0.8x - 0.004x^2 = 0.004x(200 - x),$$

which is positive for $0 < x < 200$ and negative for $x > 200$; thus the maximum amount
that will dissolve is $M = 200$ g.

(b) Since the given equation conforms to Equation (6) in the text, the solution is given
there by Equation (7), with $M = 200$, $P_0 = 50$, and $k = 0.004$:

$$x(t) = \frac{10000}{50 + 150e^{-0.8t}}.$$

Substituting $x = 100$, we solve for $t = 1.25 \ln 3 \approx 1.37$ sec.

25. **(a)** Following the suggestions (and thus taking $t = 0$ in 1925), we estimate the rate of population growth in 1925 to be

$$P'(0) = \frac{P(1) - P(-1)}{2} = \frac{25.38 - 24.63}{2} = 0.375$$

million people annually. The corresponding estimate for the year 1975, corresponding to $t = 50$, is

$$P'(50) = \frac{P(51) - P(49)}{2} = \frac{48.04 - 47.04}{2} = 0.5$$

million people annually. Substituting these values, together with $P(0) = 25$ and $P(50) = 47.54$, into the logistic equation (3) leads to the system of equations

$$0.375 = 25k(M - 25)$$
$$0.5 = 47.54k(M - 47.54)^{\cdot}$$

As in Example 3 in the text, we solve these equations to find $M = 100$ and $k = 0.0002$. Then Equation (7) gives the population function

$$P(t) = \frac{100 \cdot 25}{25 + (100 - 25)e^{-0.0002 \cdot 100t}} = \frac{2500}{25 + 75e^{-0.02t}}.$$

(b) We find that $P = 75$ when $t = 50 \ln 9 \approx 110$, that is, in 2035 A.D.

27. Our assumptions lead to the differential equation $\dfrac{dP}{dt} = kP^2 - 0.01P$ for the animal population $P(t)$. Substituting $P(0) = 200$ and $P'(0) = 2$, we find that $k = 0.0001$, so that

$$\frac{dP}{dt} = 0.0001P^2 - 0.01P = 0.0001P(P - 100).$$

Separating variables gives $\displaystyle\int \frac{1}{P(P - 100)}\, dP = \int 0.0001\, dt$. By the method of partial fractions

$$\int \frac{1}{P(P - 100)}\, dP = -\frac{1}{100}\int \frac{1}{P} - \frac{1}{P - 100}\, dP = \frac{1}{100}\left(\ln|P - 100| - \ln P\right),$$

and so the general solution of the differential equation is

$$\frac{1}{100}\ln|P - 100| - \ln P = 0.0001t + C,$$

or $\ln\dfrac{|P - 100|}{P} = \dfrac{t}{100} + C$, or $\dfrac{P - 100}{P} = Ce^{t/100}$. The initial condition $P(0) = 200$ gives $C = \dfrac{1}{2}$, and so $\dfrac{P - 100}{P} = \dfrac{1}{2}e^{t/100}$, leading to the general solution $P(t) = \dfrac{200}{2 - e^{t/100}}$.

(a) Setting $P = 1000$ gives $t = 100 \ln \dfrac{9}{5} \approx 58.78$ months.

(b) Doomsday occurs as the denominator $2 - e^{t/100}$ approaches zero, that is, as t approaches $100 \ln 2 \approx 69.31$ months, since the population P becomes infinite then.

29. Here we have the logistic equation

$$\frac{dP}{dt} = 0.03135P - 0.0001489P^2 = 0.0001489P(210.544 - P),$$

where $k = 0.0001489$ and $P = 210.544$. With $P_0 = 3.9$ as well, Eq. (7) in the text gives

$$P(t) = \frac{(210.544)(3.9)}{3.9 + (210.544 - 3.9)e^{-(0.0001489)(210.544)t}} = \frac{821.122}{3.9 + 206.644e^{-0.03135t}}.$$

(a) This solution gives $P(140) \approx 127.008$, fairly close to the actual 1930 U.S. census figure of 123.2 million.

(b) As t grows without bound, $P(t)$ approaches $\dfrac{821.122}{3.9} = 210.544$ million.

(c) Since the actual U.S. population in 2000 was about 281 million—already exceeding the maximum population predicted by the logistic equation—we see that that this model did *not* continue to hold throughout the 20th century.

31. Substituting $P(0) = 10^6$ and $P'(0) = 3 \times 10^5$ into the differential equation $P'(t) = \beta_0 e^{-\alpha t}P$ yields $\beta_0 = 0.3$. Hence the solution given in Problem 30 is

$$P(t) = P_0 \exp\left[\frac{0.3}{\alpha}\left(1 - e^{-\alpha t}\right)\right].$$ The fact that $P(6) = 2P_0$ now yields the equation

$$0.3\left(1 - e^{-6\alpha}\right) - \alpha \ln 2 = 0,$$

which we seek to solve for the constant α. We let $f(\alpha)$ denote the left-hand side $0.3\left(1 - e^{-6\alpha}\right) - \alpha \ln 2$ of this equation and apply Newton's iterative formula

$$\alpha_{n+1} = \alpha_n - \frac{f(\alpha_n)}{f'(\alpha_n)}$$

with initial guess $\alpha_0 = 1$ (suggested by a plot of $f(\alpha)$), leading quickly to $\alpha \approx 0.3915$. Therefore the limiting cell population as t grows without bound is

$$P_0 \exp\left(\frac{\beta_0}{\alpha}\right) = 10^6 \exp\left(\frac{0.3}{0.3915}\right) \approx 2.15 \times 10^6.$$

Thus the tumor does not grow much further after 6 months.

33.   **(a)** Separating variables in the extinction-explosion equation gives

$\int \dfrac{1}{P(P-M)}\,dP = \int k\,dt$. By the method of partial fractions

$$\int \frac{1}{P(P-M)}\,dP = \frac{1}{M}\int \frac{1}{P-M} - \frac{1}{P}\,dP = \frac{1}{M}\left(\ln|P-M| - \ln P\right),$$

and so the general solution is

$$\frac{1}{M}\left(\ln|P-M| - \ln P\right) = kt + C,$$

or $\ln\dfrac{|P-M|}{P} = kMt + C$, or $\dfrac{|P-M|}{P} = Ce^{kMt}$. The initial condition $P(0) = P_0$ gives

$C = \dfrac{|P_0 - M|}{P_0}$. If the initial population P_0 is less than the threshold population M, then

$C = \dfrac{M - P_0}{P_0}$. Moreover, as in Problem 32, in this case $P < M$ for all t. Thus for $P_0 < M$

the solution of the extinction-explosion initial value problem is $\dfrac{M - P}{P} = \dfrac{M - P_0}{P_0}e^{kMt}$.

Similarly, if $P_0 > M$, then $P > M$ for all t, and so the solution is $\dfrac{P - M}{P} = \dfrac{P_0 - M}{P_0}e^{kMt}$.

Solving either of these equivalent expressions for P yields

$$P_0(P - M) = P(P_0 - M)e^{kMt},$$

or

$$\left[P_0 + (M - P_0)e^{kMt}\right]P = MP_0,$$

or finally

$$P(t) = \frac{MP_0}{P_0 + (M - P_0)e^{kMt}}.$$

(b) If $P_0 < M$, then the coefficient $M - P_0$ is positive and the denominator increases
without bound, so $P(t) \to 0$ as $t \to \infty$ But if $P_0 > M$, then the denominator
$P_0 - (P_0 - M)e^{kMt}$ approaches zero—so $P(t) \to +\infty$—as t approaches the positive value
$\dfrac{1}{kM}\ln\dfrac{P_0}{P_0 - M}$ from the left. Thus the population either becomes extinct or explodes.

35. Any way you look at it, you should conclude that the larger the parameter $k > 0$, the faster the logistic population $P(t)$ approaches its limiting population M:

To examine the question *geometrically*, we will assume that $M = 10$ and that $k_1 = 1$ and $k_2 = 2$, leading to the logistic equations $\dfrac{dP}{dt} = P(10 - P)$ and $\dfrac{dP}{dt} = 2P(10 - P)$. We draw slope fields and solution curves for each of these equations, using the same initial values $P(0)$ in both cases:

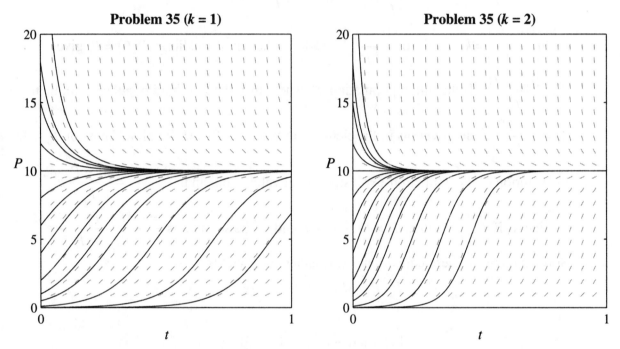

Problem 35 (k = 1)   **Problem 35 (k = 2)**

These diagrams suggest that the larger the value of k, the more rapidly the population $P(t)$ approaches the limiting population M.

To look at things *analytically*, we examine the distance between the solution (7) in the text of the logistic initial value problem and the limiting population M:

$$\left| M - \frac{MP_0}{P_0 + (M - P_0)e^{-kMt}} \right| = \left| \frac{M\left[P_0 + (M - P_0)e^{-kMt} \right] - MP_0}{P_0 + (M - P_0)e^{-kMt}} \right| = \frac{M|M - P_0|}{P_0\left(e^{kMt} - 1\right) + M}.$$

For fixed M, t, and P_0 this distance decreases as k increases; thus, the larger the value of k, the more rapidly $P(t)$ approaches M.

Finally, *numerically*, we tabulate values of $P(t)$, $t = 0, 0.1, 0.2, \ldots, 0.9, 1$, for the two solutions illustrated graphically above, using $P_0 = 0.1$ in both cases. Once again the evidence is that the larger value of k leads to the more rapid approach to M:

$k = 1$										
t	0	0.1	0.2	0.3	0.4	0.5	0.6	0.7	0.8	0.9	1.0
$P(t)$	0.1	0.267	0.695	1.687	3.555	5.999	8.030	9.172	9.679	9.879	9.955

$k = 2$										
t	0	0.1	0.2	0.3	0.4	0.5	0.6	0.7	0.8	0.9	1.0
$P(t)$	0.1	0.695	3.555	8.03	9.679	9.955	9.994	9.999	9.999	10.00	10.00

37 $k = 0.0000668717$ and $M = 338.027$, so that $P(t) = \dfrac{25761.7}{76.212 + 261.815e^{-0.0226045t}}$, which

predicts that $P = 192.525$ in the year 2000.

39.   Separating variables gives

$\displaystyle\int \frac{1}{P}\,dP = \int (k + b\cos 2\pi t)\,dt$, or

$\ln P = kt + \dfrac{b}{2\pi}\sin 2\pi t + C$. The initial

condition $P(0) = P_0$ implies that $C = \ln P_0$,

so the desired particular solution is

$P = P_0 \exp\left(kt + \dfrac{b}{2\pi}\sin 2\pi t \right)$. Of course the

natural growth equation $P' = kP$ with the
same initial condition has solution

$P(t) = P_0 e^{kt}$. The results of both growth

patterns are indicated in the graph shown

Problem 39

with the typical numerical values $P_0 = 100$, $k = 0.03$, and $b = 0.06$. Under the periodic
growth law the population oscillates about the curve representing natural growth. We see
that the two agree at the end of each full year.

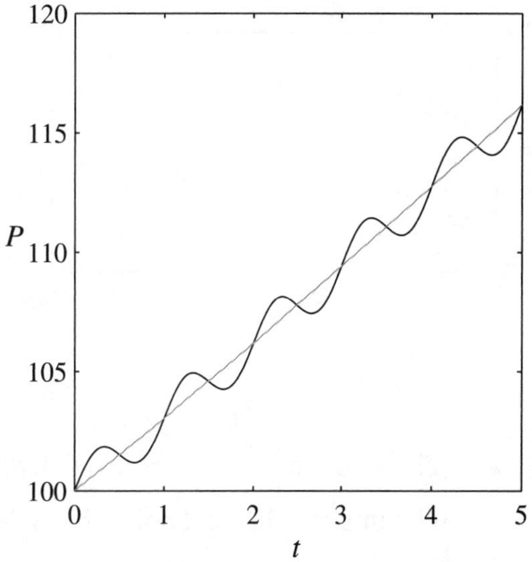

SECTION 2.2

EQUILIBRIUM SOLUTIONS AND STABILITY

In Problems 1-12 we identify the stable and unstable critical points as well as the funnels and
spouts along the equilibrium solutions. In each problem the indicated solution satisfying
$x(0) = x_0$ is derived by separation of variables, and we show typical solution curves
corresponding to different values of x_0 .

1. The unstable critical point $x = 4$ leads to a spout along the equilibrium solution $x(t) = 4$.

Separating variables gives $\int \dfrac{1}{x-4} dx = \int dt$, or $\ln|x-4| = t + C$, where C is an arbitrary constant. Thus the general solution is $x = Ce^t + 4$, where C is an arbitrary nonzero constant. The initial condition $x(0) = x_0$ then gives $x_0 = C + 4$, or $C = x_0 - 4$. Thus the solution is given by $x(t) = (x_0 - 4)e^t + 4$.

Problem 1

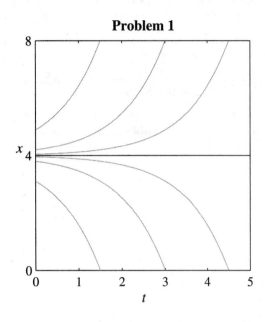

Problem 3

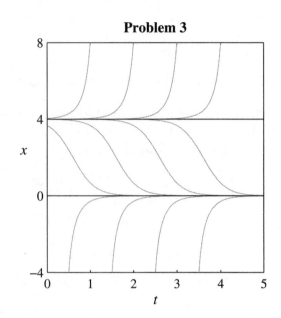

3. The stable critical point $x = 0$ leads to a funnel along the equilibrium solution $x(t) = 0$. The unstable critical point $x = 4$ leads to a spout along the equilibrium solution $x(t) = 4$.

Separating variables gives $\int \dfrac{1}{x^2 - 4x} dx = \int dt$, or $\int \dfrac{1}{x-4} - \dfrac{1}{x} dx = \int 4\, dt$. Integrating

gives $\ln|x-4| - \ln|x| = 4t + C$, or $\dfrac{x-4}{x} = Ce^{4t}$, where C is an arbitrary nonzero constant.

The initial condition $x(0) = x_0$ gives $\dfrac{x_0 - 4}{x_0} = C$, leading to $\dfrac{x-4}{x} = \dfrac{x_0 - 4}{x_0}e^{4t}$, or finally

the solution $x(t) = \dfrac{4x_0}{x_0 + (4 - x_0)e^{4t}}$.

5. The stable critical point $x = -2$ leads to a funnel along the equilibrium solution $x(t) = -2$. The unstable critical point $x = 2$ leads to a spout along the equilibrium solution $x(t) = 2$. Separating variables gives $\int \dfrac{1}{x^2 - 4} dx = \int dt$, or

$\int \dfrac{1}{x-2} - \dfrac{1}{x+2} dx = \int 4 dt$. Integrating gives $\ln|x-2| - \ln|x+2| = -4t + C$, or

$\dfrac{x-2}{x+2} = Ce^{-4t}$, where C is an arbitrary nonzero constant. The initial condition $x(0) = x_0$

gives $\dfrac{x_0 - 2}{x_0 + 2} = C$, leading to $\dfrac{x-2}{x+2} = \dfrac{x_0 - 2}{x_0 + 2} e^{-3t}$, or finally the solution

$$x(t) = 2 \dfrac{(x_0 + 2) + (x_0 - 2)e^{4t}}{(x_0 + 2) - (x_0 - 2)e^{4t}}.$$

Problem 5

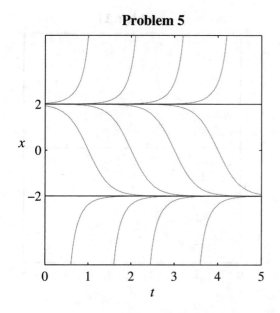

Problem 7

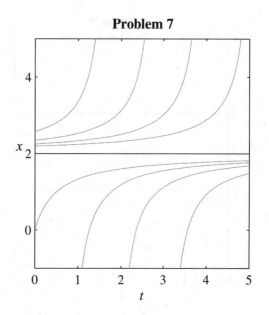

7. The lone critical point $x = 2$ is *semi-stable*; solutions with $x_0 > 2$ approach $+\infty$ as t increases, whereas those with $x_0 < 2$ approach 2 as t increases. Separating variables

gives $\int \dfrac{1}{(x-2)^2} dx = \int dt$, or $\dfrac{1}{x-2} = -t + C$, where C is an arbitrary nonzero constant.

The initial condition $x(0) = x_0$ gives $\dfrac{1}{x_0 - 2} = C$, leading to

$\dfrac{1}{x-2} = -t + \dfrac{1}{x_0 - 2} = \dfrac{1 - t(x_0 - 2)}{x_0 - 2}$, or finally the solution

$$x(t) = 2 + \dfrac{x_0 - 2}{1 - t(x_0 - 2)} = \dfrac{x_0(2t - 1) - 4t}{(x_0 - 2)t - 1}.$$

9. Factoring gives $x^2 - 5x + 4 = (x-4)(x-1)$. The stable critical point $x = 1$ leads to a funnel along the equilibrium solution $x(t) = 1$. The unstable critical point $x = 4$ leads to a spout along the equilibrium solution $x(t) = 4$. Separating variables gives

$$\int \frac{1}{(x-4)(x-1)} dx = \int dt \text{, or } \int \frac{1}{x-4} - \frac{1}{x-1} dx = \int 3 dt \text{. Integrating gives}$$

$\ln|x-4| - \ln|x-1| = 3t + C$, or $\dfrac{x-4}{x-1} = Ce^{3t}$, where C is an arbitrary nonzero constant.

The initial condition $x(0) = x_0$ gives $\dfrac{x_0 - 4}{x_0 - 1} = C$, leading to $\dfrac{x-4}{x-1} = \dfrac{x_0 - 4}{x_0 - 1} e^{3t}$, or finally

the solution $x(t) = \dfrac{4(1-x_0) + (x_0 - 4)e^{3t}}{(1-x_0) + (x_0 - 4)e^{3t}}$.

Problem 9	**Problem 11**
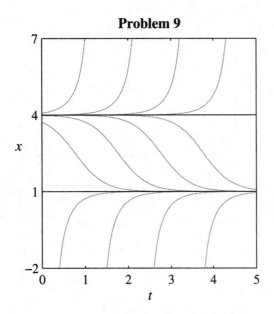	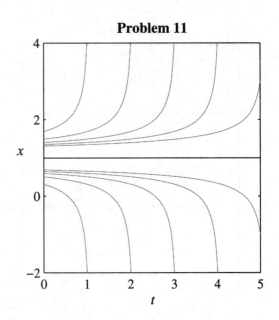

11. The unstable critical point $x = 1$ leads to a spout along the equilibrium solution $x(t) = 4$.

Separating variables gives $\int \dfrac{1}{(x-1)^3} dx = \int dt$, and integrating gives $\dfrac{1}{2(x-1)^2} = -t + C$,

where C is an arbitrary constant. The initial condition $x(0) = x_0$ gives $\dfrac{1}{2(x_0 - 1)^2} = C$,

leading to $\dfrac{1}{2(x-1)^2} = -t + \dfrac{1}{2(x_0 - 1)^2}$, or $(x-1)^2 = \dfrac{(x_0 - 1)^2}{1 - 2t(x_0 - 1)^2}$, or finally the solution

$x(t) = 1 \pm \dfrac{x_0 - 1}{\sqrt{1 - 2t(x_0 - 1)^2}}$.

In each of Problems 13–18 we present the figure showing the slope field and typical solution curves, and then record the visually apparent classification of critical points for the given differential equation.

13.   The critical points $x = 2$ and $x = -2$ are both unstable.

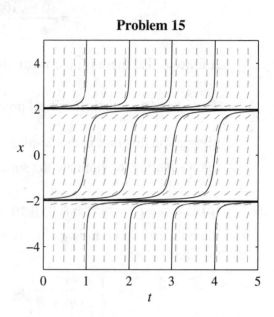

Problem 13

Problem 15

15.   The critical points $x = 2$ and $x = -2$ are both unstable.

17.   The critical points $x = 2$ and $x = 0$ are unstable, while the critical point $x = -2$ is stable.

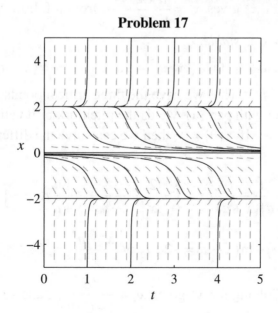

Problem 17

19.   The critical points of the given differential equation are the roots of the quadratic equation $\frac{1}{10}x(10-x)-h=0$, that is, $x^2-10x+10h=0$. Thus a critical point c is given in terms of h by

$$c = \frac{10 \pm \sqrt{100-40h}}{2} = 5 \pm \sqrt{25-10h}\,.$$

It follows that there is no critical point if $h > \frac{5}{2}$, only the single critical point $c = 0$ if $h = \frac{5}{2}$, and two distinct critical points if $h < \frac{5}{2}$, so that $10-25h > 0$. Hence the bifurcation diagram in the hc-plane is the parabola $(c-5)^2 = 25-10h$ that is obtained upon squaring to eliminate the square root above.

21.   **(a)** If $k = -a^2$, where $a \ge 0$, then $kx - x^3 = -a^2 x - x^3 = -x(a^2 + x^2) = 0$ only if $x = 0$, so the only critical point is $c = 0$. If $a > 0$, then we can solve the differential equation by writing

$$\int \frac{a^2}{x(a^2+x^2)}dx = \int \frac{1}{x} - \frac{x}{a^2+x^2}dx = -\int a^2\,dt,$$

or $\ln x - \frac{1}{2}\ln(a^2+x^2) = -a^2 t + C$, or $\frac{x^2}{a^2+x^2} = Ce^{-2a^2 t}$, where C is an arbitrary nonzero constant. Solving for x^2 gives $x^2 = \frac{a^2 Ce^{-2a^2 t}}{1-Ce^{-2a^2 t}}$, from it follows that $x \to 0$ as $t \to \infty$, so the critical point $c = 0$ is *stable*.

(b) If $k = a^2$, where $a > 0$, then $kx - x^3 = +a^2 x - x^3 = -x(x+a)(x-a) = 0$ if either $x = 0$ or $x = \pm a = \pm\sqrt{k}$. Thus we have the three critical points $c = 0$ and $c = \pm\sqrt{k}$; this observation, together with part **(a)**, yields the pitchfork bifurcation diagram shown in Fig. 2.2.13 of the textbook. If $x(0) \ne 0$, then we can solve the differential equation by writing

$$\int \frac{2a^2}{x(x-a)(x+a)}dx = \int -\frac{2}{x} + \frac{1}{x-a} + \frac{1}{x+a}dx = -\int 2a^2\,dt,$$

or $-2\ln x + \ln(x-a) + \ln(x-a) = -2a^2 t$, or $\frac{x^2-a^2}{x^2} = Ce^{-2a^2 t}$, where C is an arbitrary nonzero constant. Solving for x^2 gives $x^2 = \frac{a^2}{1-Ce^{-2a^2 t}}$, and so $x = \frac{\pm\sqrt{k}}{\sqrt{1-Ce^{-2a^2 t}}}$. It follows that if $x(0) \ne 0$, then $x \to \sqrt{k}$ if $x > 0$ and $x \to -\sqrt{k}$ if $x < 0$. This implies that the critical point $c = 0$ is *unstable*, while the critical points $c = \pm\sqrt{k}$ are *stable*.

23.   **(a)** If $h < kM$, then writing the differential equation as

$$x' = kx(M - x) - hx = kx\left[\left(M - \frac{h}{k}\right) - x\right],$$

still a logistic equation but with the *reduced* limiting population $M - \dfrac{h}{k}$.

(b) If $h > kM$, then the differential equation can be rewritten in the form $x' = -ax - bx^2$, with a and b both positive. The solution of this equation is $x(t) = \dfrac{ax_0}{(a + bx_0)e^{at} - bx_0}$, so it is clear that $x(t) \to 0$ as $t \to \infty$.

25.   In the first alternative form that is given, all of the coefficients within parentheses are positive if $H < x_0 < N$. Hence it is clear that $x(t) \to N$ as $t \to \infty$, which confirms (17).

In the second alternative form, all of the coefficients within parentheses are positive if $x_0 < H$. Hence the denominator is initially equal to $N - H > 0$, but decreases as t increases, and reaches the value 0 when $t = t_1 = \dfrac{1}{k(N - H)}\ln\dfrac{N - x_0}{H - x_0} > 0$. Meanwhile the numerator is initially $(N - H)x_0$, but approaches $(H - N)(H - x_0) < 0$ as $t \to t_1$. Conclusion (18) follows.

27.   Separation of variables in the differential equation $x' = -k\left[(x - a)^2 + b^2\right]$ yields

$$x(t) = a - b\tan\left(bkt + \tan^{-1}\frac{a - x_0}{b}\right).$$

It follows that $x(t) \to -\infty$ in a finite period of time.

29.   This is simply a matter of analyzing the signs of x' in the various cases $x < a$, $a < x < b$, $b < x < c$, and $c > x$. Alternatively, plot slope fields and typical solution curves for the two differential equations using typical numerical values such as $a = -1$, $b = 1$, and $c = 2$.

SECTION 2.3

ACCELERATION-VELOCITY MODELS

This section consists of three essentially independent subsections that can be studied separately: resistance proportional to velocity, resistance proportional to velocity-squared, and inverse-square gravitational acceleration.

1. The velocity v of the car (in km/hr) is related to the time t (in seconds) by the initial value problem $v' = k(250 - v)$, $v(0) = 0$, $v(10) = 100$. Separating variables gives

    $\int \dfrac{1}{250 - v} dv = \int k \, dt$, and integration yields $\ln|250 - v| = kt + C$, or $250 - v = Ce^{kt}$, or

    finally $v = Ce^{kt} + 250$, where C is an arbitrary nonzero constant. The initial condition

    $v(0) = 0$ gives $C = -250$, so that $v = 250(1 - e^{kt})$, and the condition $v(10) = 100$

    implies that $k = \dfrac{1}{10} \ln\left(\dfrac{250}{150}\right) \approx 0.0511$. Finally, solving the equation $v(t) = 200$ for t

    gives $t = -\dfrac{\ln 50}{250k} \approx 31.5 \sec$.

3. The velocity v of the boat (in ft/s) is related to the time t (in seconds) by the initial value problem $v' = -kv$, $v(0) = 40$, $v(10) = 20$. By Problem 2a, $v(t) = 40^{-kt}$, and the

    condition $v(10) = 20$ implies that $k = \dfrac{1}{10} \ln 2 \approx 0.0693$. By Problem 2b, then, the boat

    travels a distance of $\dfrac{v_0}{k} = \dfrac{40 \cdot 10}{\ln 2} \approx 577$ ft altogether.

5. We are assuming that the velocity v of the motorboat satisfies the initial value problem $v' = -kv^2$, $v(0) = 40$, with $v(10) = 20$ as well. We seek $x(60)$. The result of Problem

    4 gives $v = \dfrac{v_0}{1 + v_0 kt} = \dfrac{40}{1 + 40kt}$, and then the condition $v(10) = 20$ implies that

    $20 = \dfrac{40}{1 + 400k}$, or $k = \dfrac{1}{400}$. Thus $v(t) = \dfrac{400}{10 + t}$, and then

    $$x = \int v(t) \, dt = 400 \ln(t + 10) + C.$$

    The initial condition $x(0) = 0$ implies that $C = -400 \ln 10$, so that $x(t) = 400 \ln \dfrac{t + 10}{10}$. It

    follows that $x(60) = 400 \ln 7 \approx 778$ ft.

7.   The car satisfies the initial value problem $v' = 10 - 0.1v$, $v(0) = 0$. Separating variables

gives $\int \dfrac{1}{10 - 0.1v} dv = \int dt$, or $\ln(10 - 0.1v) = -\dfrac{t}{10} + C$. The initial condition $v(0) = 0$

gives $C = \ln 10$, so that $\ln(10 - 0.1v) = -\dfrac{t}{10} + \ln 10$, or $\ln(1 - 0.01v) = -\dfrac{t}{10}$, or

$v(t) = 100(1 - e^{-t/10})$. As $t \to \infty$, we find $v(t) \to 100$ ft/sec, the answer to **a**. Further,

setting $v(t) = 90$ ft/sec (that is, 90% of limiting velocity) gives $t = 10 \ln 10 \approx 23.0259$ sec.

Since $x = \int 100(1 - e^{-t/10}) dt = 100t + 1000e^{-t/10} + C'$, where the initial condition $x(0) = 0$

gives $C' = -1000$, we find that $x(23.0259) \approx 1402.59$ ft, the answer to **b**.

9.   Separating variables gives $\int \dfrac{1000}{5000 - 100v} dv = \int dt$, or $10 \ln(5000 - 100v) = -t + C$, or

$v = 50 + Ce^{-t/10}$. The initial condition $v(0) = 0$ implies that or $C = -50$, so that

$v(t) = 50(1 - e^{-t/10})$. As $t \to \infty$, $v(t) \to 50$ ft/sec ≈ 34 mph.

11.   If the paratrooper's terminal velocity was 100 mph $= \dfrac{440}{3}$ ft/sec, then Equation (7) in the

text yields $\dfrac{g}{\rho} = \dfrac{440}{3}$, or $\rho = \dfrac{3}{440} \cdot 32 = \dfrac{12}{55}$. Equation (9) then becomes

$$y(t) = -1200 + \dfrac{440}{3}t - \dfrac{55}{12} \cdot \dfrac{440}{3}(1 - e^{-12t/55}),$$

and solving the equation $y(t) = 0$ *via* technology gives $t \approx 12.5$ sec. Thus the newspaper

account is inaccurate.

Given the hints and integrals provided in the text, Problems 13–16 are fairly straightforward (and
fairly tedious) integration problems.

17.   Equation (13) from the text gives

$$v(t) = \sqrt{\frac{9.8}{0.0011}} \tan\left(C_1 - t\sqrt{0.0011 \cdot 9.8}\right) = 94.3880 \tan\left(C_1 - 0.1038267t\right),$$

where $C_1 = \tan^{-1}\left(49\sqrt{\frac{0.0011}{9.8}}\right) \approx 0.4788372$. Thus

$$v(t) = 94.3880 \tan\left(0.4788372 - 0.1038267t\right).$$

Then Equation (14) gives

$$y(t) = \frac{1}{0.0011}\ln\left|\frac{\cos\left(0.4788372 - \sqrt{0.0011 \cdot 9.8}t\right)}{\cos 0.4788372}\right| \approx 909.0909\ln\left|\frac{\cos\left(0.4788372 - 0.103827t\right)}{\cos 0.4788372}\right|.$$

Setting $v(t) = 0$ leads to $t \approx 4.612\,\text{sec}$, at which time $y \approx 108.465\,\text{m}$.

19.   The initial value problem for the velocity of the motorboat is $v' = 4 - \frac{1}{400}v^2$, $v(0) = 0$.

Separating variables gives $\int \frac{1}{4 - \frac{1}{400}v^2}\,dv = \int dt$, or $\int \frac{1/40}{1 - (v/40)^2}\,dv = \int \frac{1}{10}\,dt$, or

$\tanh^{-1}\frac{v}{40} = \frac{t}{10} + C$. The initial condition $v(0) = 0$ gives $C = 0$, so that

$v(t) = 40\tanh\frac{t}{10}$. Finally, $v(10) = \tanh 1 \approx 30.46$ ft/sec and

$$\lim_{t \to \infty} v(t) = 40\lim_{t \to \infty}\tanh\frac{t}{10} = 40 \text{ ft/sec}.$$

21.   The initial value problem for the velocity of the ball is $v' = -g - \rho v^2$, $v(0) = v_0$, with the
added condition that $y(0) = 0$, where y is the height of the ball. Separating variables

gives $\int \frac{dv}{g + \rho v^2}\,dv = -\int dt$, or $\int \frac{\sqrt{\rho/g}}{1 + \left(\sqrt{\rho/g}\,v\right)^2}\,dv = -\int \sqrt{g\rho}\,dt$, or

$\tan^{-1}\left(\sqrt{\rho/g}\,v\right) = -\sqrt{g\rho}\,t + C$. The initial condition $v(0) = v_0$ implies that

$C = \tan^{-1}\left(\sqrt{\rho/g}\,v_0\right)$, and so

$$v(t) = -\sqrt{\frac{g}{\rho}}\tan\left[t\sqrt{g\rho} - \tan^{-1}\left(v_0\sqrt{\frac{\rho}{g}}\right)\right].$$

We solve $v(t) = 0$ for $t = \frac{1}{\sqrt{g\rho}}\tan^{-1}\left(v_0\sqrt{\frac{\rho}{g}}\right)$ and substitute in Equation (17) for $y(t)$:

$$y_{max} = \frac{1}{\rho} \ln \left| \frac{\cos\left(\tan^{-1} v_0 \sqrt{\rho/g} - \tan^{-1} v_0 \sqrt{\rho/g}\right)}{\cos\left(\tan^{-1} v_0 \sqrt{\rho/g}\right)} \right| = \frac{1}{\rho} \ln \left[\sec\left(\tan^{-1} v_0 \sqrt{\rho/g}\right) \right]$$

$$= \frac{1}{\rho} \ln \sqrt{1 + \frac{\rho v_0^2}{g}} = \frac{1}{2\rho} \ln \left(1 + \frac{\rho v_0^2}{g}\right).$$

23. Before the parachute opens, the paratrooper's descent is modeled by the initial value problem $v' = -32 + 0.00075 v^2$, $v(0) = 0$, with $y(0) = 10000$. Solving gives $v(t) = -206.559 \tanh(0.154919t)$, and then $v(30) = -206.521$ ft/sec. Integrating once again gives $y(t) = 10000 - 1333.33 \ln(\cosh 0.154919t)$, with $y(30) = 4727.30$ ft. After the parachute opens, the initial value problem becomes $v' = -32 + 0.075 v^2$, $v(0) = -206.521$, with $y(0) = 4727.30$. Solving gives

$$v(t) = -20.6559 \tanh(1.54919t + 0.00519595),$$

followed by

$$y(t) = 4727.30 - 13.3333 \ln(\cosh 1.54919t + 0.00519595).$$

We find that $y = 0$ when $t = 229.304$. Thus he opens his parachute after 30 sec at a height of 4727 feet, and the total time of descent is $30 + 229.304 = 259.304$ sec, about 4 minutes and 19.3 seconds.

25. **(a)** The rocket's apex occurs when $v = 0$. We get the desired formula when we set $v = 0$ in Eq. (23), $v^2 = v_0^2 + 2GM\left(\frac{1}{r} - \frac{1}{R}\right)$, and solve for r.

(b) We substitute $v = 0$, $r = R + 10^5$ (note $100\,\text{km} = 10^5\,\text{m}$), and the mks values $G = 6.6726 \times 10^{-11}$, $M = 5.975 \times 10^{24}$, and $R = 6.378 \times 10^6$ in Eq. (23) and solve for $v_0 = 1389.21\,\text{m/s} \approx 1.389\,\text{km/s}$.

(c) When we substitute $v_0 = 0.9\sqrt{2GM/R}$ in the formula derived in part **a**, we find that $r_{max} = \frac{100}{19} R$.

27. **(a)** Substitution of $v_0^2 = \dfrac{2GM}{R} = \dfrac{k^2}{R}$ in Eq. (23) of the textbook gives

$$\frac{dr}{dt} = v = \sqrt{\frac{2GM}{r}} = \frac{k}{\sqrt{r}}.$$

We separate variables and proceed to integrate: $\int \sqrt{r}\, dr = \int k\, dt$ implies that

$\dfrac{2}{3} r^{3/2} = kt + \dfrac{2}{3} R^{3/2}$, since $r = R$ when $t = 0$. We solve for $r(t) = \left(\dfrac{2}{3} kt + R^{3/2} \right)^{2/3}$ and

note that $r(t) \to \infty$ as $t \to \infty$.

(b) If $v_0 > \dfrac{2GM}{R}$, then Eq. (23) gives

$$\frac{dr}{dt} = v = \sqrt{\frac{2GM}{r} + \left(v_0^2 - \frac{2GM}{R} \right)} = \sqrt{\frac{k^2}{r} + \alpha} > \frac{k}{\sqrt{r}}.$$

Therefore, at every instant in its ascent, the upward velocity of the projectile in this part is greater than the velocity at the same instant of the projectile of part (a). It's as though the projectile of part (a) is the fox, and the projectile of this part is a rabbit that runs faster. Since the fox goes to infinity, so does the faster rabbit.

29. Integration of $v \dfrac{dv}{dy} = -\dfrac{GM}{(y+R)^2}$, $y(0) = 0$, $v(0) = v_0$ gives

$$\frac{1}{2} v^2 = \frac{GM}{y+R} - \frac{GM}{R} + \frac{1}{2} v_0^2,$$

which simplifies to the desired formula for v^2. Then substitution of $G = 6.6726 \times 10^{-11}\,\text{N} \cdot (\text{m/kg})^2$, $M = 5.975 \times 10^{24}\,\text{kg}$, $R = 6.378 \times 10^6\,\text{m}$, $v = 0$, and $v_0 = 1$ yields an equation that we easily solve for $y = 51427.3\,\text{m}$, that is, about 51.427 km.

SECTION 2.4

NUMERICAL APPROXIMATION: EULER'S METHOD

In each of Problems 1–10 we also give first the explicit form of Euler's iterative formula for the given differential equation $y' = f(x, y)$. As we illustrate in Problem 1, the desired iterations are readily implemented, either manually or with a computer system or graphing calculator. Then we list the indicated values of $y(\tfrac{1}{2})$ rounded off to 3 decimal places.

1.   For the differential equation $y' = f(x, y)$ with $f(x, y) = -y$, the iterative formula of Euler's method is $y_{n+1} = y_n + h(-y_n)$. The TI-83 screen on the left shows a graphing calculator implementation of this iterative formula.

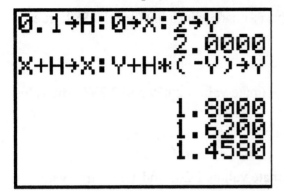   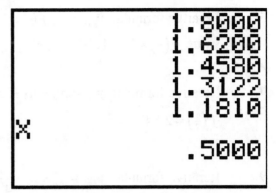

After the variables are initialized (in the first line), and the formula is entered, each press of the enter key carries out an additional step. The screen on the right shows the results of 5 steps from $x = 0$ to $x = 0.5$ with step size $h = 0.1$—winding up with $y(0.5) \approx 1.181$. Similarly, using $h = 0.25$ gives $y(0.5) \approx 1.125$. The true value is $y\left(\frac{1}{2}\right) \approx 1.213$.

The following *Mathematica* instructions produce precisely the line of data shown:

```
f[x_,y_] = -y;
g[x_] = 2*Exp[-x];
y0 = 2;
h = 0.25;
x = 0;
y1 = y0;

Do[k = f[x,y1];       (* the left-hand slope *)
   y1 = y1 + h*k;       (* Euler step to update y *)
   x = x + h,        (* update x *)
   {i,1,2}]

h = 0.1;
x = 0;
y2 = y0;

Do[k = f[x,y2];       (* the left-hand slope *)
   y2 = y2 + h*k;       (* Euler step to update y *)
   x = x + h,        (* update x *)
   {i,1,5}]

Print[x,"      ",y1,"       ",y2,"       ",g[0.5]]

0.5     1.125     1.18098     1.21306
```

3. Iterative formula: $y_{n+1} = y_n + h(y_n + 1)$; approximate values 2.125 and 2.221; true value $y\left(\frac{1}{2}\right) \approx 2.297$.

5. Iterative formula: $y_{n+1} = y_n + h(y_n - x_n - 1)$; approximate values 0.938 and 0.889; true value $y\left(\frac{1}{2}\right) \approx 0.851$.

7. Iterative formula: $y_{n+1} = y_n + h(-3x_n^2 y_n)$; approximate values 2.859 and 2.737; true value $y\left(\frac{1}{2}\right) \approx 2.647$.

9. Iterative formula: $y_{n+1} = y_n + h\dfrac{1 + y_n^2}{4}$; approximate values 1.267 and 1.278; true value $y\left(\frac{1}{2}\right) \approx 1.287$.

The tables of approximate and actual values called for in Problems 11–16 were produced using the following MATLAB script (appropriately altered for each problem).

```
% Section 2.4, Problems 11-16
x0 = 0;
y0 = 1;
% first run:
h = 0.01;
x = x0;
y = y0;
y1 = y0;
for n = 1:100
   y = y + h*(y-2);
   y1 = [y1,y];
   x = x + h;
   end
% second run:
h = 0.005;
x = x0;  y = y0;  y2 = y0;
for n = 1:200
   y = y + h*(y-2);
   y2 = [y2,y];
   x = x + h;
   end
% exact values
x = x0 : 0.2 : x0+1;
ye = 2 - exp(x);
% display table
ya = y2(1:40:201);
err = 100*(ye-ya)./ye;
```

```
[x; y1(1:20:101); ya; ye; err]
```

11.   The iterative formula of Euler's method is $y_{n+1} = y_n + h(y_n - 2)$, and the exact solution is $y(x) = 2 - e^x$. The resulting table of approximate and actual values is

x	0.0	0.2	0.4	0.6	0.8	1.0
y ($h = 0.01$)	1.0000	0.7798	0.5111	0.1833	−0.2167	−0.7048
y ($h = 0.005$)	1.0000	0.7792	0.5097	0.1806	−0.2211	−0.7115
y actual	1.0000	0.7786	0.5082	0.1779	−0.2255	−0.7183
error	0%	−0.08%	−0.29%	−1.53%	1.97%	0.94%

13.   Iterative formula: $y_{n+1} = y_n + 2h\dfrac{x_n^3}{y_n}$; exact solution: $y(x) = \left(8 + x^4\right)^{1/2}$.

x	1.0	1.2	1.4	1.6	1.8	2.0
y ($h = 0.01$)	3.0000	3.1718	3.4368	3.8084	4.2924	4.8890
y ($h = 0.005$)	3.0000	3.1729	3.4390	3.8117	4.2967	4.8940
y actual	3.0000	3.1739	3.4412	3.8149	4.3009	4.8990
error	0%	0.03%	0.06%	0.09%	0.10%	0.10%

15.   Iterative formula: $y_{n+1} = y_n + h\left(3 - \dfrac{2y_n}{x_n}\right)$; exact solution: $y(x) = x + \dfrac{4}{x^2}$.

x	2.0	2.2	2.4	2.6	2.8	3.0
y ($h = 0.01$)	3.0000	3.0253	3.0927	3.1897	3.3080	3.4422
y ($h = 0.005$)	3.0000	3.0259	3.0936	3.1907	3.3091	3.4433
y actual	3.0000	3.0264	3.0944	3.1917	3.3102	3.4444
error	0%	0.019%	0.028%	0.032%	0.033%	0.032%

The tables of approximate values called for in Problems 17–24 were produced using a MATLAB script similar to the one listed preceding the Problem 11 solution above.

17.

x	0.0	0.2	0.4	0.6	0.8	1.0
y ($h = 0.1$)	0.0000	0.0010	0.0140	0.0551	0.1413	0.2925
y ($h = 0.02$)	0.0000	0.0023	0.0198	0.0688	0.1672	0.3379
y ($h = 0.004$)	0.0000	0.0026	0.0210	0.0717	0.1727	0.3477
y ($h = 0.0008$)	0.0000	0.0027	0.0213	0.0723	0.1738	0.3497

These data indicate that $y(1) \approx 0.35$, in contrast with Example 5 in the text, where the initial condition is $y(0) = 1$.

In Problems 18–24 we give only the final approximate values of y obtained using Euler's method with step sizes $h = 0.1$, $h = 0.02$, $h = 0.004$, and $h = 0.0008$.

19. With $x_0 = 0$ and $y_0 = 1$, the approximate values of $y(2)$ obtained are:

h	0.1	0.02	0.004	0.0008
y	6.1831	6.3653	6.4022	6.4096

21. With $x_0 = 1$ and $y_0 = 2$, the approximate values of $y(2)$ obtained are:

h	0.1	0.02	0.004	0.0008
y	2.8508	2.8681	2.8716	2.8723

23. With $x_0 = 0$ and $y_0 = 0$, the approximate values of $y(1)$ obtained are:

h	0.1	0.02	0.004	0.0008
y	1.2262	1.2300	1.2306	1.2307

25. Here $f(t,v) = 32 - 1.6v$ and $t_0 = 0$, $v_0 = 0$. With $h = 0.01$, 100 iterations of $v_{n+1} = v_n + hf(t_n, v_n)$ yield $v(1) \approx 16.014$, and 200 iterations with $h = 0.005$ yield $v(1) \approx 15.998$. Thus we observe an approximate velocity of 16.0 ft/sec after 1 second — 80% of the limiting velocity of 20 ft/sec.

With $h = 0.01$, 200 iterations yield $v(2) \approx 19.2056$, and 400 iterations with $h = 0.005$ yield $v(2) \approx 19.1952$. Thus we observe an approximate velocity of 19.2 ft/sec after 2 seconds — 96% of the limiting velocity of 20 ft/sec.

27. Here $f(x,y) = x^2 + y^2 - 1$ and $x_0 = 0, y_0 = 0$. The following table gives the approximate values for the successive step sizes h and corresponding numbers n of steps. It appears likely that $y(2) = 1.00$ rounded off accurate to 2 decimal places.

h	0.1	0.01	0.001	0.0001	0.00001
n	20	200	2000	20000	200000
$y(2)$	0.7772	0.9777	1.0017	1.0042	1.0044

29.   With step sizes $h = 0.15$, $h = 0.03$, and $h = 0.006$, we get the following results:

x	y with $h = 0.15$	y with $h = 0.03$	y with $h = 0.006$
-1.0	1.0000	1.0000	1.0000
-0.7	1.0472	1.0512	1.0521
-0.4	1.1213	1.1358	1.1390
-0.1	1.2826	1.3612	1.3835
$+0.2$	0.8900	1.4711	0.8210
$+0.5$	0.7460	1.2808	0.7192

While the values for $h = 0.15$ alone are not conclusive, a comparison of the values of y for all three step sizes with $x > 0$ suggests some anomaly in the transition from negative to positive values of x.

31.   With step sizes $h = 0.1$ and $h = 0.01$ we get the following results:

x	y with $h = 0.1$	y with $h = 0.01$
0.0	1.0000	1.0000
0.1	1.2000	1.2200
0.2	1.4428	1.4967
$\vdots$	$\vdots$	$\vdots$
0.7	4.3460	6.4643
0.8	5.8670	11.8425
0.9	8.3349	39.5010

Clearly there is some difficulty near $x = 0.9$.

SECTION 2.5

A CLOSER LOOK AT THE EULER METHOD

In each of Problems 1–10 we give first the predictor formula for u_{n+1} and then the improved Euler corrector for y_{n+1}. These predictor-corrector iterations are readily implemented, either manually or with a computer system or graphing calculator (as we illustrate in Problem 1). We give in each problem a table showing the approximate values obtained, as well as the corresponding values of the exact solution.

1. $u_{n+1} = y_n + h(-y_n)$; $y_{n+1} = y_n + \dfrac{h}{2}(-y_n - u_{n+1})$

The TI-83 screen on the left above shows a graphing calculator implementation of this iteration. After the variables are initialized (in the first line), and the formulas are entered, each press of the enter key carries out an additional step. The screen on the right shows the results of 5 steps from $x = 0$ to $x = 0.5$ with step size $h = 0.1$ — winding up with $y(0.5) \approx 1.2142$ — and we see the approximate values shown in the second row of the table below.

x	0.0	0.1	0.2	0.3	0.4	0.5
y with $h = 0.1$	2.0000	1.8100	1.6381	1.4824	1.3416	1.2142
y actual	2.0000	1.8097	1.6375	1.4816	1.3406	1.2131

3. $u_{n+1} = y_n + h(y_n + 1)$; $y_{n+1} = y_n + \dfrac{h}{2}\left[(y_n + 1) + (u_{n+1} + 1)\right]$

x	0.0	0.1	0.2	0.3	0.4	0.5
y with $h = 0.1$	1.0000	1.2100	1.4421	1.6985	1.9818	2.2949
y actual	1.0000	1.2103	1.4428	1.6997	1.9837	2.2974

5. $u_{n+1} = y_n + h(y_n - x_n - 1)$; $y_{n+1} = y_n + \dfrac{h}{2}\left[(y_n - x_n - 1) + (u_{n+1} - x_n - h - 1)\right]$

x	0.0	0.1	0.2	0.3	0.4	0.5
y with $h = 0.1$	1.0000	0.9950	0.9790	0.9508	0.9091	0.8526
y actual	1.0000	0.9948	0.9786	0.9501	0.9082	0.8513

7. $u_{n+1} = y_n - 3x_n^2 y_n h$; $y_{n+1} = y_n - \dfrac{h}{2}\left[3x_n^2 y_n + 3(x_n + h)^2 u_{n+1}\right]$

x	0.0	0.1	0.2	0.3	0.4	0.5
y with $h = 0.1$	3.0000	2.9955	2.9731	2.9156	2.8082	2.6405
y actual	3.0000	2.9970	2.9761	2.9201	2.8140	2.6475

9. $u_{n+1} = y_n + h \cdot \dfrac{1 + y_n^2}{4}$; $y_{n+1} = y_n + h \cdot \dfrac{1 + y_n^2 + 1 + u_{n+1}^2}{8}$

x	0.0	0.1	0.2	0.3	0.4	0.5
y with $h = 0.1$	1.0000	1.0513	1.1053	1.1625	1.2230	1.2873
y actual	1.0000	1.0513	1.1054	1.1625	1.2231	1.2874

The results given below for Problems 11–16 were computed using the following MATLAB script.

```
% Section 2.5, Problems 11-16
x0 = 0;  y0 = 1;
% first run:
h = 0.01;
x = x0;  y = y0;  y1 = y0;
for  n = 1:100
    u = y + h*f(x,y);                    %predictor
    y = y + (h/2)*(f(x,y)+f(x+h,u));     %corrector
    y1 = [y1,y];
    x = x + h;
    end
% second run:
h = 0.005;
x = x0;  y = y0;  y2 = y0;

for  n = 1:200
    u = y + h*f(x,y);                    %predictor
    y = y + (h/2)*(f(x,y)+f(x+h,u));     %corrector
    y2 = [y2,y];
    x = x + h;
end

% exact values
x = x0 : 0.2 : x0+1;
ye = g(x);

% display table
ya = y2(1:40:201);
```

```
err = 100*(ye-ya)./ye;
x = sprintf('%10.5f',x), sprintf('\n');
y1 = sprintf('%10.5f',y1(1:20:101)), sprintf('\n');
ya = sprintf('%10.5f',ya), sprintf('\n');
ye = sprintf('%10.5f',ye), sprintf('\n');
err = sprintf('%10.5f',err), sprintf('\n');
table = [x; y1; ya; ye; err]
```

For each problem the differential equation $y' = f(x,y)$ and the known exact solution $y = g(x)$ are stored in the files **f.m** and **g.m** — for instance, the files

```
function yp = f(x,y)
yp = y-2;

function ye = g(x,y)
ye = 2-exp(x);
```

for Problem 11. (The exact solutions for Problems 11–16 here are given in the solutions for Problems 11–16 in Section 2.4.)

11.

x	0.0	0.2	0.4	0.6	0.8	1.0
$y\,(h = 0.01)$	1.00000	0.77860	0.50819	0.17790	–0.22551	–0.71824
$y\,(h = 0.005)$	1.00000	0.77860	0.50818	0.17789	–0.22553	–0.71827
y actual	1.00000	0.77860	0.50818	0.17788	–0.22554	–0.71828
error	0.000%	–0.000%	–0.000%	–0.003%	0.003%	0.002%

13.

x	1.0	1.2	1.4	1.6	1.8	2.0
$y\,(h = 0.01)$	3.00000	3.17390	3.44118	3.81494	4.30091	4.89901
$y\,(h = 0.005)$	3.00000	3.17390	3.44117	3.81492	4.30089	4.89899
y actual	3.00000	3.17389	3.44116	3.81492	4.30088	4.89898
error	0.0000%	–0.0001%	–0.0001%	-0.0001%	–0.0002%	–0.0002%

15.

x	2.0	2.2	2.4	2.6	2.8	3.0
$y\,(h = 0.01)$	3.000000	3.026448	3.094447	3.191719	3.310207	3.444448
$y\,(h = 0.005)$	3.000000	3.026447	3.094445	3.191717	3.310205	3.444445
y actual	3.000000	3.026446	3.094444	3.191716	3.310204	3.444444
error	0.00000%	–0.00002%	–0.00002%	–0.00002%	–0.00002%	–0.00002%

17.

With $h =$ 0.1:	$y(1) \approx 0.35183$
With $h =$ 0.02:	$y(1) \approx 0.35030$
With $h =$ 0.004:	$y(1) \approx 0.35023$
With $h =$ 0.0008:	$y(1) \approx 0.35023$

The table of numerical results is

x	y with $h = 0.1$	y with $h = 0.02$	y with $h = 0.004$	y with $h = 0.0008$
0.0	0.00000	0.00000	0.00000	0.00000
0.2	0.00300	0.00268	0.00267	0.00267
0.4	0.02202	0.02139	0.02136	0.02136
0.6	0.07344	0.07249	0.07245	0.07245
0.8	0.17540	0.17413	0.17408	0.17408
1.0	0.35183	0.35030	0.35023	0.35023

In Problems 18–24 we give only the final approximate values of y obtained using the improved Euler method with step sizes $h = 0.1$, $h = 0.02$, $h = 0.004$, and $h = 0.0008$.

19.

Value of h	Estimated value of $y(2)$
0.1	6.40834
0.02	6.41134
0.004	6.41147
0.0008	6.41147

21.

Value of h	Estimated value of $y(2)$
0.1	2.87204
0.02	2.87245
0.004	2.87247
0.0008	2.87247

23.

Value of h	Estimated value of $y(1)$
0.1	1.22967
0.02	1.23069
0.004	1.23073
0.0008	1.23073

25.   Here $f(t,v) = 32 - 1.6v$ and $t_0 = 0$, $v_0 = 0$. With $h = 0.01$, 100 iterations of

$$k_1 = f(t,v_n), \; k_2 = f(t + h, v_n + hk_1), \; v_{n+1} = v_n + \frac{h}{2}(k_1 + k_2)$$

yield $v(1) \approx 15.9618$, and 200 iterations with $h = 0.005$ yield $v(1) \approx 15.9620$. Thus we observe an approximate velocity of 15.962 ft/sec after 1 second — 80% of the limiting velocity of 20 ft/sec.

With $h = 0.01$, 200 iterations yield $v(2) \approx 19.1846$, and 400 iterations with $h = 0.005$ yield $v(2) \approx 19.1847$. Thus we observe an approximate velocity of 19.185 ft/sec after 2 seconds — 96% of the limiting velocity of 20 ft/sec.

27.   Here $f(x,y) = x^2 + y^2 - 1$ and $x_0 = 0$, $y_0 = 0$. The following table gives the approximate values for the successive step sizes h and corresponding numbers n of steps. It appears likely that $y(2) = 1.0045$ rounded off accurate to 4 decimal places.

h	0.1	0.01	0.001	0.0001
n	20	200	2000	20000
$y(2)$	1.01087	1.00452	1.00445	1.00445

In the solutions for Problems 29 and 30 we illustrate the following general MATLAB ode solver.

```
function  [t,y] = ode(method, yp, t0,b, y0, n)
%   [t,y] = ode(method, yp, t0,b, y0, n)
```

```
%  calls the method described by 'method' for the
%  ODE 'yp' with function header
%
%               y' = yp(t,y)
%
%  on the interval  [t0,b]  with initial (column)
%  vector  y0.  Choices for method are 'euler',
%  'impeuler', 'rk' (Runge-Kutta), 'ode23', 'ode45'.
%  Results are saved at the endPoints of n subintervals,
%  that is, in steps of length  h = (b - t0)/n.  The
%  result  t  is an (n+1)-column vector from b to t1,
%  while  y  is a matrix with  n+1  rows (one for each
%  t-value) and one column for each dependent variable.

h = (b - t0)/n;             % step size
t = t0 : h : b;
t = t';                        % col. vector of t-values
y = y0';                    % 1st row of result matrix
for  i = 2 : n+1             % for i=2 to i=n+1
   t0 = t(i-1);             % old t
   t1 = t(i);                 % new t
   y0 = y(i-1,:)';             % old y-row-vector
   [T,Y] = feval(method, yp, t0,t1, y0);
   y = [y;Y'];                 % adjoin new y-row-vector
end
```

To use the improved Euler method, we call as **'method'** the following function.

```
function [t,y] = impeuler(yp, t0,t1, y0)
%
%  [t,y] = impeuler(yp, t0,t1, y0)
%  Takes one improved Euler step for
%
%        y' = yprime( t,y ),
%
%  from t0  to  t1  with initial value  the
%  column vector  y0.

h = t1 - t0;
k1 = feval( yp, t0, y0          );
k2 = feval( yp, t1, y0 + h*k1 );
k  = (k1 + k2)/2;
t = t1;
y = y0 + h*k;
```

29.   Here our differential equation is described by the MATLAB function

```
function  vp = vpbolt1(t,v)
vp = -0.04*v - 9.8;
```

Then the commands

```
n = 50;
[t1,v1] = ode('impeuler','vpbolt1',0,10,49,n);
n = 100;
[t2,v2] = ode('impeuler','vpbolt1',0,10,49,n);
t = (0:10)';
ve = 294*exp(-t/25)-245;
[t, v1(1:5:51), v2(1:10:101), ve]
```

generate the table

t	with $n = 50$	with $n = 100$	actual v
0	49.0000	49.0000	49.0000
1	37.4722	37.4721	37.4721
2	26.3964	26.3963	26.3962
3	15.7549	15.7547	15.7546
4	5.5307	5.5304	5.5303
5	-4.2926	-4.2930	-4.2932
6	-13.7308	-13.7313	-13.7314
7	-22.7989	-22.7994	-22.7996
8	-31.5115	-31.5120	-31.5122
9	-39.8824	-39.8830	-39.8832
10	-47.9251	-47.9257	-47.9259

We notice first that the final two columns agree to 3 decimal places (each difference being less than 0.0005). Scanning the $n = 100$ column for sign changes, we suspect that $v = 100$ (at the bolt's apex) occurs just after $t = 4.5 \, \text{sec}$. Then interpolation between $t = 4.5$ and $t = 4.6$ in the table

```
[t2(40:51),v2(40:51)]
```

3.9000	6.5345
4.0000	5.5304
4.1000	4.5303
4.2000	3.5341
4.3000	2.5420
4.4000	1.5538
4.5000	0.5696
4.6000	-0.4108
4.7000	-1.3872
4.8000	-2.3597
4.9000	-3.3283
5.0000	-4.2930

indicates that $t = 4.56$ at the bolt's apex. Finally, interpolation in

```
[t2(95:96),v2(95:96)]
```

9.4000	-43.1387
9.5000	-43.9445

gives the impact velocity $v(9.41) \approx -43.22\,\text{m/s}$.

SECTION 2.6

THE RUNGE-KUTTA METHOD

Each problem can be solved with a "template" of computations like those listed in Problem 1. We include a table showing the slope values k_1, k_2, k_3, k_4 and the xy-values at the ends of two successive steps of size $h = 0.25$.

1. To make the first step of size $h = 0.25$ we start with the function defined by

    ```
    f[x_, y_] := -y
    ```

    and the initial values

    ```
    x = 0;      y = 2;      h = 0.25;
    ```

    and then perform the calculations

    ```
    k1 = f[x, y]
    k2 = f[x + h/2, y + h*k1/2]
    k3 = f[x + h/2, y + h*k2/2]
    k4 = f[x + h, y + h*k3]
    y  = y + (h/6)*(k1 + 2*k2 + 2*k3 + k4)
    x  = x + h
    ```

    in turn. Here we are using *Mathematica* notation that translates transparently to standard mathematical notation describing the corresponding manual computations. A repetition of this same block of calculations carries out a second step of size $h = 0.25$. The following table lists the intermediate and final results obtained in these two steps.

k_1	k_2	k_3	k_4	x	Approx. y	Actual y
−2	−1/75	−1.78125	−1.55469	0.25	1.55762	1.55760
−1.55762	−1.36292	−1.38725	−1.2108	0.5	1.21309	1.21306

3.

k_1	k_2	k_3	k_4	x	Approx. y	Actual y
2	2.25	2.28125	2.57031	0.25	1.56803	1.56805
2.56803	2.88904	2.92916	3.30032	0.5	2.29740	2.29744

5.

k_1	k_2	k_3	k_4	x	Approx. y	Actual y
0	−0.125	−0.14063	−0.28516	0.25	0.96598	0.96597
−28402	−0.44452	−0.46458	−0.65016	0.5	0.85130	0.85128

7.

k_1	k_2	k_3	k_4	x	Approx. y	Actual y
0	−0.14063	−0.13980	−0.55595	0.25	2.95347	2.95349
−0.55378	−1.21679	−1.18183	−1.99351	0.5	2.6475	2.64749

9.

k_1	k_2	k_3	k_4	x	Approx. y	Actual y
0.5	0.53223	0.53437	0.57126	0.25	1.13352	1.13352
0.57122	0.61296	0.61611	0.66444	0.5	1.28743	1.28743

The results given below for Problems 11–16 were computed using the following MATLAB script.

```
% Section 2.6, Problems 11-16
x0 = 0;  y0 = 1;

% first run:
h = 0.2;
x = x0;  y = y0;  y1 = y0;
for  n = 1:5
   k1 = f(x,y);
   k2 = f(x+h/2,y+h*k1/2);
   k3 = f(x+h/2,y+h*k2/2);
   k4 = f(x+h,y+h*k3);
   y = y +(h/6)*(k1+2*k2+2*k3+k4);
   y1 = [y1,y];
   x = x + h;
   end

% second run:
```

```
h = 0.1;
x = x0;  y = y0;   y2 = y0;
for  n = 1:10
    k1 = f(x,y);
    k2 = f(x+h/2,y+h*k1/2);
    k3 = f(x+h/2,y+h*k2/2);
    k4 = f(x+h,y+h*k3);
    y = y +(h/6)*(k1+2*k2+2*k3+k4);
    y2 = [y2,y];
    x = x + h;
end

% exact values
x = x0 : 0.2 : x0+1;
ye = g(x);

% display table
y2 = y2(1:2:11);
err = 100*(ye-y2)./ye;
x = sprintf('%10.6f',x), sprintf('\n');
y1 = sprintf('%10.6f',y1), sprintf('\n');
y2 = sprintf('%10.6f',y2), sprintf('\n');
ye = sprintf('%10.6f',ye), sprintf('\n');
err = sprintf('%10.6f',err), sprintf('\n');
table = [x;y1;y2;ye;err]
```

For each problem the differential equation $y' = f(x,y)$ and the known exact solution $y = g(x)$ are stored in the files **f.m** and **g.m** — for instance, the files

```
function  yp = f(x,y)
yp = y-2;
```

and

```
function ye = g(x,y)
ye = 2-exp(x);
```

for Problem 11.

11.

x	0.0	0.2	0.4	0.6	0.8	1.0
$y\,(h = 0.2)$	1.000000	0.778600	0.508182	0.177894	−0.225521	−0.718251
$y\,(h = 0.1)$	1.000000	0.778597	0.508176	0.177882	−0.225540	−0.718280
y actual	1.000000	0.778597	0.508175	0.177881	−0.225541	−0.718282
error	0.00000%	−0.00002%	−0.00009%	−0.00047%	−0.00061%	−0.00029%

13.

x	1.0	1.2	1.4	1.6	1.8	2.0
$y\,(h=0.2)$	3.000000	3.173896	3.441170	3.814932	4.300904	4.899004
$y\,(h=0.1)$	3.000000	3.173894	3.441163	3.814919	4.300885	4.898981
y actual	3.000000	3.173894	3.441163	3.814918	4.300884	4.898979
error	0.00000%	−0.00001%	−0.00001%	−0.00002%	−0.00003%	−0.00003%

15.

x	2.0	2.2	2.4	2.6	2.9	3.0
$y\,(h=0.2)$	3.000000	3.026448	3.094447	3.191719	3.310207	3.444447
$y\,(h=0.1)$	3.000000	3.026446	3.094445	3.191716	3.310204	3.444445
y actual	3.000000	3.026446	3.094444	3.191716	3.310204	3.444444
error	−0.000000%	−0.000004%	−0.000005%	−0.000005%	−0.000005%	−0.000004%

17.

Value of h	Estimated value of $y(1)$
0.2	0.350258
0.1	0.350234
0.05	0.350232
0.025	0.350232

The table of numerical results is

x	y with $h=0.2$	y with $h=0.1$	y with $h=0.05$	y with $h=0.025$
0.0	0.000000	0.000000	0.000000	0.000000
0.2	0.002667	0.002667	0.002667	0.002667
0.4	0.021360	0.021359	0.021359	0.021359
0.6	0.072451	0.072448	0.072448	0.072448
0.8	0.174090	0.174081	0.174080	0.174080
1.0	0.350258	0.350234	0.350232	0.350232

In Problems 18−24 we give only the final approximate values of y obtained using the Runge-Kutta method with step sizes $h=0.2$, $h=0.1$, $h=0.05$, and $h=0.025$.

19.

Value of h	Estimated value of $y(2)$
0.2	1.679459
0.1	6.411474
0.05	6.411474
0.025	6.411474
0.025	−1.259993

21.

Value of h	Estimated value of $y(2)$
0.2	2.872467
0.1	2.872468
0.05	2.872468
0.025	2.872468

23.

Value of h	Estimated value of $y(1)$
0.2	1.230735
0.1	1.230731
0.05	1.230731
0.025	1.230731

25.   Here $f(t,v) = 32 - 1.6v$ and $t_0 = 0$, $v_0 = 0$. With $h = 0.1$, 10 iterations of

$$k_1 = f(t_n, v_n) \qquad k_2 = f\left(t_n + \frac{1}{2}h, v_n + \frac{1}{2}hk_1\right)$$

$$k_3 = f\left(t_n + \frac{1}{2}h, v_n + \frac{1}{2}hk_2\right) \qquad k_4 = f(t_n + h, v_n + hk_3)$$

$$k = \frac{1}{6}(k_1 + 2k_2 + 2k_3 + k_4) \qquad v_{n+1} = v_n + hk$$

yield $v(1) \approx 15.9620$, and 20 iterations with $h = 0.05$ yield $v(1) \approx 15.9621$. Thus we observe an approximate velocity of 15.962 ft/sec after 1 second — 80% of the limiting velocity of 20 ft/sec.

With $h = 0.1$, 20 iterations yield $v(2) \approx 19.1847$, and 40 iterations with $h = 0.05$ yield $v(2) \approx 19.1848$. Thus we observe an approximate velocity of 19.185 ft/sec after 2 seconds — 96% of the limiting velocity of 20 ft/sec.

27. Here $f(x,y) = x^2 + y^2 - 1$ and $x_0 = 0$, $y_0 = 0$. The following table gives the approximate values for the successive step sizes h and corresponding numbers n of steps. It appears likely that $y(2) = 1.00445$ rounded off accurate to 5 decimal places.

h	1	0.1	0.01	0.001
n	2	20	200	2000
$y(2)$	1.05722	1.00447	1.00445	1.00445

In the solutions for Problems 29 and 30 we use the general MATLAB solver **ode** that was listed prior to the Problem 29 solution in Section 2.5. To use the Runge-Kutta method, we call as **'method'** the following function.

```
function [t,y] = rk(yp, t0,t1, y0)

%   [t, y] = rk(yp, t0, t1, y0)
%   Takes one Runge-Kutta step for
%
%       y' = yp( t,y ),
%
%   from t0  to  t1  with initial value  the
%   column vector  y0.

h = t1 - t0;
k1 = feval(yp, t0        , y0             );
k2 = feval(yp, t0 + h/2, y0 + (h/2)*k1 );
k3 = feval(yp, t0 + h/2, y0 + (h/2)*k2 );
k4 = feval(yp, t0 + h   ,y0 +    h *k3 );
k  = (1/6)*(k1 + 2*k2 + 2*k3 + k4);
t = t1;
y = y0 + h*k;
```

29. Here our differential equation is described by the MATLAB function

```
function  vp = vpbolt1(t,v)
vp = -0.04*v - 9.8;
```

Then the commands

```
n = 100;
[t1,v1] = ode('rk','vpbolt1',0,10,49,n);
n = 200;
[t2,v] = ode('rk','vpbolt1',0,10,49,n);
t = (0:10)';
ve = 294*exp(-t/25)-245;
[t, v1(1:n/20:1+n/2), v(1:n/10:n+1), ve]
```

generate the table

t	with $n = 100$	with $n = 200$	actual v
0	49.0000	49.0000	49.0000
1	37.4721	37.4721	37.4721
2	26.3962	26.3962	26.3962
3	15.7546	15.7546	15.7546
4	5.5303	5.5303	5.5303
5	−4.2932	−4.2932	−4.2932
6	−13.7314	−13.7314	−13.7314
7	−22.7996	−22.7996	−22.7996
8	−31.5122	−31.5122	−31.5122
9	−39.8832	−39.8832	−39.8832
10	−47.9259	−47.9259	−47.9259

We notice first that the final three columns agree to the 4 displayed decimal places. Scanning the last column for sign changes in v, we suspect that $v = 0$ (at the bolt's apex) occurs just after $t = 4.5\,\text{sec}$. Then interpolation between $t = 4.55$ and $t = 4.60$ in the table

```
[t2(91:95),v(91:95)]
```

4.5000	0.5694
4.5500	0.0788
4.6000	−0.4109
4.6500	−0.8996
4.7000	−1.3873

indicates that $t = 4.56$ at the bolt's apex. Now the commands

```
y = zeros(n+1,1);
h = 10/n;


for j = 2:n+1
   y(j) = y(j-1) + v(j-1)*h + 0.5*(-.04*v(j-1) - 9.8)*h^2;
end
ye = 7350*(1 - exp(-t/25)) - 245*t;
[t, y(1:n/10:n+1), ye]
```

generate the table

t	approx. y	actual y
0	0	0
1	43.1974	43.1976
2	75.0945	75.0949
3	96.1342	96.1348
4	106.7424	106.7432
5	107.3281	107.3290
6	98.2842	98.2852
7	79.9883	79.9895
8	52.8032	52.8046
9	17.0775	17.0790
10	−26.8540	−26.8523

We see at least 2-decimal place agreement between approximate and actual values of y. Finally, interpolation between $t = 9$ and $t = 10$ here suggests that $y = 0$ just after $t = 9.4$. Then interpolation between $t = 9.40$ and $t = 9.45$ in the table

`[t2(187:191),y(187:191)]`

9.3000	4.7448
9.3500	2.6182
9.4000	0.4713
9.4500	−1.6957
9.5000	−3.8829

indicates that the bolt is aloft for about 9.41 seconds.

CHAPTER 3

LINEAR EQUATIONS OF HIGHER ORDER

SECTION 3.1

INTRODUCTION: SECOND-ORDER LINEAR EQUATIONS

In this section the central ideas of the theory of linear differential equations are introduced and illustrated concretely in the context of **second-order** equations. These key concepts include superposition of solutions (Theorem 1), existence and uniqueness of solutions (Theorem 2), linear independence, the Wronskian (Theorem 3), and general solutions (Theorem 4). This discussion of second-order equations serves as preparation for the treatment of nth order linear equations in Section 3.2. Although the concepts in this section may seem somewhat abstract to students, the problems set is quite tangible and largely computational.

In each of Problems 1–16 the verification that y_1 and y_2 satisfy the given differential equation is a routine matter. As in Example 2, we then impose the given initial conditions on the general solution $y = c_1 y_1 + c_2 y_2$. This yields two linear equations that determine the values of the constants c_1 and c_2.

1. Imposition of the initial conditions $y(0) = 0$, $y'(0) = 5$ on the general solution

$y(x) = c_1 e^x + c_2 e^{-x}$ yields the two equations $c_1 + c_2 = 0$, $c_1 - c_2 = 5$ with solution $c_1 = \dfrac{5}{2}$,

$c_2 = -\dfrac{5}{2}$. Hence the desired particular solution is $y(x) = \dfrac{5}{2}\left(e^x - e^{-x}\right)$.

3. Imposition of the initial conditions $y(0) = 3$, $y'(0) = 8$ on the general solution

$y(x) = c_1 \cos 2x + c_2 \sin 2x$ yields the two equations $c_1 = 3$, $2c_2 = 8$ with solution $c_1 = 3$,

$c_2 = 4$. Hence the desired particular solution is $y(x) = 3\cos 2x + 4\sin 2x$.

5. Imposition of the initial conditions $y(0) = 1$, $y'(0) = 0$ on the general solution

$y(x) = c_1 e^x + c_2 e^{2x}$ yields the two equations $c_1 + c_2 = 1$, $c_1 + 2c_2 = 0$ with solution

$c_1 = 2$, $c_2 = -1$. Hence the desired particular solution is $y(x) = 2e^x - e^{2x}$.

7. Imposition of the initial conditions $y(0) = -2$, $y'(0) = 8$ on the general solution

$y(x) = c_1 + c_2 e^{-x}$ yields the two equations $c_1 + c_2 = -2$, $-c_2 = 8$ with solution $c_1 = 6$,

$c_2 = -8$. Hence the desired particular solution is $y(x) = 6 - 8e^{-x}$.

9. Imposition of the initial conditions $y(0) = 2$, $y'(0) = -1$ on the general solution $y(x) = c_1 e^{-x} + c_2 x e^{-x}$ yields the two equations $c_1 = 2$, $-c_1 + c_2 = -1$ with solution $c_1 = 2$ $c_2 = 1$. Hence the desired particular solution is $y(x) = 2e^{-x} + xe^{-x}$.

11. Imposition of the initial conditions $y(0) = 0$, $y'(0) = 5$ on the general solution $y(x) = c_1 e^x \cos x + c_2 e^x \sin x$ yields the two equations $c_1 = 0$, $c_1 + c_2 = 5$ with solution $c_1 = 0$, $c_2 = 5$. Hence the desired particular solution is $y(x) = 5e^x \sin x$.

13. Imposition of the initial conditions $y(1) = 3$, $y'(1) = 1$ on the general solution $y(x) = c_1 x + c_2 x^2$ yields the two equations $c_1 + c_2 = 3$, $c_1 + 2c_2 = 1$ with solution $c_1 = 5$, $c_2 = -2$. Hence the desired particular solution is $y(x) = 5x - 2x^2$.

15. Imposition of the initial conditions $y(1) = 7$, $y'(1) = 2$ on the general solution $y(x) = c_1 x + c_2 x \ln x$ yields the two equations $c_1 = 7$, $c_1 + c_2 = 2$ with solution $c_1 = 7$, $c_2 = -5$. Hence the desired particular solution is $y(x) = 7x - 5x \ln x$.

17. If $y = \dfrac{c}{x}$, then $y' + y^2 = -\dfrac{c}{x^2} + \dfrac{c^2}{x^2} = \dfrac{c(c-1)}{x^2} \neq 0$ unless either $c = 0$ or $c = 1$.

19. If $y = 1 + \sqrt{x}$, then $yy'' + (y')^2 = \left(1 + \sqrt{x}\right)\left(-\dfrac{x^{-3/2}}{4}\right) + \left(\dfrac{x^{-1/2}}{2}\right)^2 = -\dfrac{x^{-3/2}}{4} \neq 0$.

21. Linearly independent, because $x^3 = +x^2 |x|$ if $x > 0$, whereas $x^3 = -x^2 |x|$ if $x < 0$.

23. Linearly independent, because $f(x) = +g(x)$ if $x > 0$, whereas $f(x) = -g(x)$ if $x < 0$.

25. $f(x) = e^x \sin x$ and $g(x) = e^x \cos x$ are linearly independent, because $f(x) = kg(x)$ would imply that $\sin x = k \cos x$, whereas $\sin x$ and $\cos x$ are linearly independent, as noted in Example 3.

27. The operator notation used elsewhere in this chapter is convenient here. Let $L[y]$ denote $y'' + py' + qy$. Then $L[y_c] = 0$ and $L[y_p] = f$, so $L[y_c + y_p] = 0 + f = f$.

29.   There is no contradiction, because if the given differential equation is divided by x^2 to get the form in Equation (8) in the text, then the resulting functions $p(x) = -\dfrac{4}{x}$ and $q(x) = \dfrac{6}{x^2}$ are not continuous at $x = 0$.

31. $W(y_1, y_2) = -2x$ vanishes at $x = 0$, whereas if y_1 and y_2 were (linearly independent) solutions of an equation $y'' + py' + qy = 0$ with p and q both continuous on an open interval I containing $x = 0$, then Theorem 3 would imply that $W \neq 0$ on I.
(c) Because the exponential factor is never zero.

In Problems 33–42 we give the characteristic equation, its roots, and the corresponding general solution.

33. $r^2 - 3r + 2 = 0$; $r = 1, 2$; $y(x) = c_1 e^x + c_2 e^{2x}$

35. $r^2 + 5r = 0$; $r = 0, -5$; $y(x) = c_1 + c_2 e^{-5x}$

37. $2r^2 - r - 1 = 0$; $r = 1, -\dfrac{1}{2}$; $y(x) = c_1 e^{-x/2} + c_2 e^x$

39. $4r^2 + 4r + 1 = 0$; $r = -\dfrac{1}{2}$ (repeated); $y(x) = (c_1 + c_2 x) e^{-x/2}$

41. $6r^2 - 7r - 20 = 0$; $r = -\dfrac{4}{3}, \dfrac{5}{2}$; $y(x) = c_1 e^{-4x/3} + c_2 e^{5x/2}$

In Problems 43–48 we first write and simplify the equation with the indicated characteristic roots, and then write the corresponding differential equation.

43. $(r - 0)(r + 10) = r^2 + 10r = 0$; $y'' + 10y' = 0$

45. $(r + 10)(r + 10) = r^2 + 20r + 100 = 0$; $y'' + 20y' + 100y = 0$

47. $(r - 0)(r - 0) = r^2 = 0$; $y'' = 0$

49.   The solution curve with $y(0) = 1$, $y'(0) = 6$ is $y(x) = 8e^{-x} - 7e^{-2x}$. We find that $y'(x) = 0$ when $x = \ln\dfrac{7}{4}$, so that $e^{-x} = \dfrac{4}{7}$ and $e^{-2x} = \dfrac{16}{49}$. It follows that $y\left(\ln\dfrac{7}{4}\right) = \dfrac{16}{7}$, so the high point on the curve is $\left(\ln \dfrac{7}{4}, \dfrac{16}{7} \right) \approx (0.56, 2.29)$, which looks consistent with Fig. 3.1.6.

51.   **(a)** The substitution $v = \ln x$ gives

$$y' = \frac{dy}{dx} = \frac{dy}{dv}\frac{dv}{dx} = \frac{1}{x}\frac{dy}{dv}.$$

Then another differentiation using the chain and product rules gives

$$\begin{aligned}
y'' &= \frac{d^2 y}{dx^2} \\
&= \frac{d}{dx}\left(\frac{dy}{dx} \right) \\
&= \frac{d}{dx}\left(\frac{1}{x} \cdot \frac{dy}{dv} \right) \\
&= -\frac{1}{x^2} \cdot \frac{dy}{dv} + \frac{1}{x} \cdot \frac{d}{dx}\left(\frac{dy}{dv} \right) \\
&= -\frac{1}{x^2} \cdot \frac{dy}{dv} + \frac{1}{x} \cdot \frac{d}{dv}\left(\frac{dy}{dv} \right) \cdot \frac{dv}{dx} \\
&= -\frac{1}{x^2} \cdot \frac{dy}{dv} + \frac{1}{x^2} \cdot \frac{d^2 y}{dv^2}.
\end{aligned}$$

Substitution of these expressions for y' and y'' into Eq. (21) in the text then yields immediately the desired Eq. (23):

$$a \frac{d^2 y}{dv^2} + (b - a)\frac{dy}{dv} + cy = 0.$$

(b) If the roots r_1 and r_2 of the characteristic equation of Eq. (23) are real and distinct, then a general solution of the original Euler equation is

$$y(x) = c_1 e^{r_1 v} + c_2 e^{r_2 v} = c_1 \left(e^v \right)^{r_1} + c_2 \left(e^v \right)^{r_2} = c_1 x^{r_1} + c_2 x^{r_2}.$$

53.   The substitution $v = \ln x$ yields the converted equation $\dfrac{dy^2}{dv^2} + \dfrac{dy}{dv} - 12y = 0$, whose characteristic equation $r^2 + r - 12 = 0$ has roots $r_1 = -4$ and $r_2 = 3$. Because $e^v = x$, the corresponding general solution is $y = c_1 e^{-4v} + c_2 e^{3v} = c_1 x^{-4} + c_2 x^3$.

55.   The substitution $v = \ln x$ yields the converted equation $\dfrac{dy^2}{dv^2} = 0$, whose characteristic

equation $r^2 = 0$ has repeated roots $r_1, r_2 = 0$. Because $v = \ln x$, the corresponding general

solution is $y = c_1 + c_2 v = c_1 + c_2 \ln x$.

SECTION 3.2

GENERAL SOLUTIONS OF LINEAR EQUATIONS

Students should check each of Theorems 1 through 4 in this section to see that, in the case $n = 2$, it reduces to the corresponding theorem in Section 3.1. Similarly, the computational problems for this section largely parallel those for the previous section. By the end of Section 3.2 students should understand that, although we do not prove the existence-uniqueness theorem now, it provides the basis for everything we do with linear differential equations.

The linear combinations listed in Problems 1–6 were discovered "by inspection"—that is, by trial and error.

1. $\dfrac{5}{2} \cdot 2x + \left(-\dfrac{8}{3} \right) \cdot 3x^2 + (-1)\left(5x - 8x^2 \right) = 0$ for all x.

3. $1 \cdot 0 + 0 \cdot \sin x + 0 \cdot e^x = 0$ for all x.

5. $1 \cdot 17 + (-34) \cdot \cos^2 x + 17 \cdot \cos 2x = 0$ for all x, because $2\cos^2 x = 1 + \cos 2x$.

7. $W = \begin{vmatrix} 1 & x & x^2 \\ 0 & 1 & 2x \\ 0 & 0 & 2 \end{vmatrix} = 2$ is nonzero everywhere.

9. $W = e^x \left(\cos^2 x + \sin^2 x \right) = e^x$ is never zero.

11. $W = x^3 e^{2x}$ is nonzero if $x \neq 0$.

In each of Problems 13-20 we first form the general solution
$y(x) = c_1 y_1(x) + c_2 y_2(x) + c_3 y_3(x)$, then calculate $y'(x)$ and $y''(x)$, and finally impose the given initial conditions to determine the values of the coefficients c_1, c_2, c_3.

13.   Imposition of the initial conditions $y(0) = 1$, $y'(0) = 2$, $y''(0) = 0$ on the general solu-

tion $y(x) = c_1 e^x + c_2 e^{-x} + c_3 e^{-2x}$ yields the three equations

$$c_1 + c_2 + c_3 = 1, \quad c_1 - c_2 - 2c_3 = 2, \quad c_1 + c_2 + 4c_3 = 0,$$

with solution $c_1 = \dfrac{4}{3}$, $c_2 = 0$, $c_3 = -\dfrac{1}{3}$. Hence the desired particular solution is given by

$$y(x) = \frac{1}{3}\left(4e^x - e^{-2x}\right).$$

15. Imposition of the initial conditions $y(0) = 2$, $y'(0) = 0$, $y''(0) = 0$ on the general solution $y(x) = c_1 e^x + c_2 x e^x + c_3 x^2 e^x$ yields the three equations

$$c_1 = 2, \quad c_1 + c_2 = 0, \quad c_1 + 2c_2 + 2c_3 = 0,$$

with solution $c_1 = 2$, $c_2 = -2$, $c_3 = 1$. Hence the desired particular solution is given by
$$y(x) = \left(2 - 2x + x^2\right)e^x.$$

17. Imposition of the initial conditions $y(0) = 3$, $y'(0) = 1$, $y''(0) = 2$ on the general solution $y(x) = c_1 + c_2 \cos 3x + c_3 \sin 3x$ yields the three equations

$$c_1 + c_3 = 3, \quad 3c_3 = -1, \quad -9c_2 = 2$$

with solution $c_1 = \dfrac{29}{9}$, $c_2 = -\dfrac{2}{9}$, $c_3 = -\dfrac{1}{3}$. Hence the desired particular solution is given

by $y(x) = \dfrac{29}{9} - \dfrac{2}{9}\cos 3x - \dfrac{1}{3}\sin 3x$.

19. Imposition of the initial conditions $y(1) = 6$, $y'(1) = 14$, $y''(1) = 22$ on the general solution $y(x) = c_1 x + c_2 x^2 + c_3 x^3$ yields the three equations

$$c_1 + c_2 + c_3 = 6, \quad c_1 + 2c_2 + 3c_3 = 14, \quad 2c_2 + 6c_3 = 22,$$

with solution $c_1 = 1$, $c_2 = 2$, $c_3 = 3$. Hence the desired particular solution is given by
$$y(x) = x + 2x^2 + 3x^3.$$

In each of Problems 21-24 we first form the general solution

$$y(x) = y_c(x) + y_p(x) = c_1 y_1(x) + c_2 y_2(x) + y_p(x),$$

then calculate $y'(x)$, and finally impose the given initial conditions to determine the values of the coefficients c_1 and c_2.

21. Imposition of the initial conditions $y(0) = 2$, $y'(0) = -2$ on the general solution
$y(x) = c_1 \cos x + c_2 \sin x + 3x$ yields the two equations $c_1 = 2$, $c_2 + 3 = -2$ with solution

$c_1 = 2$, $c_2 = -5$. Hence the desired particular solution is given by

$$y(x) = 2\cos x - 5\sin x + 3x.$$

23.   Imposition of the initial conditions $y(0) = 3$, $y'(0) = 11$ on the general solution

$y(x) = c_1 e^{-x} + c_2 e^{3x} - 2$ yields the two equations $c_1 + c_2 - 2 = 3$, $-c_1 + 3c_2 = 11$ with solution $c_1 = 1$, $c_2 = 4$. Hence the desired particular solution is given by

$$y(x) = e^{-x} + 4e^{3x} - 2.$$

25. $Ly = L[y_1 + y_2] = Ly_1 + Ly_2 = f + g$

27.   The equations

$$c_1 + c_2 x + c_3 x^2 = 10, \quad c_2 + 2c_3 x = 0, \quad 2c_3 = 0$$

(the latter two obtained by successive differentiation of the first one) evidently imply that $c_1 = c_2 = c_3 = 0$.

29.   If $c_0 e^{rx} + c_1 x e^{rx} + \cdots + c_n x^n e^{rx} = 0$, then division by e^{rx} yields $c_0 + c_1 x + \cdots + c_n x^n = 0$, so the result of Problem 28 applies.

31.   **(a)** Substitution of $x = a$ in the differential equation gives $y''(a) = -py'(a) - q(a)$.

(b) If $y(0) = 1$ and $y'(0) = 0$, then the equation $y'' - 2y' - 5y = 0$ implies that

$$y''(0) = 2y'(0) + 5y(0) = 5.$$

33.   This follows from the fact that

$$\begin{vmatrix} 1 & 1 & 1 \\ a & b & c \\ a^2 & b^2 & c^2 \end{vmatrix} = (b-a)(c-b)(c-a)$$

when a, b, and c are distinct, which can be verified by expanding both sides of the equation.

37.   When we substitute $y = vx^3$ in the given differential equation and simplify, we get the

separable equation $xv'' + v' = 0$, which we write as $\dfrac{v''}{v'} = -\dfrac{1}{x}$. Integrating gives

$\ln v' = -\ln x + \ln A$, and then solving for v' leads to $v' = \dfrac{A}{x}$, or finally $v(x) = A\ln x + B$.

With $A = 1$ and $B = 0$ we get $v(x) = \ln x$, and thus $y_2(x) = x^3 \ln x$.

39. When we substitute $y = v e^{x/2}$ in the given differential equation and simplify, we eventually get the simple equation $v'' = 0$, with general solution $v(x) = Ax + B$. With $A = 1$ and $B = 0$ we get $v(x) = x$, and hence $y_2(x) = x e^{x/2}$.

41. When we substitute $y = v e^x$ in the given differential equation and simplify, we get the separable equation $(1 + x) v'' + x v' = 0$, which we write as $\dfrac{v''}{v'} = -\dfrac{x}{1+x} = -1 + \dfrac{1}{1+x}$. Integrating gives $\ln v' = -x + \ln(1+x) + \ln A$, and then solving for v' leads to $v' = A(1+x) e^{-x}$, or finally $v(x) = A \int (1+x) e^{-x} \, dx = -A(2+x) e^{-x} + B$. With $A = -1$ and $B = 0$ we get $v(x) = (2+x) e^{-x}$, and hence $y_2(x) = 2 + x$.

43. When we substitute $y = vx$ in the given differential equation and simplify, we get the separable equation $x(x^2 - 1) v'' = (2 - 4x^2) v'$, which we write using the method of partial fractions as

$$\frac{v''}{v'} = \frac{2 - 4x^2}{x(x^2 - 1)} = -\frac{2}{x} - \frac{1}{1+x} + \frac{1}{1-x}.$$

Integrating gives

$$\ln v' = -2 \ln x - \ln(1+x) - \ln(1-x) + \ln A,$$

and then solving for v' leads to

$$v' = \frac{A}{x^2(1-x^2)} = A\left[\frac{1}{x^2} + \frac{1}{2(1+x)} + \frac{1}{2(1-x)} \right],$$

or finally

$$v(x) = A\left[-\frac{1}{x} + \frac{1}{2} \ln(1+x) - \frac{1}{2} \ln(1-x) \right] + B.$$

With $A = -1$ and $B = 0$ we get $v(x) = \dfrac{1}{x} - \dfrac{1}{2} \ln(1+x) + \dfrac{1}{2} \ln(1-x)$, and hence

$$y_2(x) = 1 - \frac{x}{2} \ln \frac{1+x}{1-x}.$$

SECTION 3.3

HOMOGENEOUS EQUATIONS WITH CONSTANT COEFFICIENTS

This is a purely computational section devoted to the single most widely applicable type of higher order differential equations—linear ones with constant coefficients. In Problems 1–20, we first write the characteristic equation and list its roots, then give the corresponding general solution of the given differential equation. Explanatory comments are included only when the solution of the characteristic equation is not routine.

1. $r^2 - 4 = (r-2)(r+2) = 0$; $r = -2, 2$; $y(x) = c_1 e^{2x} + c_2 e^{-2x}$

3. $r^2 + 3r - 10 = (r+5)(r-2) = 0$; $r = -5, 2$; $y(x) = c_1 e^{2x} + c_2 e^{-5x}$

5. $r^2 + 6r + 9 = (r+3)^2 = 0$; $r = -3$ (repeated); $y(x) = c_1 e^{-3x} + c_2 x e^{-3x}$

7. $4r^2 - 12r + 9 = (2r-3)^2 = 0$; $r = \dfrac{3}{2}$ (repeated); $y(x) = c_1 e^{3x/2} + c_2 x e^{3x/2}$

9. $r^2 + 8r + 25 = 0$; $r = \dfrac{-8 \pm \sqrt{-36}}{2} = -4 \pm 3i$; $y(x) = e^{-4x}(c_1 \cos 3x + c_2 \sin 3x)$

11. $r^4 - 8r^3 + 16r^2 = r^2(r-4)^2 = 0$; $r = 0, 0, 4, 4$; $y(x) = c_1 + c_2 x + c_3 e^{4x} + c_4 x e^{4x}$

13. $9r^3 + 12r^2 + 4r = r(3r+2)^2 = 0$; $r = 0, -\dfrac{2}{3}, -\dfrac{2}{3}$; $y(x) = c_1 + c_2 e^{-2x/3} + c_3 x e^{-2x/3}$

15. $4r^4 - 8r^2 + 16 = (r^2 - 4)^2 = (r-2)^2(r+2)^2 = 0$; $r = 2, 2, -2, -2$;

$$y(x) = c_1 e^{2x} + c_2 x e^{2x} + c_3 e^{-2x} + c_4 x e^{-2x}$$

17. $6r^4 + 11r^2 + 4 = (2r^2 + 1)(3r^2 + 4) = 0$; $r = \pm\dfrac{i}{\sqrt{2}}, \pm\dfrac{2i}{\sqrt{3}}$;

$$y(x) = c_1 \cos\frac{x}{\sqrt{2}} + c_2 \sin\frac{x}{\sqrt{2}} + c_3 \cos\frac{2x}{\sqrt{3}} + c_4 \sin\frac{2x}{\sqrt{3}}$$

19.   Factoring by grouping gives $r^3 + r^2 - r - 1 = r(r^2 - 1) + (r^2 - 1) = (r-1)(r+1)^2 = 0$;
$r = 1, -1, -1$; $y(x) = c_1 e^x + c_2 e^{-x} + c_3 x e^{-x}$.

21.   Imposition of the initial conditions $y(0) = 7$, $y'(0) = 11$ on the general solution
$y(x) = c_1 e^x + c_2 e^{3x}$ yields the two equations $c_1 + c_2 = 7$, $c_1 + 3c_2 = 11$ with solution
$c_1 = 5$, $c_2 = 2$. Hence the desired particular solution is $y(x) = 5e^x + 2e^{3x}$.

23.   Imposition of the initial conditions $y(0) = 3$, $y'(0) = 1$ on the general solution
$y(x) = e^{3x}(c_1 \cos 4x + c_2 \sin 4x)$ yields the two equations $c_1 = 3$, $3c_1 + 4c_2 = 1$ with solution $c_1 = 3$, $c_2 = -2$. Hence the desired particular solution is
$y(x) = e^{3x}(3\cos 4x - 2\sin 4x)$.

25.   Imposition of the initial conditions $y(0) = -1$, $y'(0) = 0$, $y''(0) = 1$ on the general solution $y(x) = c_1 + c_2 x + c_3 e^{-2x/3}$ yields the three equations

$$c_1 + c_3 = -1, \quad c_2 - \frac{2c_3}{3} = 0, \quad \frac{4c_3}{9} = 1,$$

with solution $c_1 = -\dfrac{13}{4}$, $c_2 = \dfrac{3}{2}$, $c_3 = \dfrac{9}{4}$. Hence the desired particular solution is

$$y(x) = -\frac{13}{4} + \frac{3}{2}x + \frac{9}{4}e^{-2x/3}.$$

27.   First we spot the root $r = 1$. Then long division of the polynomial $r^3 + 3r^2 - 4$ by $r - 1$ yields the quadratic factor $r^2 + 4r + 4 = (r+2)^2$, with roots $r = -2, -2$. Hence the general solution is $y(x) = c_1 e^x + c_2 e^{-2x} + c_3 x e^{-2x}$.

29.   First we spot the root $r = -3$. Then long division of the polynomial $r^3 + 27$ by $r + 3$ yields the quadratic factor $r^2 - 3r + 9$, with roots $r = \dfrac{3}{2} \pm i\dfrac{3\sqrt{3}}{2}$. Hence the general solution is $y(x) = c_1 e^{-3x} + e^{3x/2}\left(c_2 \cos\dfrac{3\sqrt{3}}{2}x + c_3 \sin\dfrac{3\sqrt{3}}{2}x\right)$.

31.   The characteristic equation $r^3 + 3r^2 + 4r - 8 = 0$ has the evident root $r = 1$, and long division then yields the quadratic factor $r^2 + 4r + 8 = (r+2)^2 + 4$, corresponding to the complex conjugate roots $-2 \pm 2i$. Hence the general solution is
$y(x) = c_1 e^x + e^{-2x}(c_2 \cos 2x + c_3 \sin 2x)$.

33.   Knowing that $y = e^{3x}$ is one solution, we divide the characteristic polynomial $r^3 + 3r^2 - 54$ by $r - 3$ and get the quadratic factor $r^2 + 6r + 18 = (r+3)^2 + 9$. Hence the general solution is $y(x) = c_1 e^{3x} + e^{-3x}(c_2 \cos 3x + c_3 \sin 3x)$.

35.   The fact that $y = \cos 2x$ is one solution tells us that $r^2 + 4$ is a factor of the characteristic polynomial $6r^4 + 5r^3 + 25r^2 + 20r + 4$. Then long division yields the quadratic factor

$6r^2 + 5r + 1 = (3r + 1)(2r + 1)$, with roots $r = -\dfrac{1}{2}, -\dfrac{1}{3}$. Hence the general solution is

$$y(x) = c_1 e^{-x/2} + c_2 e^{-x/3} + c_3 \cos 2x + c_4 \sin 2x.$$

37.   The characteristic equation is $r^4 - r^3 = r^3(r - 1) = 0$, so the general solution is
$y(x) = A + Bx + Cx^2 + De^x$. Imposition of the given initial conditions yields the equations

$$A + D = 18, \quad B + D = 12, \quad 2C + D = 13, \quad D = 7$$

with solution $A = 11$, $B = 5$, $C = 3$, $D = 7$. Hence the desired particular solution is
$y(x) = 11 + 5x + 3x^2 + 7e^x$.

39.   The characteristic polynomial is $(r - 2)^3 = r^3 - 6r^2 + 12r - 8$, so the differential equation
is $y''' - 6y'' + 12y' - 8y = 0$.

41.   The characteristic polynomial is $(r^2 + 4)(r^2 - 4) = r^4 - 16$, so the differential equation is
$y^{(4)} - 16y = 0$.

45.   The characteristic polynomial is the quadratic polynomial of Problem 44(b). Hence the
general solution is

$$y(x) = c_1 e^{-ix} + c_2 e^{3ix} = c_1(\cos x - i \sin x) + c_2(\cos 3x + i \sin 3x).$$

47.   The characteristic roots are $r = \pm\sqrt{-2 + 2i\sqrt{3}} = \pm(1 + i\sqrt{3})$, so the general solution is

$$y(x) = c_1 e^{(1+i\sqrt{3})x} + c_2 e^{-(1+i\sqrt{3})x} = c_1 e^x\left(\cos\sqrt{3}x + i\sin\sqrt{3}x\right) + c_2 e^{-x}\left(\cos\sqrt{3}x - i\sin\sqrt{3}x\right).$$

49.   We adopt the same strategy as was used in Problem 48. The general solution is
$y(x) = Ae^{2x} + Be^{-x} + C\cos x + D\sin x$. Imposition of the given initial conditions yields
the equations

$$
\begin{aligned}
A + B + C \quad\quad &= 0 \\
2A - B \quad + D &= 0 \\
4A + B - C \quad\quad &= 0 \\
8A - B \quad - D &= 30
\end{aligned}
$$

that we solve for $A = 2$, $B = -5$, $C = 3$, and $D = -9$. Thus

$$y(x) = 2e^{2x} - 5e^{-x} + 3\cos x - 9\sin x.$$

51.   In the solution of Problem 51 in Section 3.1 we showed that the substitution $v = \ln x$

gives $y' = \dfrac{dy}{dx} = \dfrac{1}{x}\dfrac{dy}{dv}$ and $y'' = \dfrac{d^2 y}{dx^2} = -\dfrac{1}{x^2}\cdot\dfrac{dy}{dv} + \dfrac{1}{x^2}\cdot\dfrac{d^2 y}{dv^2}$. A further differentiation using the chain rule gives

$$y''' = \frac{d^3 y}{dx^3} = \frac{2}{x^3}\cdot\frac{dy}{dv} - \frac{3}{x^3}\cdot\frac{d^2 y}{dv^2} + \frac{1}{x^3}\cdot\frac{d^3 y}{dv^3}.$$

Substitution of these expressions for y', y'', and y''' into the third-order Euler equation $ax^3 y''' + bx^2 y'' + cxy' + dy = 0$, together with collection of coefficients, yields the desired constant-coefficient equation

$$a\frac{d^3 y}{dv^3} + (b - 3a)\frac{d^2 y}{dv^2} + (c - b + 2a)\frac{dy}{dv} + d\cdot y = 0.$$

In Problems 52 through 58 we list first the transformed constant-coefficient equation, then its characteristic equation and roots, and finally the corresponding general solution with $v = \ln x$ and $e^v = x$.

53. $\dfrac{d^2 y}{dv^2} + 6\dfrac{dy}{dv} + 25y = 0$; $r^2 + 6r + 25 = 0$; $r = -3 \pm 4i$;

$$y(x) = e^{-3v}\left(c_1 \cos 4v + c_2 \sin 4v\right) = x^{-3}\left[c_1 \cos\left(4\ln x\right) + c_2 \sin\left(4\ln x\right)\right]$$

55. $\dfrac{d^3 y}{dv^3} - 4\dfrac{d^2 y}{dv^2} + 4\dfrac{dy}{dv} = 0$; $r^3 - 4r^2 + 4r = 0$; $r = 0, 2, 2$;

$$y(x) = c_1 + c_2 e^{2v} + c_3 v e^{2v} = c_1 + x^2\left(c_2 + c_3 \ln x\right)$$

57. $\dfrac{d^3 y}{dv^3} - 5\dfrac{d^2 y}{dv^2} + 5\dfrac{dy}{dv} = 0$; $r^3 - 4r^2 + 4r = 0$; $r = 0, 3 \pm \sqrt{3}$;

$$y(x) = c_1 + c_2 e^{(3 - \sqrt{3})v} + c_3 v e^{(3 + \sqrt{3})v} = c_1 + x^3\left(c_2 x^{-\sqrt{3}} + c_3 x^{+\sqrt{3}}\right)$$

SECTION 3.4

MECHANICAL VIBRATIONS

In this section we discuss four types of free motion of a mass on a spring—undamped, underdamped, critically damped, and overdamped. However, the undamped and underdamped cases—in which actual oscillations occur—are emphasized because they are both the most interesting and the most important cases for applications.

1.   Frequency: $\omega_0 = \sqrt{\dfrac{k}{m}} = \sqrt{\dfrac{16}{4}} = 2\,\text{rad/sec} = \dfrac{1}{\pi}\,\text{Hz}$; period: $P = \dfrac{2\pi}{\omega_0} = \dfrac{2\pi}{2} = \pi\,\text{sec}$

3.   The spring constant is $k = \dfrac{15\,N}{0.20\,\text{m}} = 75\,\text{N/m}$. The solution of $3x'' + 75x = 0$ with

$x(0) = 0$ and $x'(0) = -10$ is $x(t) = -2\sin 5t$. Thus the amplitude is 2 m, the frequency

is $\omega_0 = \sqrt{\dfrac{k}{m}} = \sqrt{\dfrac{75}{3}} = 5\,\text{rad/sec} = \dfrac{2.5}{\pi}\,\text{Hz}$, and the period is $\dfrac{2\pi}{5}\,\text{sec}$.

5.   The gravitational acceleration at distance R from the center of the earth is $g = \dfrac{GM}{R^2}$. Ac-

cording to Equation (6) in the text, the (circular) frequency ω of a (linearized) pendulum

is given by $\omega^2 = \dfrac{g}{L} = \dfrac{GM}{R^2 L}$, so its period is $p = \dfrac{2\pi}{\omega} = 2\pi R\sqrt{\dfrac{L}{GM}}$.

7.   The period equation $p = 3960\sqrt{100.10} = (3960 + x)\sqrt{100}$ yields

$x \approx 1.9795\,\text{mi} \approx 10.450\,\text{ft}$ for the altitude of the mountain.

9.   Designating $x(t)$ as in the suggestion, we see that the mass is subject to a restorative

force $F_S = -kx$ together with the force of gravity $W = mg$. We also assume that the

mass is subject to a damping force $F_R = -cx'$. Applying Newton's law then gives

$mx'' = -kx + mg - cx'$, or $mx'' + cx' + kx = mg$. Finally, substituting $y = x - s_0$, so that

$x = y + s_0$ and thus $x' = y'$ and $x'' = y''$, yields $my'' + cy' + k(y + s_0) = mg$, or

$my'' + cy' + ky = mg - ks_0$, which is Equation (5) with $F(t)$ assuming the constant value

$mg - ks_0$.

11.   The differential equation from Problem 10 must be modified to reflect the fact that the

weight density of water is $62.4\,\text{lb/ft}^3$ (as opposed to $1\,\text{g/cm}^3$ in the cgs system). Thus

the weight of water displaced by the buoy is given by $62.4\pi r^2 \cdot x$. Moreover, the mass

and weight of the buoy are given to be 3.125 slugs and 100 lb, respectively. Applying

$ma = F$ then gives $3.125x'' = 100 - 62.4\pi r^2 \cdot x$, or $x'' + \dfrac{62.4\pi}{3.125} r^2 \cdot x = 32$. The frequency

of the oscillations of the buoy is therefore $\dfrac{\omega_0}{2\pi}$, where $\omega_0 = \sqrt{\dfrac{62.4\pi}{3.125}} \cdot r$. Since the fre-

quency of the buoy's motion is observed to be $\dfrac{4\,\text{cycles}}{10\,\text{sec}} = 0.4\,\text{cycles/sec}$, we can equate the two to conclude that $\dfrac{1}{2\pi}\sqrt{\dfrac{62.4\pi}{3.125}} \cdot r = 0.4$, which gives

$$r = 0.8\sqrt{\dfrac{3.125\pi}{62.4}} \approx 0.3173\,\text{ft} \approx 3.8\,\text{in}.$$

13. **(a)** The characteristic equation $10r^2 + 9r + 2 = (5r+2)(2r+1) = 0$ has roots $r = -\dfrac{2}{5}, -\dfrac{1}{2}$.
When we impose the initial conditions $x(0) = 0$, $x'(0) = 5$ on the general solution $x(t) = c_1 e^{-2t/5} + c_2 e^{-t/2}$ we get the particular solution $x(t) = 50\left(e^{-2t/5} - e^{-t/2}\right)$.

(b) The derivative $x'(t) = 25e^{-t/2} - 20e^{-2t/5} = 5e^{-2t/5}\left(5e^{-t/10} - 4\right) = 0$ when

$t = 10\ln\dfrac{5}{4} \approx 2.23144$. Hence the mass's farthest distance to the right is given by

$$x\left(10\ln\dfrac{5}{4}\right) = \dfrac{512}{125} = 4.096.$$

In Problems 15-21 the graph of the damped motion $x(t)$, that is, with the dashpot attached, is shown as a solid line; the graph of the corresponding undamped motion $u(t)$ is dashed.

15. **With damping:** The characteristic equation $\dfrac{1}{2}r^2 + 3r + 4 = 0$ has roots $r = -2, -4$. When we impose the initial conditions $x(0) = 2$, $x'(0) = 0$ on the general solution $x(t) = c_1 e^{-2t} + c_2 e^{-4t}$ we get the particular solution $x(t) = 4e^{-2t} - 2e^{-4t}$ that describes overdamped motion.

Without damping: The characteristic equation $\dfrac{1}{2}r^2 + 4 = 0$ has roots $r = \pm 2i\sqrt{2}$. When we impose the initial conditions $x(0) = 2$, $x'(0) = 0$ on the general solution $u(t) = A\cos\left(2\sqrt{2}t\right) + B\sin\left(2\sqrt{2}t\right)$ we get the particular solution $u(t) = 2\cos\left(2\sqrt{2}t\right)$.

Problem 15   **Problem 17**

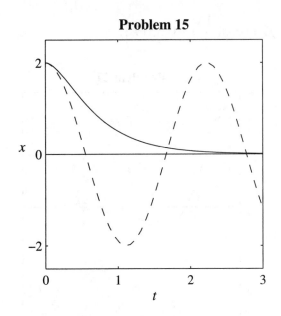   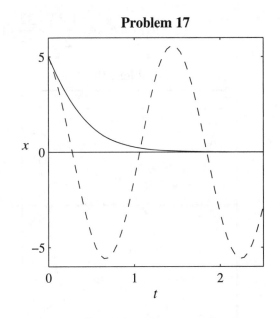

17.   **With damping:** The characteristic equation $r^2 + 8r + 16 = 0$ has roots $r = -4, -4$. When we impose the initial conditions $x(0) = 5$, $x'(0) = -10$ on the general solution

$x(t) = (c_1 + c_2 t)e^{-4t}$ we get the particular solution $x(t) = 5e^{-4t}(2t + 1)$ that describes critically damped motion.

Without damping: The characteristic equation $r^2 + 16 = 0$ has roots $r = \pm 4i$. When we impose the initial conditions $x(0) = 5$, $x'(0) = -10$ on the general solution

$u(t) = A\cos 4t + B\sin 4t$ we get the particular solution

$$u(t) = 5\cos 4t - \frac{5}{2}\sin 4t \approx \frac{5}{2}\sqrt{5}\cos(4t - 5.8195).$$

19.   The characteristic equation $4r^2 + 20r + 169 = 0$ has roots $r = -\frac{5}{2} \pm 6i$. When we impose the initial conditions $x(0) = 4$, $x'(0) = 16$ on the general solution

$x(t) = e^{-5t/2}(A\cos 6t + B\sin 6t)$ we get the particular solution

$$x(t) = e^{-5t/2}\left(4\cos 6t + \frac{13}{3}\sin 6t\right) \approx \frac{1}{3}\sqrt{313}\, e^{-5t/2}\cos(6t - 0.8254)$$

that describes underdamped motion.

Without damping: The characteristic equation $4r^2 + 169 = 0$ has roots $r = \pm\frac{13}{2}i$. When we impose the initial conditions $x(0) = 4$, $x'(0) = 16$ on the general solution

$u(t) = A\cos\left(\frac{13}{2}t\right) + B\sin\left(\frac{13}{2}t\right)$ we get the particular solution

$$u(t) = 4\cos\left(\frac{13t}{2}\right) + \frac{32}{13}\sin\left(\frac{13t}{2}\right) \approx \frac{4}{13}\sqrt{233}\cos\left(\frac{13}{2}t - 0.5517\right).$$

Problem 19   **Problem 21**

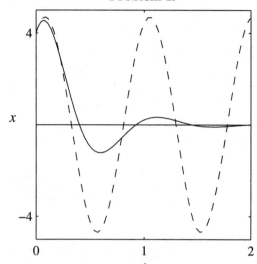   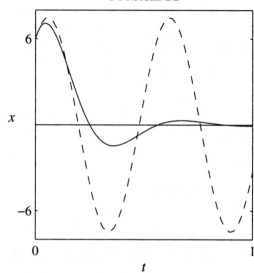

21.   **With damping:** The characteristic equation $r^2 + 10r + 125 = 0$ has roots $r = -5 \pm 10i$. When we impose the initial conditions $x(0) = 6$, $x'(0) = 50$ on the general solution $x(t) = e^{-5t}(A\cos 10t + B\sin 10t)$ we get the particular solution

$$x(t) = e^{-5t}(6\cos 10t + 8\sin 10t) \approx 10e^{-5t}\cos(10t - 0.9273)$$

that describes underdamped motion.

Without damping: The characteristic equation $r^2 + 125 = 0$ has roots $r = \pm 5\sqrt{5}\,i$. When we impose the initial conditions $x(0) = 6$, $x'(0) = 50$ on the general solution $u(t) = A\cos\left(5\sqrt{5}\,t\right) + B\sin\left(5\sqrt{5}\,t\right)$ we get the particular solution

$$u(t) = 6\cos\left(5\sqrt{5}\,t\right) + 2\sqrt{5}\sin\left(5\sqrt{5}\,t\right) \approx 2\sqrt{14}\cos\left(5\sqrt{5}\,t - 0.6405\right).$$

23.   **(a)** With $m = 100\,\text{slug}$ we get $\omega = \sqrt{\dfrac{k}{100}}$. But we are given that

$$\omega = (80\,\text{cycles/min}) \cdot 2\pi \cdot (1\,\text{min}/60\,\text{sec}) = \frac{8\pi}{3},$$

and equating the two values yields $k \approx 7018\,\text{lb/ft}$.

(b) With $\omega_1 = 2\pi \cdot \dfrac{78}{60}\,\text{cycles/sec}$, Equation (21) in the text yields $c \approx 372.31\,\text{lb}/(\text{ft/sec})$.

Hence $p = \dfrac{c}{2m} \approx 1.8615$. Finally $e^{-pt} = 0.01$ gives $t \approx 2.47\,\text{sec}$.

31.   The binomial series

$$(1+x)^\alpha = 1 + \alpha x + \frac{\alpha(\alpha-1)}{2!}x^2 + \frac{\alpha(\alpha-1)(\alpha-2)}{3!}x^3 + \cdots$$

converges if $|x| < 1$. (See, for instance, Section 10.8 of Edwards and Penney, Calculus: Early Transcendentals, 7th edition, Pearson, 2008.) With $\alpha = \dfrac{1}{2}$ and $x = -\dfrac{c^2}{4mk}$ in Eq. (22) of Section 3.4 in the differential equations text, the binomial series gives

$$\omega_1 = \sqrt{\omega_0^2 - p^2} = \sqrt{\frac{k}{m} - \frac{c^2}{4m^2}} = \sqrt{\frac{k}{m}}\sqrt{1 - \frac{c^2}{4mk}} = \sqrt{\frac{k}{m}}\left(1 - \frac{c^2}{8mk} - \frac{c^4}{128m^2k^2} - \cdots\right) \approx \omega_0\left(1 - \frac{c^2}{8mk}\right).$$

33.   If $x_1 = x(t_1)$ and $x_2 = x(t_2)$ are two successive local maxima, then $\omega_1 t_2 = \omega_1 t_1 + 2\pi$, and so $x_1 = Ce^{-pt_1}\cos(\omega_1 t_1 - \alpha)$ and $x_2 = Ce^{-pt_2}\cos(\omega_1 t_2 - \alpha) = Ce^{-pt_2}\cos(\omega_1 t_1 - \alpha)$. Hence

$$\frac{x_1}{x_2} = e^{-p(t_1-t_2)} \text{ and therefore } \ln\left(\frac{x_1}{x_2}\right) = -p(t_1 - t_2) = \frac{2\pi p}{\omega_1}.$$

35.   The characteristic equation $r^2 + 2r + 1 = 0$ has roots $r = -1, -1$. When we impose the initial conditions $x(0) = 0$, $x'(1) = 0$ on the general solution $x(t) = (c_1 + c_2 t)e^{-t}$ we get the particular solution $x_1(t) = te^{-t}$.

37.   The characteristic equation $r^2 + 2r + (1 + 10^{-2n}) = 0$ has roots $r = -1 \pm 10^{-n}i$. When we impose the initial conditions $x(0) = 0$, $x'(1) = 0$ on the general solution

$$x(t) = e^{-t}\left[A\cos(10^{-n}t) + B\sin(10^{-n}t)\right]$$

we get the equations $c_1 = 0$, $-c_1 + 10^{-n}c_2 = 1$ with solution $c_1 = 0$, $c_2 = 10^n$. This gives the particular solution $x_3(t) = 10^n e^{-t}\sin(10^{-n}t)$.

SECTION 3.5

NONHOMOGENEOUS EQUATIONS AND UNDETERMINED COEFFICIENTS

The method of undetermined coefficients is based on "educated guessing". If we can guess correctly the **form** of a particular solution of a nonhomogeneous linear equation with constant coefficients, then we can determine the particular solution explicitly by substitution in the given differential equation. It is pointed out at the end of Section 3.5 that this simple approach is not always successful—in which case the method of variation of parameters is available if a complementary function is known. However, undetermined coefficients *does* turn out to work well with a surprisingly large number of the nonhomogeneous linear differential equations that arise in elementary scientific applications.

In each of Problems 1-20 we give first the form of the trial solution y_{trial}, then the equations in the coefficients we get when we substitute y_{trial} into the differential equation and collect like terms, and finally the resulting particular solution y_p.

1. $y_{\text{trial}} = Ae^{3x}$; $25A = 1$; $y_p = \dfrac{1}{25}e^{3x}$.

3. $y_{\text{trial}} = A\cos 3x + B\sin 3x$; $-15A - 3B = 0$, $3A - 15B = 2$; $y_p = \dfrac{1}{39}\cos 3x - \dfrac{5}{39}\sin 3x$.

5.   First we substitute $\dfrac{1 - \cos 2x}{2}$ for $\sin^2 x$ on the right-hand side of the differential equation, leading to $y_{\text{trial}} = A + B\cos 2x + C\sin 2x$, and then

$$A = \frac{1}{2}, \quad -3B + 2C = -\frac{1}{2}, \quad -2B - 3C = 0;$$

$$y_p = \frac{1}{2} + \frac{3}{26}\cos 2x - \frac{1}{13}\sin 2x.$$

7.   First we substitute $\dfrac{e^x - e^{-x}}{2}$ for $\sinh x$ on the right-hand side of the differential equatio,

leading to $y_{\text{trial}} = Ae^x + Be^{-x}$; $-3A = \dfrac{1}{2}$, $-3B = -\dfrac{1}{2}$; $y_p = \dfrac{1}{6}e^{-x} - \dfrac{1}{6}e^x = -\dfrac{1}{3}\sinh x$. (Note

that according to Rule 1 in the text, we could also have started with
$y_{\text{trial}} = A\cosh x + B\sinh x$.)

9.   First we note that e^x is part of the complementary function $y_c = c_1 e^x + c_2 e^{-3x}$. Then
$y_{\text{trial}} = A + x(B + Cx)e^x$, and then

$$-3A = 1 \quad 4B + 2C = 0 \quad 8C = 1;$$

$$y_p = -\frac{1}{3} + x\left(-\frac{1}{16} + \frac{1}{8}x\right)e^x.$$

11.   First we note the duplication with the complementary function
$y_c = c_1 + c_2 \cos 2x + c_3 \sin 2x$. Then $y_{\text{trial}} = x(A + Bx)$; $4A = -1$, $8B = 3$;

$$y_p = x\left(-\frac{1}{4} + \frac{3}{8}x\right) = \frac{1}{8}(3x^2 - 2x).$$

13. $y_{\text{trial}} = e^x \left(A\cos x + B\sin x \right)$; $7A + 4B = 0$, $-4A + 7B = 1$; $y_p = \dfrac{1}{65} e^x \left(7\sin x - 4\cos x \right)$.

15. This is something of a trick problem. We cannot solve the characteristic equation $r^5 + 5r^4 - 1 = 0$ to find the complementary function, but we can see that the complementary function contains no constant term (why?). Hence we can take $y_{\text{trial}} = A$, leading immediately to the particular solution $y_p = -17$.

17. First we note the duplication with the complementary function $y_c = c_1 \cos x + c_2 \sin x$. Then $y_{\text{trial}} = x\left[(A + Bx)\cos x + (C + Dx)\sin x \right]$;

$$2B + 2C = 0 \quad 4D = 1 \quad -2A + 2D = 1 \quad -4B = 0;$$

$$y_p = x\left(-\frac{1}{4}\cos x + \frac{1}{4}x\sin x \right) = \frac{1}{4}\left(x^2 \sin x - x\cos x \right).$$

19. First we note the duplication with the part $c_1 + c_2 x$ of the complementary function (which corresponds to the factor r^2 of the characteristic polynomial). Then $y_{\text{trial}} = x^2 \left(A + Bx + Cx^2 \right)$;

$$4A + 12B = -1, \quad 12B + 48C = 0, \quad 24C = 3;$$

$$y_p = x^2 \left(\frac{5}{4} - \frac{1}{2}x + \frac{1}{8}x^2 \right) = \frac{1}{8}\left(10x^2 - 4x^3 + x^4 \right).$$

In Problems 21-30 we list first the complementary function y_c, then the initially proposed trial function y_i, and finally the actual trial function y_p, in which duplication with the complementary function has been eliminated.

21. $y_c = e^x \left(c_1 \cos x + c_2 \sin x \right)$; $y_i = e^x \left(A\cos x + B\sin x \right)$; $y_p = x \cdot e^x \left(A\cos x + B\sin x \right)$

23. $y_c = c_1 \cos 2x + c_2 \sin 2x$; $y_i = (A + Bx)\cos 2x + (C + Dx)\sin 2x$;
$y_p = x \cdot \left[(A + Bx)\cos 2x + (C + Dx)\sin 2x \right]$

25. $y_c = c_1 e^{-x} + c_2 e^{-2x}$; $y_i = (A + Bx)e^{-x} + (C + Dx)e^{-2x}$;
$y_p = x \cdot (A + Bx)e^{-x} + x \cdot (C + Dx)e^{-2x}$

27. $y_c = (c_1 \cos x + c_2 \sin x) + (c_3 \cos 2x + c_4 \sin 2x)$;

$y_i = (A \cos x + B \sin x) + (C \cos 2x + D \sin 2x)$;

$y_p = x \cdot \left[(A \cos x + B \sin x) + (C \cos 2x + D \sin 2x) \right]$

29. $y_c = (c_1 + c_2 x + c_3 x^2) e^x + c_4 e^{2x} + c_5 e^{-2x}$; $y_i = (A + Bx) e^x + C e^{2x} + D e^{-2x}$;

$y_p = x^3 \cdot (A + Bx) e^x + x \cdot (C e^{2x}) + x \cdot (D e^{-2x})$

In Problems 31-40 we list first the complementary function y_c, the trial solution y_{tr} for the method of undetermined coefficients, and the corresponding general solution $y_g = y_c + y_p$, where y_p results from determining the coefficients in y_{tr} so as to satisfy the given nonhomogeneous differential equation. Then we list the linear equations obtained by imposing the given initial conditions, and finally the resulting particular solution $y(x)$.

31. $y_c = c_1 \cos 2x + c_2 \sin 2x$; $y_{tr} = A + Bx$; $y_g = c_1 \cos 2x + c_2 \sin 2x + \dfrac{x}{2}$; $c_1 = 1$, $2c_2 + \dfrac{1}{2} = 2$;

$y(x) = \cos 2x + (3/4)\sin 2x + x/2$

33. $y_c = c_1 \cos 3x + c_2 \sin 3x$; $y_{tr} = A \cos 2x + B \sin 2x$; $y_g = c_1 \cos 3x + c_2 \sin 3x + \dfrac{1}{5} \sin 2x$;

$c_1 = 1$, $3c_2 + \dfrac{2}{5} = 0$, $y(x) = \cos 3x - \dfrac{2}{15} \sin 3x + \dfrac{1}{5} \sin 2x$

35. $y_c = e^x (c_1 \cos x + c_2 \sin x)$; $y_{tr} = A + Bx$; $y_g = e^x (c_1 \cos x + c_2 \sin x) + 1 + \dfrac{x}{2}$; $c_1 + 1 = 3$,

$c_1 + c_2 + \dfrac{1}{2} = 0$; $y(x) = e^x \left(2 \cos x - \dfrac{5}{2} \sin x \right) + 1 + \dfrac{x}{2}$

37. $y_c = c_1 + c_2 e^x + c_3 x e^x$; $y_{tr} = x \cdot (A) + x^2 \cdot (B + Cx) e^x$

$$y_g = c_1 + c_2 e^x + c_3 x e^x + x - \dfrac{1}{2} x^2 e^x + \dfrac{1}{6} x^3 e^x$$

$$c_1 + c_2 = 0, \quad c_2 + c_3 + 1 = 0, \quad c_2 + 2c_3 - 1 = 1$$

$$y(x) = 4 + x + e^x \left(-4 + 3x - \dfrac{1}{2} x^2 + \dfrac{1}{6} x^3 \right)$$

39. $y_c = c_1 + c_2 x + c_3 e^{-x}$; $y_{tr} = x^2 \cdot (A + Bx) + x \cdot (C e^{-x})$

$$y_g = c_1 + c_2 x + c_3 e^{-x} - \frac{x^2}{2} + \frac{x^3}{6} + xe^{-x}$$

$$c_1 + c_3 = 1, \quad c_2 - c_3 + 1 = 0, \quad c_3 - 3 = 1$$

$$y(x) = \frac{1}{6}\left(-18 + 18x - 3x^2 + x^3\right) + (4 + x)e^{-x}$$

41.　The trial solution $y_{tr} = A + Bx + Cx^2 + Dx^3 + Ex^4 + Fx^5$ leads to the equations

$$
\begin{array}{rcl}
2A - B - 2C - 6D + 24E & = & 0 \\
-2B - 2C - 6D - 24E + 120F & = & 0 \\
-2C - 3D - 12E - 60F & = & 0 \\
-2D - 4E - 20F & = & 0 \\
-2E - 5F & = & 0 \\
-2F & = & 8
\end{array}
$$

that are readily solved by back-substitution. The resulting particular solution is

$$y(x) = -255 - 450x + 30x^2 + 20x^3 + 10x^4 - 4x^5.$$

43.　**(a)** Applying Euler's formula gives

$$\cos 3x + i \sin 3x = (\cos x + i \sin x)^3 = \cos^3 x + 3i \cos^2 x \sin x - 3 \cos x \sin^2 x - i \sin^3 x.$$

When we equate real parts we get the equation

$$\cos^3 x - 3(\cos x)(1 - \cos^2 x) = 4 \cos^3 x - 3 \cos x$$

and readily solve for $\cos^3 x = \frac{3}{4}\cos x + \frac{1}{4}\cos 3x$. The formula for $\sin^3 x$ is derived similarly by equating imaginary parts in the first equation above.

(b) Upon substituting the trial solution $y_p = A \cos x + B \sin x + C \cos 3x + D \sin 3x$ in the differential equation $y'' + 4y = \frac{3}{4}\cos x + \frac{1}{4}\cos 3x$, we find that

$$A = \frac{1}{4}, \quad B = 0, \quad C = -\frac{1}{20}, \quad D = 0.$$

The resulting general solution is

$$y(x) = c_1 \cos 2x + c_2 \sin 2x + \frac{1}{4}\cos x - \frac{1}{20}\cos 3x.$$

45.　We substitute

$$\sin^4 x = \frac{1}{4}(1 - \cos 2x)^2 = \frac{1}{4}\left(1 - 2\cos 2x + \cos^2 2x\right) = \frac{1}{8}\left(3 - 4\cos 2x + \cos 4x\right)$$

on the right-hand side of the differential equation, and then substitute the trial solution

$$y_p = A\cos 2x + B\sin 2x + C\cos 4x + D\sin 4x + E.$$

We find that

$$A = -\frac{1}{10}, \quad B = 0, \quad C = -\frac{1}{56}, \quad D = 0, \quad E = \frac{1}{24}.$$

The resulting general solution is

$$y = c_1\cos 3x + c_2\sin 3x + \frac{1}{24} - \frac{1}{10}\cos 2x - \frac{1}{56}\cos 4x.$$

In Problems 47–49 we list the independent solutions y_1 and y_2 of the associated homogeneous equation, their Wronskian $W = W(y_1, y_2)$, the coefficient functions

$$u_1(x) = -\int \frac{y_2(x)f(x)}{W(x)}dx \quad \text{and} \quad u_2(x) = \int \frac{y_1(x)f(x)}{W(x)}dx$$

in the particular solution $y_p = u_1 y_1 + u_2 y_2$ of Eq. (32) in the text, and finally y_p itself.

47. $y_1 = e^{-2x}$, $y_2 = e^{-x}$, $W = e^{-3x}$, $u_1 = -\frac{4}{3}e^{3x}$, $u_2 = 2e^{2x}$, $y_p = \frac{2}{3}e^x$

49. $y_1 = e^{2x}$, $y_2 = xe^{2x}$, $W = e^{4x}$, $u_1 = -x^2$, $u_2 = 2x$, $y_p = x^2 e^{2x}$.

51. $y_1 = \cos 2x$, $y_2 = \sin 2x$, $W = 2$. Liberal use of trigonometric sum and product identities yields

$$u_1 = \frac{1}{20}(\cos 5x - 5\cos x) \quad \text{and} \quad u_2 = \frac{1}{20}(\sin 5x + 5\sin x).$$

Thus

$$\begin{aligned}
y_p &= \frac{1}{20}(\cos 5x - 5\cos x)\cos 2x + \frac{1}{20}(\sin 5x + 5\sin x)\sin 2x \\
&= \frac{1}{20}\left[(\cos 5x\cos 2x + \sin 5x\sin 2x) - 5(\cos x\cos 2x - \sin x\sin 2x)\right] \\
&= \frac{1}{20}(\cos 3x - 5\cos 3x) \\
&= -\frac{1}{5}\cos 3x \ (!)
\end{aligned}$$

53. $y_1 = \cos 3x$, $y_2 = \sin 3x$, $W = 3$; $u_1' = -\frac{2}{3}\tan 3x$, so $u_1 = \frac{2}{9}\ln|\cos 3x|$; $u_2' = \frac{2}{3}$, so $u_2 = \frac{2}{3}x$. Thus

$$y_p = (\cos 3x) \cdot \frac{2}{9} \ln |\cos 3x| + (\sin 3x) \cdot \frac{2}{3} x$$

$$= \frac{2}{9} \left[3x \sin 3x + (\cos 3x) \ln |\cos 3x| \right].$$

55. $y_1 = \cos 2x$, $y_2 = \sin 2x$, $W = 2$;

$$u_1' = -\frac{1}{2} \sin^2 x \sin 2x = -\frac{1}{2} \cdot \frac{1 - \cos 2x}{2} \cdot \sin 2x = -\frac{1}{4} \left(\sin 2x - \cos 2x \sin 2x \right),$$

so $u_1 = \frac{1}{16} \left(2 \cos 2x - \cos^2 2x \right)$;

$$u_2' = \frac{1}{2} \sin^2 x \cos 2x = \frac{1}{2} \cdot \frac{1 - \cos 2x}{2} \cdot \cos 2x = \frac{1}{4} \left(\cos 2x - \cos^2 2x \right) = \frac{1}{8} \left[2 \cos 2x - (1 + \cos 4x) \right],$$

so $u_2 = \frac{1}{8} \left(\sin 2x - x - \frac{\sin 4x}{4} \right)$. Thus

$$y_p = \frac{1}{16} \left(2 \cos 2x - \cos^2 2x \right) \cos 2x + \frac{1}{8} \left(\sin 2x - x - \frac{\sin 4x}{4} \right) \sin 2x$$

$$= \frac{1}{16} \left(\underline{2 \cos^2 2x} - \cos^3 2x + \underline{2 \sin^2 2x} - 2x \sin 2x - \frac{1}{2} \sin 4x \sin 2x \right)$$

$$= \frac{1}{16} \left(2 - 2x \sin 2x \underline{\underline{- \cos^3 2x - \frac{1}{2} \sin 4x \sin 2x}} \right),$$

because the single-underlined terms sum to 2. The double-underlined terms reduce to

$$- \cos^3 2x - \frac{1}{2} \sin 4x \sin 2x = - \cos^3 2x - \frac{1}{2} \left(2 \sin 2x \cos 2x \right) \sin 2x$$

$$= - \cos^3 2x - \sin^2 2x \cos 2x$$

$$= - \cos^3 2x - \left(1 - \cos^2 2x \right) \cos 2x$$

$$= - \cos 2x.$$

Therefore we can take $y_p = \frac{1}{16} \left(2 - 2x \sin 2x - \cos 2x \right)$. However, because $\cos 2x$ is a solution of the associated homogeneous equation $y'' + y = 0$—that is, because $\cos 2x$ is part of the complimentary function y_c—we can in fact omit the $\cos 2x$ term from y_p, leading to the simpler version $y_p = \frac{1}{8} \left(1 - x \sin 2x \right)$.

57.   With $y_1 = x$, $y_2 = x^{-1}$, and $f(x) = 72x^3$, Equations (31) in the text take the form

$$xu_1' + x^{-1}u_2' = 0,$$
$$u_1' - x^{-2}u_2' = 72x^3.$$

Upon multiplying the second equation by x and then adding, we readily solve first for $u_1' = 36x^3$, so $u_1 = 9x^4$; then $u_2' = -x^2u_1'$, so $u_2 = -6x^6$. It follows that

$$y_p = y_1u_1 + y_2u_2 = x(9x^4) + (x^{-1})(-6x^6) = 3x^5.$$

59. $y_1 = x^2$, $y_2 = x^2\ln x$, $W = x^3$, $f(x) = x^2$; $u_1' = -x\ln x$, $u_2' = x$; $y_p = \dfrac{1}{4}x^4$.

61. $y_1 = \cos(\ln x)$, $y_2 = \sin(\ln x)$, $W = \dfrac{1}{x}$, $f(x) = \dfrac{\ln x}{x^2}$; from $u_1' = -\dfrac{(\ln x)\sin(\ln x)}{x}$ and

$u_2' = \dfrac{(\ln x)\cos(\ln x)}{x}$ integration by parts yields

$$u_1 = -\int \frac{(\ln x)\sin(\ln x)}{x}\,dx = -\int \frac{\sin(\ln x)}{x}\cdot \ln x\,dx$$

$$= \cos(\ln x)\cdot \ln x - \int \frac{\cos(\ln x)}{x}\,dx = \cos(\ln x)\cdot \ln x - \sin(\ln x)$$

and

$$u_2 = \int \frac{(\ln x)\cos(\ln x)}{x}\,dx = \int \frac{\cos(\ln x)}{x}\cdot \ln x\,dx$$

$$= \sin(\ln x)\cdot \ln x - \int \frac{\sin(\ln x)}{x}\,dx = \sin(\ln x)\cdot \ln x + \cos(\ln x).$$

Thus

$$y_p = u_1y_1 + u_2y_2$$

$$= \left[\cos(\ln x)\cdot \ln x - \sin(\ln x)\right]\cos(\ln x) + \left[\sin(\ln x)\cdot \ln x + \cos(\ln x)\right]\sin(\ln x)$$

$$= \left[\cos^2(\ln x) + \sin^2(\ln x)\right]\ln x - \sin(\ln x)\cos(\ln x) + \cos(\ln x)\sin(\ln x)$$

$$= \ln x \;(!)$$

63. This is simply a matter of solving the equations in (31) for the derivatives

$$u_1' = -\frac{y_2(x)f(x)}{W(x)} \quad \text{and} \quad u_2' = \frac{y_1(x)f(x)}{W(x)},$$

integrating each, and then substituting the results in (32).

SECTION 3.6

FORCED OSCILLATIONS AND RESONANCE

1. Trial of $x = A\cos 2t$ yields the particular solution $x_p = 2\cos 2t$. (Can you see that because the differential equation contains no first-derivative term, there is no need to include a $\sin 2t$ term in the trial solution?) Hence the general solution is

$$x(t) = c_1 \cos 3t + c_2 \sin 3t + 2\cos 2t.$$

The initial conditions imply that $c_1 = -2$ and $c_2 = 0$, so $x(t) = 2\cos 2t - 2\cos 3t$. The figure shows the graph of $x(t)$.

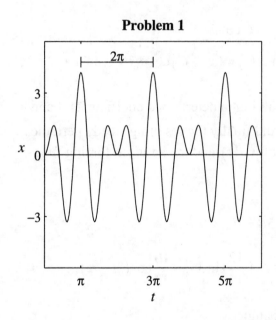

Problem 1

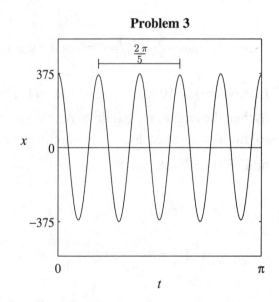

Problem 3

3. First we apply the method of undetermined coefficients with trial solution $x = A\cos 5t + B\sin 5t$ to find the particular solution

$$x_p = 3\cos 5t + 4\sin 5t = 5\left(\frac{3}{5}\cos 5t + \frac{4}{5}\sin 5t\right) = 5\cos(5t - \beta),$$

where $\beta = \tan^{-1}\dfrac{4}{3} \approx 0.9273$. Hence the general solution is

$$x(t) = c_1 \cos 10t + c_2 \sin 10t + 3\cos 5t + 4\sin 5t.$$

The initial conditions $x(0) = 375$, $x'(0) = 0$ now yield $c_1 = 372$ and $c_2 = -2$, so the part of the solution with frequency $\omega = 10$ is

$$x_c = 372\cos 10t - 2\sin 10t$$

$$= \sqrt{138388}\left(\frac{372}{\sqrt{138388}}\cos 10t - \frac{2}{\sqrt{138388}}\sin 10t\right)$$

$$= \sqrt{138388}\cos(10t - \alpha),$$

where $\alpha = 2\pi - \tan^{-1}\frac{1}{186} \approx 6.2778$ is a fourth-quadrant angle. The figure shows the graph of $x(t)$.

5. Substitution of the trial solution $x = C\cos\omega t$ gives $C = \dfrac{F_0}{k - m\omega^2}$. Then imposition of the initial conditions $x(0) = x_0$, $x'(0) = 0$ on the general solution

$$x(t) = c_1\cos\omega_0 t + c_2\sin\omega_0 t + C\cos\omega t$$

(where $\omega_0 = \sqrt{k/m}$) gives the particular solution $x(t) = (x_0 - C)\cos\omega_0 t + C\cos\omega t$.

In Problems 7–10 we give first the trial solution x_p involving undetermined coefficients A and B, then the equations that determine these coefficients, and finally the resulting steady periodic solution x_{sp}. In each case the figure shows the graphs of $x_{sp}(t)$ and the adjusted forcing function $F_1(t) = F(t)/m\omega$.

7. $x_p = A\cos 3t + B\sin 3t$; $-5A + 12B = 10$, $12A + 5B = 0$.

$$x_{sp}(t) = -\frac{50}{169}\cos 3t + \frac{120}{169}\sin 3t = \frac{10}{13}\left(-\frac{5}{13}\cos 3t + \frac{12}{13}\sin 3t\right) = \frac{10}{13}\cos(3t - \alpha),$$

where $\alpha = \pi - \tan^{-1}\frac{12}{5} \approx 1.9656$, a 2nd-quadrant angle.

Problem 7

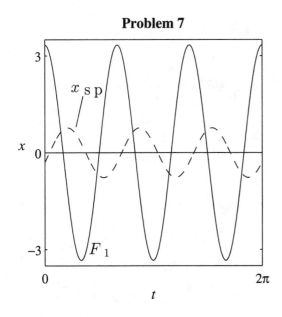

Problem 9

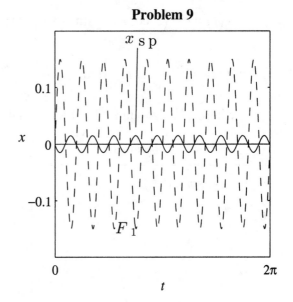

9. $x_p = A\cos 10t + B\sin 5t$; $-199A + 20B = 0$, $20A + 199B = -3$.

$$x_{sp}(t) = -\frac{60}{40001}\cos 10t - \frac{597}{40001}\sin 10t$$

$$= \frac{3}{\sqrt{40001}}\left(-\frac{20}{\sqrt{40001}}\cos 10t - \frac{199}{\sqrt{40001}}\sin 10t\right)$$

$$= \frac{3}{\sqrt{40001}}\cos(10t - \alpha),$$

where $\alpha = \pi + \tan^{-1}\dfrac{199}{20} \approx 4.6122$, a 3rd-quadrant angle.

Each solution in Problems 11–14 has two parts. For the first part, we give first the trial solution x_p involving undetermined coefficients A and B, then the equations that determine these coefficients, and finally the resulting steady periodic solution x_{sp}. For the second part, we give first the general solution $x(t)$ involving the coefficients c_1 and c_2 in the transient solution, then the equations that determine these coefficients, and finally the resulting transient solution x_{tr}, so that $x(t) = x_{tr}(t) + x_{sp}(t)$ satisfies the given initial conditions. For each problem, the graph shows the graphs of both $x(t)$ and $x_{sp}(t)$.

11. $x_p = A\cos 3t + B\sin 3t$; $-4A + 12B = 10$, $12A + 4B = 0$.

$$x_{sp}(t) = -\frac{1}{4}\cos 3t + \frac{3}{4}\sin 3t = \frac{\sqrt{10}}{4}\left(-\frac{1}{\sqrt{10}}\cos 3t + \frac{3}{\sqrt{10}}\sin 3t\right) = \frac{\sqrt{10}}{4}\cos(3t - \alpha),$$

where $\alpha = \pi - \tan^{-1}3 \approx 1.8925$, a 2nd-quadrant angle.

$$x(t) = e^{-2t}(c_1 \cos t + c_2 \sin t) + x_{sp}(t); \ c_1 - \frac{1}{4} = 0, \ -2c_1 + c_2 + \frac{9}{4} = 0.$$

$$x_{tr}(t) = e^{-2t}\left(\frac{1}{4}\cos t - \frac{7}{4}\sin t\right) = \frac{\sqrt{50}}{4}e^{-2t}\left(\frac{1}{\sqrt{50}}\cos t - \frac{7}{\sqrt{50}}\sin t\right) = \frac{5}{4}\sqrt{2}e^{-2t}\cos(t-\beta),$$

where $\beta = 2\pi - \tan^{-1}7 \approx 4.8543$, a 4th-quadrant angle.

Problem 11

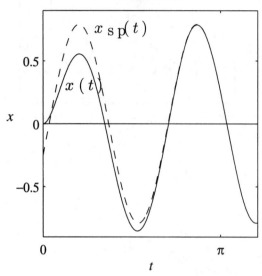

Problem 13

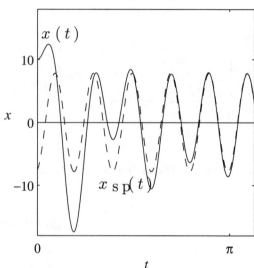

13. $x_p = A\cos 10t + B\sin 10t$; $74A + 20B = 600$, $20A + 74B = 0$.

$$x_{sp}(t) = -\frac{11100}{1469}\cos 10t + \frac{3000}{1469}\sin 10t$$

$$= \frac{300}{\sqrt{1469}}\left(-\frac{37}{\sqrt{1469}}\cos 10t + \frac{10}{\sqrt{1469}}\sin 10t\right)$$

$$= \frac{300}{\sqrt{1469}}\cos(10t - \alpha),$$

where $\alpha = \pi - \tan^{-1}\frac{10}{37} \approx 2.8776$, a 2nd-quadrant angle.

$$x(t) = e^{-t}(c_1 \cos 5t + c_2 \sin 5t) + x_{sp}(t); \ c_1 - \frac{11100}{1469} = 10, \ -c_1 + 5c_2 = -\frac{30000}{1469}.$$

$$x_{tr}(t) = \frac{e^{-t}}{1469}(25790\cos 5t - 842\sin 5t)$$

$$= \frac{2\sqrt{166458266}}{1469}e^{-t}\left(\frac{12895}{\sqrt{166458266}}\cos 5t - \frac{421}{\sqrt{166458266}}\sin 5t\right)$$

$$= 2\sqrt{\frac{113314}{1469}}e^{-t}\cos(5t - \beta),$$

where $\beta = 2\pi - \tan^{-1} \dfrac{421}{12895} \approx 6.2505$, a 4th-quadrant angle.

In Problems 15-18 we substitute $x(t) = A(\omega)\cos\omega t + B(\omega)\sin\omega t$ into the differential equation $mx'' + cx' + kx = F_0 \cos\omega t$ with the given numerical values of m, c, k, and F_0. We give first the equations in A and B that result upon collection of coefficients of $\cos\omega t$ and $\sin\omega t$, followed by the values of $A(\omega)$ and $B(\omega)$ that we get by solving these equations. Finally, $C = \sqrt{A^2 + B^2}$ gives the amplitude of the resulting forced oscillations as a function of the forcing frequency ω, and we show the graph of the function $C(\omega)$.

15. $(2 - \omega^2)A + 2\omega B = 2$, $-2\omega A + (2 - \omega^2)B = 0$; $A = \dfrac{2(2 - \omega^2)}{4 + \omega^4}$, $B = \dfrac{4\omega}{4 + \omega^4}$;

$C(\omega) = \dfrac{2}{\sqrt{4 + \omega^4}}$ begins with $C(0) = 1$ and steadily decreases as ω increases. Hence there is no practical resonance frequency.

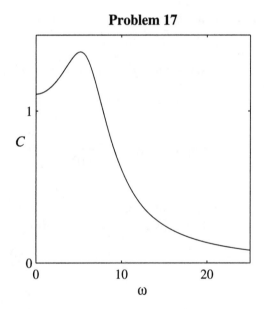

Problem 15

Problem 17

17. $(45 - \omega^2)A + 6\omega B = 50$, $-6\omega A + (45 - \omega^2)B = 0$; $A = \dfrac{50(45 - \omega^2)}{2025 - 54\omega^2 + \omega^4}$,

$B = \dfrac{300\omega}{2025 - 54\omega^2 + \omega^4}$; $C(\omega) = \dfrac{50}{\sqrt{2025 - 54\omega^2 + \omega^4}}$. So, to find the maximum value of $C(\omega)$, we calculate its derivative

$$C'(\omega) = \dfrac{-100\omega(-27 + \omega^2)}{\left(2025 - 54\omega^2 + \omega^4\right)^{3/2}}.$$

Hence the practical resonance frequency (where the derivative vanishes) is $\omega = \sqrt{27} = 3\sqrt{3}$.

19. $m = 100/32$ slugs and $k = 1200$ lb/ft, so the critical frequency is

$$\omega_0 = \sqrt{\frac{k}{m}} = \sqrt{384} \text{ rad/sec} = \frac{\sqrt{384}}{2\pi} \text{ Hz} \approx 3.12 \text{ Hz}.$$

21.   If θ is the angular displacement from the vertical, then the (essentially horizontal) displacement of the mass is $x = L\theta$, so twice its total energy (KE + PE) is

$$m(x')^2 + kx^2 + 2mgh = mL^2(\theta')^2 + kL^2\theta^2 + 2mgL(1 - \cos\theta) = C.$$

Differentiation, substitution of $\theta \approx \sin\theta$, and simplification yield $\theta'' + \left(\dfrac{k}{m} + \dfrac{g}{L}\right)\theta = 0$, so

$$\omega_0 = \sqrt{\frac{k}{m} + \frac{g}{L}}.$$

23.   **(a)** In units of ft-lb-sec we have $m = 1000$ and $k = 10000$, so $\omega_0 = \sqrt{10}$ rad/sec ≈ 0.50 Hz.

(b) We are given that $\omega = 2\pi/2.25 \approx 2.79$ rad/sec, and the equation $mx'' + kx = F(t)$ simplifies to $x'' + 10x = \dfrac{1}{4}\omega^2 \sin\omega t$. When we substitute $x(t) = A\sin\omega t$ we find that the amplitude is $A = \dfrac{\omega^2}{4(10 - \omega^2)} \approx 0.8854$ ft ≈ 10.63 in.

25.   Substitution of the trial solution $x = A\cos\omega t + B\sin\omega t$ in the differential equation followed by collection of coefficients as usual yields the equations

$$(k - m\omega^2)A + (c\omega)B = 0, \quad -(c\omega)A + (k - m\omega^2)B = F_0$$

with coefficient determinant $\Delta = (k - m\omega^2)^2 + (c\omega)^2$ and solution

$$A = \frac{1}{\Delta}(-c\omega)F_0, \quad B = \frac{1}{\Delta}(k - m\omega^2)F_0.$$

Hence

$$x(t) = \frac{F_0}{\sqrt{\Delta}}\left(\frac{k - m\omega^2}{\sqrt{\Delta}}\sin\omega t - \frac{c\omega}{\sqrt{\Delta}}\cos\omega t\right) = C\sin(\omega t - \alpha),$$

where $C = \dfrac{1}{\sqrt{\Delta}}F_0$, $\sin\alpha = \dfrac{c\omega}{\sqrt{\Delta}}$, and $\cos\alpha = \dfrac{1}{\sqrt{\Delta}}(k - m\omega^2)$.

27.   The derivative of $C(\omega) = \dfrac{F_0}{\sqrt{\left(k - m\omega^2\right)^2 + \left(c\omega\right)^2}}$ is given by

$$C'(\omega) = -\frac{\omega F_0}{2} \cdot \frac{\left(c^2 - 2km\right) + 2\left(m\omega\right)^2}{\left[\left(k - m\omega^2\right)^2 + \left(c\omega\right)^2\right]^{3/2}}.$$

(a) Therefore, if $c^2 \geq \left(c_{cr}/\sqrt{2}\right)^2 = 2km$, it is clear from the numerator that $C'(\omega) < 0$ for all ω, so that $C(\omega)$ steadily decreases as ω increases.

(b) If instead $c^2 < 2km$, however, then the numerator (and hence $C'(\omega)$) vanishes when

$$\omega = \omega_m = \sqrt{\frac{k}{m} - \frac{c^2}{2m^2}} < \sqrt{\frac{k}{m}} = \omega_0.$$

Calculation then shows that

$$C''(\omega_m) = \frac{16 F_0 m^3 \left(c^2 - 2km\right)}{c^3 \left(4km - c^2\right)^{3/2}} < 0,$$

so it follows from the second-derivative test that $C(\omega_m)$ is a local maximum value.

29.   We need only substitute $E_0 = ac\omega$ and $F_0 = ak$ in the result of Problem 26.

SECTION 3.7

ELECTRICAL CIRCUITS

1.   With $E(t) \equiv 0$ we have the simple exponential equation $5I' + 25I = 0$, whose solution with $I(0) = 4$ is $I(t) = 4e^{-5t}$.

3.   Now the differential equation is $5I' + 25I = 100\cos 60t$. Substitution of the trial solution $I_p = A\cos 60t + B\sin 60t$ yields

$$I_p = \frac{4}{145}\left(\cos 60t + 12\sin 60t\right).$$

The complementary function is $I_c = Ce^{-5t}$; the solution with $I(0) = 0$ is

$$I(t) = \frac{4}{145}\left(\cos 60t + 12\sin 60t - e^{-5t}\right).$$

5.   The linear equation $I' + 10I = 50e^{-10t}\cos 60t$ has integrating factor $\rho = e^{10t}$. The resulting general solution is $I(t) = e^{-10t}\left(\dfrac{5}{6}\sin 60t + C\right)$. To satisfy the initial condition $I(0) = 0$, we take $C = 0$ and get $I(t) = \dfrac{5}{6}e^{-10t}\sin 60t$.

7.   **(a)** The linear differential equation $RQ' + \dfrac{1}{C}Q = E_0$ has integrating factor $\rho = e^{\frac{t}{RC}}$. The resulting solution with $Q(0) = 0$ is $Q(t) = E_0 C\left(1 - e^{-\frac{t}{RC}}\right)$. Then $I(t) = Q'(t) = \dfrac{E_0}{R}e^{-\frac{t}{RC}}$.

   (b) These solutions make it obvious that $\lim\limits_{t\to\infty} Q(t) = E_0 C$ and $\lim\limits_{t\to\infty} I(t) = 0$.

9.   Substitution of the trial solution $Q = A\cos 120t + B\sin 120t$ into the differential equation $200Q' + 4000Q = 100\cos 120t$ yields the equations

$$4000A + 24000B = 100, \quad -24000A + 4000B = 0$$

with solution $A = \dfrac{1}{1480}$, $B = \dfrac{3}{740}$. The complementary function is $Q_c = ce^{-20t}$, and imposition of the initial condition $Q(0) = 0$ yields the solution

$$Q(t) = \frac{1}{1480}\left(\cos 120t + 6\sin 120t - e^{-20t}\right).$$

The current function is then

$$I(t) = Q'(t) = \frac{1}{74}\left(36\cos 120t - 6\sin 120t + e^{-20t}\right).$$

Thus the steady-periodic current is

$$I_{sp} = \frac{6}{74}\left(6\cos 120t - \sin 120t\right) = \frac{6\sqrt{37}}{74}\left(\frac{6}{\sqrt{37}}\cos 120t - \frac{1}{\sqrt{37}}\sin 120t\right) = \frac{3}{\sqrt{37}}\cos(120t - \alpha)$$

(with $\alpha = 2\pi - \tan^{-1}\dfrac{1}{6}$), so the steady-state amplitude is $\dfrac{3}{\sqrt{37}}$.

In Problems 11-16, we give first the trial solution $I_p = A\cos\omega t + B\sin\omega t$, then the equations in A and B that we get upon substituting this trial solution into the RLC equation $LI'' + RI' + \dfrac{1}{C}I = E'(t)$, and finally the resulting steady periodic solution.

11. $I_p = A\cos 2t + B\sin 2t$; $A + 6B = 10$, $-6A + B = 0$;

$$I_{sp}(t) = \frac{10}{37}\cos 2t + \frac{60}{37}\sin 2t = \frac{10}{\sqrt{37}}\left(\frac{1}{\sqrt{37}}\cos 2t + \frac{6}{\sqrt{37}}\sin 2t\right) = \frac{10}{\sqrt{37}}\sin(2t - \delta),$$

where $\delta = 2\pi - \tan^{-1}\frac{1}{6} \approx 6.1180$, a 4th-quadrant angle.

13. $I_p = A\cos 5t + B\sin 5t$; $3A - 2B = 0$, $2A + 3B = 20$;

$$I_{sp}(t) = \frac{40}{13}\cos 5t + \frac{60}{13}\sin 5t = \frac{20}{\sqrt{13}}\left(\frac{2}{\sqrt{13}}\cos 5t + \frac{3}{\sqrt{13}}\sin 5t\right) = \frac{20}{\sqrt{13}}\sin(5t - \delta),$$

where $\delta = 2\pi - \tan^{-1}\frac{2}{3} \approx 5.6952$, a 4th-quadrant angle.

15. $I_p = A\cos 60\pi t + B\sin 60\pi t$;

$$\left(1000 - 36\pi^2\right)A + 30\pi B = 33\pi, \quad 15\pi A - \left(500 - 18\pi^2\right)B = 0;$$

$$A = \frac{33\pi\left(250 - 9\pi^2\right)}{250000 - 17775\pi^2 + 324\pi^4}, \quad B = \frac{495\pi^2}{2\left(250000 - 17775\pi^2 + 324\pi^4\right)};$$

$$I_{sp}(t) \approx I_0 \sin(60\pi t - \delta), \text{ where } I_0 = \frac{33\pi}{2\sqrt{250000 - 17775\pi^2 + 324\pi^4}} \approx 0.1591 \text{ and}$$

$$\delta = 2\pi - \tan^{-1}\frac{500 - 18\pi^2}{15\pi} \approx 4.8576.$$

In each of Problems 17–22, the first step is to substitute the given *RLC* parameters, the initial values $I(0)$ and $Q(0)$, and the voltage $E(t)$ into Eq. (16) and solve for the remaining initial value

$$I'(0) = \frac{1}{L}\left[E(0) - RI(0) - \frac{1}{C}Q(0)\right]. \qquad (*)$$

17. With $I(0) = 0$ and $Q(0) = 5$, Equation (*) gives $I'(0) = -75$. The solution of the *RLC* equation $2I'' + 16I' + 50I = 0$ with these initial conditions is $I(t) = -25e^{-4t}\sin 3t$.

19. Now our differential equation to solve is

$$2I'' + 60I' + 400I = -1000e^{-10t}.$$

We find the particular solution $I_p = -50te^{-10t}$ by substituting the trial solution Ate^{-10t}; the general solution is

$$I(t) = c_1 e^{-10t} + c_2 e^{-20t} - 50te^{-10t}.$$

The initial conditions are $I(0) = 0$ and $I'(0) = -150$, the latter found by substituting $L = 2$, $R = 60$, $\frac{1}{C} = 400$, $I(0) = 0$, $Q(0) = -1$, and $E(0) = 100$ into Equation (*). Imposition of these initial values on the general solution above yields the equations

$$c_1 + c_2 = 0, \quad -10c_1 - 20c_2 - 50 = -150,$$

with solution $c_1 = -10$, $c_2 = 10$. Thus we get the solution

$$I(t) = 10e^{-20t} - 10e^{-10t} - 50te^{-10t}.$$

Problem 21

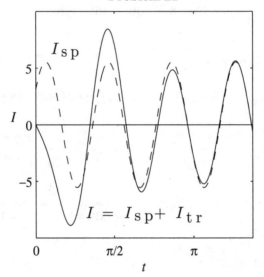

21. The differential equation $10I'' + 20I' + 100I = -1000\sin 5t$ has transient solution $I_{tr}(t) = e^{-t}(c_1 \cos 3t + c_2 \sin 3t)$, and in Problem 13 we found the steady periodic solution $I_{sp}(t) = \frac{20}{13}(2\cos 5t + 3\sin 5t)$. When we impose the initial conditions $I(0) = 0$, $I'(0) = -10$ on the general solution $I(t) = I_{tr}(t) + I_{sp}(t)$, we get the equations

$$c_1 + \frac{40}{13} = 0, \quad -c_1 + 3c_2 + \frac{300}{13} = -10,$$

with solution $c_1 = -\frac{40}{13}$, $c_2 = -\frac{470}{39}$. The figure shows the graphs of $I(t)$ and $I_{sp}(t)$.

23. The LC equation $LI'' + \frac{1}{C}I = 0$ has general solution $I(t) = c_1 \cos \omega_0 t + c_2 \sin \omega_0 t$ with critical frequency $\omega_0 = \frac{1}{\sqrt{LC}}$.

25. According to Eq. (8) in the text, the amplitude of the steady periodic current is

$\dfrac{E_0}{\sqrt{R^2 + \left(\omega L - \dfrac{1}{\omega C}\right)^2}}$. Because the radicand in the denominator is a sum of squares, it is

obvious that the denominator is least when $\omega L - \dfrac{1}{\omega C} = 0$, that is, when $\omega = \dfrac{1}{\sqrt{LC}}$.

SECTION 3.8

ENDPOINT PROBLEMS AND EIGENVALUES

The material on eigenvalues and endpoint problems in Section 3.8 can be considered optional at this point in a first course. It will not be needed until we discuss boundary value problems in the last three sections of Chapter 9 and in Chapter 10. However, after the concentration thus far on initial value problems, the inclusion of this section can give students a view of a new class of problems that have diverse and important applications (as illustrated by the subsection on the whirling string). If Section 3.8 is not covered at this point in the course, then it can be inserted just prior to Section 9.5.

1. If $\lambda = 0$, then $y'' = 0$ implies that $y(x) = A + Bx$. The endpoint conditions $y'(0) = 0$ and $y(1) = 0$ yield $B = 0$ and $A = 0$, respectively. Hence $\lambda = 0$ is *not* an eigenvalue.

If $\lambda = \alpha^2 > 0$, then the general solution of $y'' + \alpha^2 y = 0$ is

$$y(x) = A\cos\alpha x + B\sin\alpha x,$$

so

$$y'(x) = -A\alpha\sin\alpha x + B\alpha\cos\alpha x.$$

Then $y'(0) = 0$ yields $B = 0$, so $y(x) = A\cos\alpha x$. Next $y(1) = 0$ implies that $\cos\alpha = 0$, so α is an odd multiple of $\dfrac{\pi}{2}$. Hence the positive eigenvalues are

$\dfrac{(2n-1)^2\pi^2}{4}$, with associated eigenfunctions $\cos\dfrac{(2n-1)\pi x}{2}$, for $n = 1,2,3,\dots$.

3. Much as in Problem 1 we see that $\lambda = 0$ is not an eigenvalue. Suppose that $\lambda = \alpha^2 > 0$, so

$$y(x) = A\cos\alpha x + B\sin\alpha x.$$

Then the conditions $y(-\pi) = y(\pi) = 0$ yield

$$A\cos\alpha\pi + B\sin\alpha\pi = 0, \quad A\cos\alpha\pi - B\sin\alpha\pi = 0.$$

It follows that $A\cos\alpha\pi = 0 = B\sin\alpha\pi$. Hence either $A = 0$ and $B \neq 0$ with $\alpha\pi$ an even multiple of $\dfrac{\pi}{2}$, or $A \neq 0$ and $B = 0$ with $\alpha\pi$ an odd multiple of $\dfrac{\pi}{2}$. Thus the eigenvalues are $\dfrac{n^2}{4}$, $n = 1, 2, 3, \ldots$, and the nth eigenfunction is $y_n(x) = \cos\dfrac{nx}{2}$ if n is even, and $y_n(x) = \sin\dfrac{nx}{2}$ if n is odd.

5.   If $\lambda = \alpha^2 > 0$ and

$$y(x) = A\cos\alpha x + B\sin\alpha x,$$

so that

$$y'(x) = -A\alpha\sin\alpha x + B\alpha\cos\alpha x,$$

then the conditions $y(-2) = y'(2) = 0$ yield

$$A\cos 2\alpha - B\sin 2\alpha = 0, \quad A\sin 2\alpha + B\cos 2\alpha = 0.$$

It follows either that $A = B$ and $\cos 2\alpha = \sin 2\alpha$, or that $A = -B$ and $\cos 2\alpha = -\sin 2\alpha$. The former occurs if

$$2\alpha = \frac{\pi}{4}, \frac{5\pi}{4}, \frac{9\pi}{4}, \ldots$$

and the latter if

$$2\alpha = \frac{3\pi}{4}, \frac{7\pi}{4}, \frac{11\pi}{4}, \ldots$$

Hence the nth eigenvalue is

$$\lambda_n = \alpha_n^2 = \frac{(2n-1)^2 \pi^2}{64}, n = 1, 2, 3, \ldots,$$

and the associated eigenfunction is

$$y_n(x) = \begin{cases} \cos\alpha_n x + \sin\alpha_n x, & n \text{ odd} \\ \cos\alpha_n x - \sin\alpha_n x, & n \text{ even} \end{cases}.$$

7.   **(a)** If $\lambda = 0$ and $y(x) = A + Bx$, then $y(0) = A = 0$, so $y(x) = Bx$. But then $y(1) + y'(1) = 2B = 0$, so $A = B = 0$ and $\lambda = 0$ is not an eigenvalue.

(b) If $\lambda = \alpha^2 > 0$ and

$$y(x) = A\cos\alpha x + B\sin\alpha x,$$

then $y(0) = A = 0$, so $y(x) = B\sin\alpha x$. Hence

$$y(1) + y'(1) = B(\sin\alpha + \alpha\cos\alpha) = 0,$$

so α must be a positive root of the equation $\tan \alpha = -\alpha$, and hence the abscissa of a point of intersection of the lines $y = \tan z$ and $y = -z$. We see from the figure below that α_n lies just to the right of the vertical line $z = \dfrac{(2n-1)\pi}{2}$ and lies closer and closer to this line as n gets larger and larger.

Problem 7

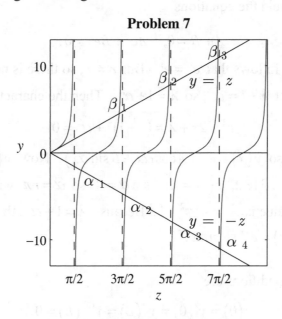

9.   If $y'' + \lambda y = 0$ and $\lambda = -\alpha^2 < 0$, then $y(x) = A e^{\alpha x} + B e^{-\alpha x}$. Then $y(0) = A + B = 0$, so $B = -A$ and therefore $y(x) = A\left(e^{\alpha x} - e^{-\alpha x}\right)$. Hence $y'(L) = A\alpha\left(e^{\alpha L} + e^{-\alpha L}\right) = 0$. But $\alpha \neq 0$ and $e^{\alpha L} + e^{-\alpha L} > 0$, so $A = 0$. Thus $\lambda = -\alpha^2$ is not an eigenvalue.

11.   If $\lambda = -\alpha^2 < 0$, then the general solution of $y'' + \lambda y = 0$ is $y(x) = A\cosh \alpha x + B \sinh \alpha x$. Then $y'(0) = 0$ implies that $B = 0$, $y(x) = \cosh \alpha x$ so (or a nonzero multiple thereof). Next,

$$y(1) + y'(1) = \cosh \alpha + \alpha \sinh \alpha = 0$$

implies that $\tanh \alpha = -\dfrac{1}{\alpha}$. But the graph of $y = \tanh \alpha$ lies in the first and third quadrants, while the graph of $y = -\dfrac{1}{\alpha}$ lies in the second and fourth quadrants. It follows that the only solution of $\tanh \alpha = -\dfrac{1}{\alpha}$ is $\alpha = 0$, and hence that our eigenvalue problem has no negative eigenvalues.

13.   **(a)** With $\lambda = 1$, the general solution of $y'' + 2y' + y = 0$ is $y(x) = A e^{-x} + B x e^{-x}$. But then $y(0) = A = 0$ and $y(1) = e^{-1}(A + B) = 0$. Hence $\lambda = 1$ is not an eigenvalue.

(b) If $\lambda < 1$, then the equation $y'' + 2y' + \lambda y = 0$ has characteristic equation $r^2 + 2r + \lambda = 0$. This equation has the two distinct real roots $\dfrac{-2 \pm \sqrt{4 - 4\lambda}}{2}$; call them r and s. Then the general solution is $y(x) = Ae^{rx} + Be^{sx}$, and the conditions $y(0) = y(1) = 0$ yield the equations

$$A + B = 0, \quad Ae^r + Be^s = 0.$$

If $A, B \neq 0$, then it follows that $e^r = e^s$. But $r \neq s$, so there is no eigenvalue $\lambda < 1$.

(c) If $\lambda > 1$, then let $\lambda - 1 = \alpha^2$, so $\lambda = 1 + \alpha^2$. Then the characteristic equation

$$r^2 + 2r + \lambda = (r + 1)^2 + \alpha^2 = 0$$

has roots $-1 \pm \alpha i$, so $y(x) = e^{-x}(A \cos \alpha x + B \sin \alpha x)$. Now $y(0) = A = 0$, so $y(x) = Ae^{-x} \sin \alpha x$. Next, $y(1) = Ae^{-1} \sin \alpha = 0$, so $\alpha = n\pi$ with n an integer. Thus the nth positive eigenvalue is $\lambda_n = n^2 \pi^2 + 1$. Because $\lambda = 1 + \alpha^2$, the eigenfunction associated with λ_n is $y_n(x) = e^{-x} \sin n\pi x$.

15.   **(a)** The endpoint conditions are

$$y(0) = y'(0) = y''(L) = y^{(3)}(L) = 0.$$

With these conditions, four successive integrations as in Example 5 yield the indicated shape function $y(x)$.

(b) The maximum value $y_{\max}$ of $y(x)$ on the closed interval $[0.L]$ must occur either at an interior point where $y'(x) = 0$ or at one of the endpoints $x = 0$ and $x = L$. Now

$$y'(x) = k\left(4x^3 - 12Lx^2 + 12L^2 x\right) = 4kx\left(x^2 - 3Lx + 3L^2\right),$$

where $k = \dfrac{w}{24EI}$ and the quadratic factor has no real zero. Hence $x = 0$ is the only zero of $y'(x)$. But $y(0) = 0$, so it follows that $y_{\max} = y(L)$.

17.   If $y(x) = k\left(x^4 - 2Lx^3 + L^3 x\right)$ with $k = \dfrac{w}{24EI}$, then $y'(x) = k\left(4x^3 - 6Lx^2 + L^3\right) = 0$ has the solution $x = \dfrac{L}{2}$ that we can verify by inspection. Now long division of the cubic $4x^3 - 6Lx^2 + L^3$ by $2x - L$ yields the quadratic factor $2x^2 - 2Lx - L^2$ whose zeros $\dfrac{2L \pm \sqrt{12L^2}}{4} = \dfrac{\left(1 \pm \sqrt{3}\right)L}{2}$ both lie outside the interval $[0, L]$. Thus $x = \dfrac{L}{2}$ is, indeed, the only zero of $y'(x) = 0$ in this interval.

CHAPTER 4

INTRODUCTION TO SYSTEMS OF DIFFERENTIAL EQUATIONS

This chapter bridges the gap between the treatment of a single differential equation in Chapters 1-3 and the comprehensive treatment of linear and nonlinear systems in Chapters 5-6. It also is designed to offer some flexibility in the treatment of linear systems, depending on the background in linear algebra that students are assumed to have—Sections 4.1 and 4.2 can stand alone as a very brief introduction to linear systems without the use of linear algebra and matrices. The final Section 4.3 of this chapter extends to systems the numerical approximation techniques of Chapter 2.

SECTION 4.1

FIRST-ORDER SYSTEMS AND APPLICATIONS

1. Let $x_1 = x$ and $x_2 = x_1' = x'$, so that $x_2' = x'' = -7x - 3x' + t^2$. Equivalent system:

$$x_1' = x_2, \quad x_2' = -7x_1 - 3x_2 + t^2 .$$

3. Let $x_1 = x$ and $x_2 = x_1' = x'$, so that $x_2' = x'' = -26x - 2x' + 82\cos 4t$. Equivalent system:

$$x_1' = x_2, \quad x_2' = -26x_1 - 2x_2 + 82\cos 4t .$$

5. Let $x_1 = x$, $x_2 = x_1' = x'$, $x_3 = x_2' = x''$, and $x_4 = x_3' = x'''$, so that $x_4' = x^{(4)} = -3x'' - x + e^t \sin 2t$. Equivalent system:

$$x_1' = x_2, \quad x_2' = x_3, \quad x_3' = x_4, \quad x_4' = -3x_2 - x_1 + e^t \sin 2t .$$

7. Let $x_1 = x$ and $x_2 = x_1' = x'$, so that $x_2' = x'' = \dfrac{\left(1 - t^2\right)x - tx'}{t^2}$. Equivalent system:

$$x_1' = x_2, \quad t^2 x_2' = \left(1 - t^2\right)x_1 - tx_2 .$$

9. Let $x_1 = x$, $x_2 = x_1' = x'$, and $x_3 = x_2' = x''$, so that $x_3' = x''' = \left(x'\right)^2 + \cos x$. Equivalent system:

$$x_1' = x_2, \quad x_2' = x_3, \quad x_3' = x_2^{\,2} + \cos x_1 .$$

11. Let $x_1 = x$, $x_2 = x_1' = x'$, $y_1 = y$, and $y_2 = y_1' = y'$, so that $x_2' = x'' = -\dfrac{kx}{\left(x^2 + y^2\right)^{3/2}}$ and

$y_2' = y'' = -\dfrac{ky}{\left(x^2 + y^2\right)^{3/2}}$. Equivalent system:

$$x_1' = x_2, \quad x_2' = -\frac{kx_1}{\left(x_1^2 + y_1^2\right)^{3/2}}, \quad y_1' = y_2, \quad y_2' = -\frac{ky_1}{\left(x_1^2 + y_1^2\right)^{3/2}}.$$

13. Let $x_1 = x$, $x_2 = x_1' = x'$, $y_1 = y$, and $y_2 = y_1' = y'$, so that $x_2' = x'' = -75x + 25y$ and
$y_2' = y'' = 50x - 50y + 50\cos 5t$. Equivalent system:

$$x_1' = x_2, \quad x_2' = -75x_1 + 25y_1, \quad y_1' = y_2, \quad y_2' = 50x_1 - 50y_1 + 50\cos 5t.$$

15. Let $x_1 = x$, $x_2 = x_1' = x'$, $y_1 = y$, $y_2 = y_1' = y'$, $z_1 = z$, and $z_2 = z_1' = z'$, so that
$x_2' = x'' = 3x - y + 2z$, $y_2' = y'' = x + y - 4z$, and $z_2' = z'' = 5x - y - z$. Equivalent system:

$$x_1' = x_2, \quad x_2' = 3x_1 - y_1 + 2z_1$$
$$y_1' = y_2, \quad y_2' = x_1 + y_1 - 4z_1$$
$$z_1' = z_2, \quad z_2' = 5x_1 - y_1 - z_1$$

17. The computation $x'' = y' = -x$ yields the single linear second-order equation $x'' + x = 0$ with characteristic equation $r^2 + 1 = 0$ and general solution $x(t) = A\cos t + B\sin t$. Then the original first equation $y = x'$ gives $y(t) = B\cos t - A\sin t$. The figure shows a direction field and typical solution curves (obviously circles?) for the given system.

Problem 17

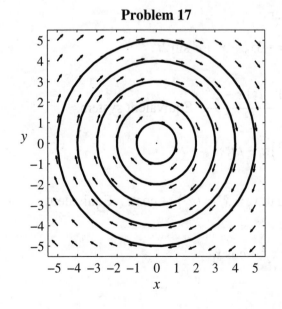

Problem 19

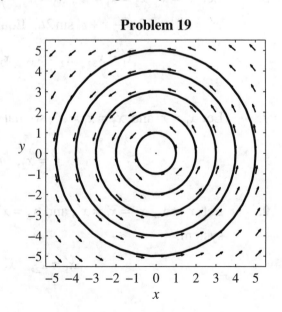

19.   The computation $x'' = -2y' = -4x$ yields the single linear second-order equation

$x'' + 4x = 0$ with characteristic equation $r^2 + 4 = 0$ and general solution

$x(t) = A\cos 2t + B\sin 2t$. Then the original first equation $y = -\dfrac{1}{2}x'$ gives

$y(t) = -B\cos 2t + A\sin 2t$. Finally, the condition $x(0) = 1$ implies that $A = 1$, and then

the condition $y(0) = 0$ gives $B = 0$. Hence the desired particular solution is given by

$$x(t) = \cos 2t, \quad y(t) = \sin 2t.$$

The figure shows a direction field and some typical circular solution curves for the given
system.

21.   The computation $x'' = \dfrac{1}{2}y' = -4x$ yields the single linear second-order equation

$x'' + 4x = 0$ with characteristic equation $r^2 + 4 = 0$ and general solution
$x(t) = A\cos 2t + B\sin 2t$. Then the original first equation $y = 2x'$ gives

$y(t) = 4B\cos 2t - 4A\sin 2t$. The figure shows a direction field and some typical ellipti-
cal solution curves.

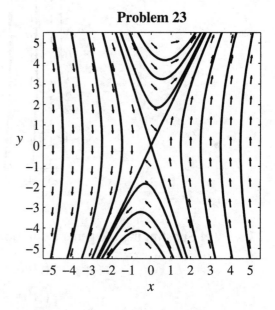

Problem 21   **Problem 23**

23.   The computation $x'' = y' = 6x - y = 6x - x'$ yields the single linear second-order equation

$x'' + x' - 6x = 0$ with characteristic equation $r^2 + r - 6 = 0$, characteristic roots $r = -3$
and 2, and general solution $x(t) = Ae^{-3t} + Be^{2t}$. Then the original first equation $y = x'$

gives $y(t) = -3Ae^{-3t} + 2Be^{2t}$. Finally, the initial conditions

$$x(0) = A + B = 1, \quad y(0) = -3A + 2B = 2$$

imply that $A = 0$ and $B = 1$, so the desired particular solution is given by

$$x(t) = e^{2t}, \quad y(t) = 2e^{2t}.$$

The figure shows a direction field and some typical solution curves.

25.   The computation $x'' = -y' = -13x - 4y = -13x + 4x'$ yields the single linear second-order equation $x'' - 4x' + 13x = 0$ with characteristic equation $r^2 - 4r + 13 = 0$ and characteristic roots $r = 2 \pm 3i$; hence the general solution is $x(t) = e^{2t}(A\cos 3t + B\sin 3t)$. The initial condition $x(0) = 0$ then gives $A = 0$, so $x(t) = Be^{2t}\sin 3t$. Then the original first equation $y = -x'$ gives $y(t) = -e^{2t}(3B\cos 3t + 2B\sin 3t)$. Finally, the initial condition $y(0) = 3$ gives $B = -1$, so the desired particular solution is given by

$$x(t) = -e^{2t}\sin 3t, \quad y(t) = e^{2t}(3\cos 3t + 2\sin 3t).$$

The figure shows a direction field and some typical solution curves.

Problem 25

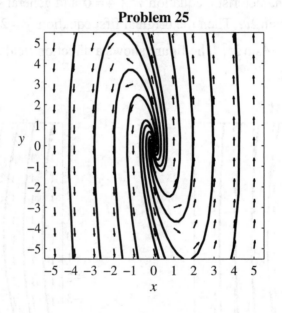

27.   **(a)** Substituting the general solution found in Problem 17 we get

$$
\begin{aligned}
x^2 + y^2 &= (A\cos t + B\sin t)^2 + (B\cos t - A\sin t)^2 \\
&= (A^2 + B^2)(\cos^2 t + \sin^2 t) \\
&= (A^2 + B^2),
\end{aligned}
$$

or $x^2 + y^2 = C^2$, the equation of a circle of radius $C = \sqrt{A^2 + B^2}$.

(b) Substituting the general solution found in Problem 18, we get

$$x^2 - y^2 = (Ae^t + Be^{-t})^2 - (Ae^t - Be^{-t})^2 = 4AB,$$

the equation of a hyperbola.

29. When we solve Equations (20) and (21) in the text for e^{-t} and e^{2t} we get $2x - y = 3Ae^{-t}$ and $x + y = 3Be^{2t}$. Hence

$$(2x - y)^2 (x + y) = (3Ae^{-t})^2 \cdot 3Be^{2t} = 27A^2B = C.$$

Clearly $y = 2x$ or $y = -x$ if $C = 0$, and expansion gives the equation $4x^3 - 3xy^2 + y^3 = C$.

31. Looking at Fig. 4.1.12 in the text, we see that

$$my_1'' = -T\sin\theta_1 + T\sin\theta_2 \approx -T\tan\theta_1 + T\tan\theta_2 = -\frac{Ty_1}{L} + \frac{T(y_2 - y_1)}{L}$$

and

$$my_2'' = -T\sin\theta_2 - T\sin\theta_3 \approx -T\tan\theta_2 - T\tan\theta_3 = -\frac{T(y_2 - y_1)}{L} - \frac{Ty_2}{L}.$$

We get the desired equations when we multiply each of these equations by $\dfrac{L}{T}$ and set

$$k = \frac{mL}{T}.$$

33. We apply Kirchhoff's law to each loop in Figure 4.1.14 in the text, and immediately get the equations

$$2(I_1' - I_2') + 50I_1 = 100\sin 60t, \quad 2(I_2' - I_1') + 25I_2 = 0.$$

35. If θ is the polar angular coordinate of the point (x, y) and we write $F = \dfrac{k}{x^2 + y^2} = \dfrac{k}{r^2}$,

then Newton's second law gives

$$mx'' = -F\cos\theta = -\frac{k}{r^2} \cdot \frac{x}{r} = -\frac{kx}{r^3}, \quad my'' = -F\sin\theta = -\frac{k}{r^2} \cdot \frac{y}{r} = -\frac{ky}{r^3}.$$

37. If $\mathbf{r} = (x, y, z)$ is the particle's position vector, then Newton's law $m\mathbf{r}'' = \mathbf{F}$ gives

$$m\mathbf{r}'' = q\mathbf{v} \times \mathbf{B} = q\begin{vmatrix} \mathbf{i} & \mathbf{j} & \mathbf{k} \\ x' & y' & z' \\ 0 & 0 & B \end{vmatrix} = +qBy'\mathbf{i} - qBx'\mathbf{j} = qB(-y', x', 0).$$

SECTION 4.2

THE METHOD OF ELIMINATION

1. The second differential equation $y' = 2y$ has the exponential solution

$$y(t) = c_2 e^{2t}.$$

Substitution in the first differential equation gives the linear first-order equation $x' + x = 3c_2 e^{2t}$ with integrating factor $\rho = e^t$. Solution of this equation in the usual way gives

$$x(t) = e^{-t}\left(c_1 + c_2 e^{3t}\right) = c_1 e^{-t} + c_2 e^{2t}.$$

The figure shows a direction field and some typical solution curves.

Problem 1

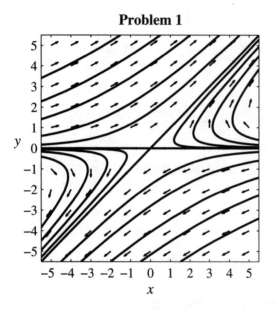

Problem 3

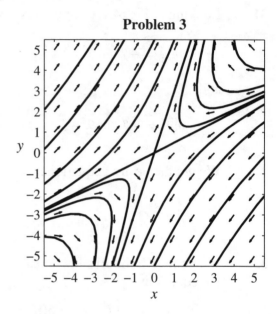

3.   From the first differential equation we get $y = \dfrac{1}{2}(3x + x')$, so $y = \dfrac{1}{2}(3x' + x'')$. Substitution of these expressions for y and y' into the second differential equation yields the second-order equation $x'' - x' - 6x = 0$ with general solution

$$x(t) = c_1 e^{-2t} + c_2 e^{3t}.$$

Substitution in $y = \dfrac{1}{2}(3x + x')$ now yields

$$y(t) = \frac{1}{2}c_1 e^{-2t} + 3c_2 e^{3t}.$$

Imposition of the initial conditions $x(0) = 0$, $y(0) = 2$ now gives the equations

$$c_1 + c_2 = 0, \quad \frac{1}{2}c_1 + 3c_2 = 2$$

with solution $c_1 = -\dfrac{4}{5}$, $c_2 = \dfrac{4}{5}$. These coefficients give the desired particular solution

$$x(t) = \frac{4}{5}\left(e^{3t} - e^{-2t}\right), \quad y(t) = \frac{2}{5}\left(6e^{3t} - e^{-2t}\right).$$

The figure shows a direction field and some typical solution curves.

5.   Substitution of $y = -\frac{1}{4}(x' + 3x)$ and $y' = -\frac{1}{4}(x'' + 3x')$ —from the first equation—into the second equation yields the second-order equation $x'' + 2x' + 5x = 0$ with general solution

$$x(t) = e^{-t}\left(c_1 \cos 2t + c_2 \sin 2t\right).$$

Substitution of this solution in $y = -\frac{1}{4}(x' + 3x)$ gives

$$y(t) = \frac{1}{2}e^{-t}\left[-(c_1 + c_2)\cos 2t + (c_1 - c_2)\sin 2t\right].$$

The figure shows a direction field and some typical solution curves.

Problem 5

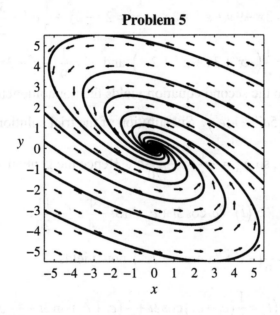

7.   Substitution of $y = x' - 4x - 2t$ and $y' = x'' - 4x' - 2$ —from the first equation—into the second equation yields the nonhomogeneous second-order equation

$x'' - 5x' + 6x = 2 - 2t$. Substitution of the trial solution $x_p = A + Bt$ yields $A = \frac{1}{18}$,

$B = -\frac{1}{3}$, so $x_p = \frac{1}{18} - \frac{t}{3}$. Hence the general solution for x is

$$x(t) = c_1 e^{2t} + c_2 e^{3t} - \frac{t}{3} + \frac{1}{18}.$$

Substitution in $y = x' - 4x - 2t$ now yields

$$y(t) = -2c_1 e^{2t} - c_2 e^{3t} - \frac{2}{3}t - \frac{5}{9}.$$

9. Substitution of $y = \frac{1}{3}(-x' + 2x + 2\sin 2t)$ and $y' = \frac{1}{3}(-x'' + 2x' + 4\cos 2t)$ —from the first equation—into the second equation yields the second-order equation $x'' - x = 7\cos 2t + 4\sin 2t$ with general solution

$$x(t) = c_1 e^{-t} + c_2 e^t - \frac{1}{5}(7\cos 2t + 4\sin 2t).$$

Substitution in $y = \frac{1}{3}(-x' + 2x + 2\sin 2t)$ now yields

$$y(t) = c_1 e^{-t} + \frac{1}{3}c_2 e^t - \frac{1}{5}(2\cos 2t + 4\sin 2t).$$

11. First we solve the given equations for the normal-form first-order equations

$$x' = 3x - 9y + e^{-t} + 2e^t, \quad y' = 2x - 3y + \frac{1}{2}e^{-t} + \frac{3}{2}e^t.$$

Substitution of $y = \frac{1}{9}(-x' + 3x + e^{-t} + 2e^t)$ and $y' = \frac{1}{9}(-x'' + 3x' - e^{-t} + 2e^t)$ —from the first equation—into the second equation yields the nonhomogeneous second-order equation $x'' + 9x = -\frac{1}{2}(5e^{-t} + 11e^t)$. Substitution of the trial solution $x_p = Ae^{-t} + Be^t$ yields $A = -\frac{1}{4}$, $B = -\frac{11}{20}$, so $x_p = -\frac{1}{4}e^{-t} - \frac{11}{20}e^t$. Hence the general solution for x is

$$x(t) = c_1 \cos 3t + c_2 \sin 3t - \frac{1}{4}e^{-t} - \frac{11}{20}e^t.$$

Substitution in $y = \frac{1}{9}(-x' + 3x + e^{-t} + 2e^t)$ now yields

$$y(t) = \frac{1}{3}(c_1 - c_2)\cos 3t + \frac{1}{3}(c_1 + c_2)\sin 3t + \frac{1}{10}e^t.$$

13. The first equation yields $y = \frac{1}{2}(x'' + 5x)$, so $y'' = \frac{1}{2}(x^{(4)} + 5x'')$. Substitution in the second equation yields

$$x^{(4)} + 13x'' + 36x = 0.$$

The characteristic equation is $r^4 + 13r^2 + 36 = (r^2 + 4)(r^2 + 9) = 0$, so the general solution for x is

$$x(t) = a_1 \cos 2t + a_2 \sin 2t + b_1 \cos 3t + b_2 \sin 3t.$$

Substitution in $y = \frac{1}{2}(x'' + 5x)$ now gives

$$y(t) = \frac{1}{2}a_1 \cos 2t + \frac{1}{2}a_2 \cos 2t - 2b_1 \cos 3t - 2b_2 \sin 3t .$$

15.   If we write the given differential equations in operator notation as

$$(D^2 - 2)x - 3Dy = 0$$
$$3Dx + (D^2 - 2)y = 0,$$

then we see that the system has operational determinant

$$(D^2 - 2)^2 + 9D^2 = D^4 + 5D^2 + 4 = (D^2 + 1)(D^2 + 4).$$

Therefore (as in Example 3) we see that x satisfies the fourth-order differential equation $(D^2 + 1)(D^2 + 4)x = 0$ with characteristic equation $(r^2 + 1)(r^2 + 4) = 0$ and general solution

$$x(t) = a_1 \cos t + a_2 \sin t + b_1 \cos 2t + b_2 \sin 2t .$$

Similarly, the general solution for y is of the form

$$y(t) = c_1 \cos t + c_2 \sin t + d_1 \cos 2t + d_2 \sin 2t .$$

Now, substitution of these two general solutions in the first equation $x'' - 3y' - 2x = 0$ and collection of coefficients gives

$$(-3a_1 - 3c_2)\cos t + (3c_1 - 3a_2)\sin t + (-6b_1 - 6d_2)\cos 2t + (3d_1 - 3b_2)\sin 2t = 0 .$$

Thus we see finally that $c_1 = a_2$, $c_2 = -a_1$, $d_1 = b_2$, $d_2 = -b_1$. Hence

$$y(t) = a_2 \cos t - a_1 \sin t + b_2 \cos 2t - b_1 \sin 2t .$$

17.   If we write the given equations in the operational form

$$(D^2 - 3D - 2)x \;+\; (D^2 - D + 2)y \;=\; 0,$$
$$(2D^2 - 9D - 4)x \;+\; (3D^2 - 2D + 6)y \;=\; 0,$$

we see (thinking of the operational determinant) that x satisfies a homogeneous fourth-order equation with characteristic equation

$$(r^2 - 3r - 2)(3r^2 - 2r + 6) - (r^2 - r + 2)(2r^2 - 9r - 4) = r^4 - 3r^2 - 4$$
$$= (r^2 + 1)(r^2 - 4)$$
$$= (r^2 + 1)(r + 2)(r - 2)$$
$$= 0.$$

Hence the general solution for x is

$$x(t) = a_1 \cos t + a_2 \sin t + b_1 e^{-2t} + b_2 e^{2t} ;$$

similarly, the general solution for y is

$$y(t) = c_1 \cos t + c_2 \sin t + d_1 e^{-2t} + d_2 e^{2t}.$$

To determine the relations between the arbitrary constants in these two general solutions, we substitute them in the first of the original differential equations and get

$$\left(-a_1 \cos t - a_2 \sin t + 4b_1 e^{-2t} + 4b_2 e^{2t}\right) + \left(-c_1 \cos t - c_2 \sin t + 4d_1 e^{-2t} + 4d_2 e^{2t}\right) +$$

$$\left(3a_1 \sin t - 3a_2 \cos t + 6b_1 e^{-2t} - 6b_2 e^{2t}\right) + \left(c_1 \sin t - c_2 \cos t + 2d_1 e^{-2t} - 2d_2 e^{2t}\right) +$$

$$\left(-2a_1 \cos t - 2a_2 \sin t - 2b_1 e^{-2t} - 2b_2 e^{2t}\right) + \left(+2c_1 \cos t + 2c_2 \sin t + 2d_1 e^{-2t} + 2d_2 e^{2t}\right) = 0.$$

If we collect coefficients of the trigonometric and exponential terms we get the equations

$$\begin{array}{ccc} -3a_1 - 3a_2 + c_1 - c_2 = 0, & & 8b_1 + 8d_1 = 0, \\ & \text{and} & \\ 3a_1 - 3a_2 + c_1 + c_2 = 0 & & -4b_2 + 4d_2 = 0. \end{array}$$

The first two of these equations imply that $c_1 = -3a_2$ and $c_2 = 3a_1$, while the latter two give $d_1 = -b_1$ and $d_2 = b_2$. We therefore see finally that

$$y(t) = -3a_2 \cos t + 3a_1 \sin t - b_1 e^{-2t} + b_2 e^{2t}.$$

19.   The operational determinant of the given system is

$$L = \begin{vmatrix} D-4 & 2 & 0 \\ 4 & D-4 & 2 \\ 0 & 4 & D-4 \end{vmatrix} = D^3 - 12D^2 + 32D,$$

so x, y, and z all satisfy a third-order homogeneous linear differential equation with characteristic equation $r^3 - 12r^2 + 32r = r(r-4)(r-8) = 0$. The corresponding general solutions are

$$x(t) = a_1 + a_2 e^{4t} + a_3 e^{8t}, \quad y(t) = b_1 + b_2 e^{4t} + b_3 e^{8t}, \quad z(t) = c_1 + c_2 e^{4t} + c_3 e^{8t}.$$

If we substitute $x(t)$ and $y(t)$ in the first differential equation $x' = 4x - 2y$ and collect coefficients of like terms, we find quickly that $b_1 = 2a_1$, $b_2 = 0$, and $b_3 = -2a_3$. Similarly, we find by substitution in the other two equations that $c_1 = 2a_1$, $c_2 = -2a_2$, and $c_3 = 2a_3$. Thus y and z are given by

$$y(t) = 2a_1 - 2a_3 e^{8t}, \quad z(t) = 2a_1 - 2a_2 e^{4t} + 2a_3 e^{8t}.$$

21. $L_1 L_2 = L_2 L_1$ because both sides simplify to the same thing upon multiplying out and collecting terms in the usual fashion of polynomial algebra. This "works" because different powers of D commute—that is, $D^i D^j = D^j D^i$ because $D^i\left(D^j x\right) = D^{i+j} x = D^j\left(D^i x\right)$.

23. Subtraction of the two equations yields $x + y = e^{-2t} - e^{-3t}$. We then verify readily that any two differentiable functions $x(t)$ and $y(t)$ satisfying this condition will constitute a solution of the given system, which thus has infinitely many solutions.

25. Infinitely many solutions, because any solution of the second equation also satisfies the first equation (because it is $D + 2$ times the second one).

27. Subtraction of the second equation from the first one gives $x + y = e^{-t}$. Then substitution in the second equation yields

$$x(t) = D^2(x + y) = e^{-t}.$$

It follows that $y(t) \equiv 0$, so there are *no* arbitrary constants.

29. Addition of the two given equations yields $D^2 x = e^{-t}$, so $x(t) = e^{-t} + a_1 t + a_2$. Then the second equation gives $D^2 y = a_1 t + a_2$, so

$$y(t) = \frac{1}{6}a_1 t^3 + \frac{1}{2}a_2 t^2 + a_3 t + a_4.$$

Thus there are *four* arbitrary constants.

31. Substitution of $I_2 = \frac{1}{25}(I_1' + 25I_1 - 50)$ and $I_2' = \frac{1}{25}(I_1'' + 25I_1')$ —from the first equation—into the second equation yields the second-order equation $3I_1'' + 30I_1' + 125I_1 = 250$ with general solution

$$I_1(t) = 2 + e^{-5t}\left[c_1 \cos\left(\frac{5\sqrt{6}}{3}t\right) + c_2 \sin\left(\frac{5\sqrt{6}}{3}t\right)\right].$$

Substitution in $I_2 = \frac{1}{25}(I_1' + 25I_1 - 50)$ now yields

$$I_2(t) = \frac{1}{15}e^{-5t}\left[\left(12c_1 + \sqrt{6}c_2\right)\cos\left(\frac{5\sqrt{6}}{3}t\right) + \left(12c_2 - \sqrt{6}c_1\right)\sin\left(\frac{5\sqrt{6}}{3}t\right)\right].$$

Imposition of the initial conditions $I_1(0) = 0$, $I_2(0) = 0$ now gives the equations

$$c_1 + 2 = 0, \quad \frac{4}{5}c_1 + \frac{\sqrt{6}}{15}c_2 = 0$$

with solution $c_1 = -2$, $c_2 = 4\sqrt{6}$. These coefficients give the desired particular solution

$$I_1(t) = 2 + e^{-5t} \left[-2\cos\left(\frac{5\sqrt{6}}{3} t \right) + 4\sqrt{6} \sin\left(\frac{5\sqrt{6}}{3} t \right) \right],$$

$$I_2(t) = \frac{20}{\sqrt{6}} e^{-5t} \sin\left(\frac{5\sqrt{6}}{3} t \right).$$

33. To solve the system

$$I_1' = -20(I_1 - I_2), \quad I_2' = 40(I_1 - I_2)$$

we first note that $I_2' = -2I_1'$, so $I_2 = -2I_1 + K$. Then $K = 2I_1(0) + I_2(0) = 2 \cdot 2 + 0 = 4$, so $I_2' = -2I_1'$, so $I_2 = -2I_1 + 4$. Substitution of this into the first equation gives the simple first-order linear equation $I_1' + 60I_1 = 80$ with general solution $I_1(t) = \frac{4}{3} + ce^{-60t}$. The initial condition $I_1(0) = 2$ gives $c = \frac{2}{3}$, so

$$I_1(t) = \frac{2}{3}(2 + e^{-60t}), \quad I_2(t) = \frac{4}{3}(1 - e^{-60t}).$$

35. The two given equations yield

$$mx^{(3)} = qBy'' = -\frac{q^2 B^2}{m} x',$$

so $x^{(3)} + \omega^2 x' = 0$. The general solution is

$$x(t) = A\cos \omega t + B\sin \omega t + C.$$

Now $x'(0) = 0$ implies $B = 0$, and then $x(0) = r_0$ gives $A + C = r_0$. Next,

$$\omega y' = x'' = -A\omega^2 \cos \omega t,$$

so $y'(0) = -\omega r_0$ implies that $A = r_0$, and hence that $C = 0$. It now follows readily that the trajectory is the circle

$$x(t) = r_0 \cos \omega t, \quad y(t) = -r_0 \sin \omega t.$$

37. **(a)** If we set $m_1 = 2$, $m_2 = \frac{1}{2}$, $k_1 = 75$, and $k_2 = 25$ in Eqs. (3) of Section 4.1, we get the system

$$2x'' = -100x + 25y, \quad \frac{1}{2}y'' = 25x - 25y$$

with operational determinant $D^4 + 100D^2 + 1875 = (D^2 + 25)(D^2 + 75)$. Hence the general form of the solution is

$$x(t) = a_1 \cos 5t + a_2 \sin 5t + b_1 \cos 5\sqrt{3}t + b_2 \sin 5\sqrt{3}t,$$
$$y(t) = c_1 \cos 5t + c_2 \sin 5t + d_1 \cos 5\sqrt{3}t + d_2 \sin 5\sqrt{3}t.$$

Upon substitution in either differential equation we see that $c_1 = 2a_1$ and $c_2 = 2a_2$, and $d_1 = -2b_1$ and $d_2 = -2b_2$. This gives

$$x(t) = a_1 \cos 5t + a_2 \sin 5t + b_1 \cos 5\sqrt{3}t + b_2 \sin 5\sqrt{3}t,$$
$$y(t) = 2a_1 \cos 5t + 2a_2 \sin 5t - 2b_1 \cos 5\sqrt{3}t - 2b_2 \sin 5\sqrt{3}t.$$

(b) In the natural mode with frequency $\omega_1 = 5$ the masses move in the same direction, while in the natural mode with frequency $\omega_2 = 5\sqrt{3}$ they move in opposite directions. In each case the amplitude of the motion of m_2 is twice that of m_1.

39.   The system has operational determinant $8D^4 + 40D^2 + 32 = 8(D^2 + 1)(D^2 + 4)$. Hence the general form of the solution is

$$x(t) = a_1 \cos t + a_2 \sin t + b_1 \cos 2t + b_2 \sin 2t,$$
$$y(t) = c_1 \cos t + c_2 \sin t + d_1 \cos 2t + d_2 \sin 2t.$$

Upon substitution in either differential equation we see that $c_1 = 2a_1$ and $c_2 = 2a_2$, and $d_1 = -b_1$ and $d_2 = -b_2$. This gives

$$x(t) = a_1 \cos t + a_2 \sin t + b_1 \cos 2t + b_2 \sin 2t,$$
$$y(t) = 2a_1 \cos t + 2a_2 \sin t - b_1 \cos 2t - b_2 \sin 2t.$$

In the natural mode with frequency $\omega_1 = 1$ the masses move in the same direction, with the amplitude of motion of the second mass twice that of the first mass (figure **a**). In the natural mode with frequency $\omega_2 = 2$ they move in opposite directions with the same amplitude of motion (figure **b**).

Problem 39a

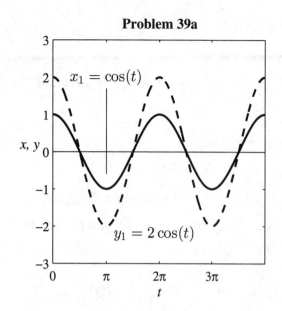

Problem 39b

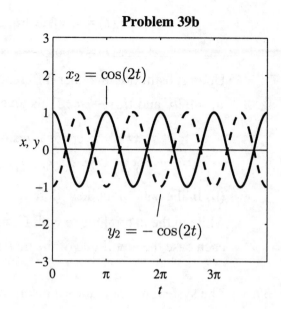

41.   The system has operational determinant $D^4 + 10D^2 + 9 = (D^2 + 1)(D^2 + 9)$. Hence the general form of the solution is

$$x(t) = a_1 \cos t + a_2 \sin t + b_1 \cos 3t + b_2 \sin 3t,$$
$$y(t) = c_1 \cos t + c_2 \sin t + d_1 \cos 3t + d_2 \sin 3t.$$

Upon substitution in either differential equation we see that $c_1 = a_1$ and $c_2 = a_2$, and $d_1 = -b_1$ and $d_2 = -b_2$. This gives

$$x(t) = a_1 \cos t + a_2 \sin t + b_1 \cos 3t + b_2 \sin 3t,$$
$$y(t) = a_1 \cos t + a_2 \sin t - b_1 \cos 3t - b_2 \sin 3t.$$

In the natural mode with frequency $\omega_1 = 1$ the masses move in the same direction (figure **a**), while in the natural mode with frequency $\omega_2 = 3$ they move in opposite directions (figure **b**). In each case the amplitudes of motion of the two masses are equal.

Problem 41a

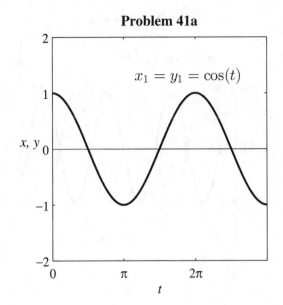

Problem 41b

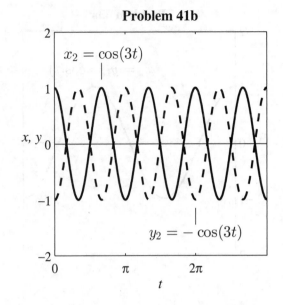

43. The system has operational determinant $D^4 + 6D^2 + 5 = (D^2 + 1)(D^2 + 5)$. Hence the general form of the solution is

$$x(t) = a_1 \cos t + a_2 \sin t + b_1 \cos\left(\sqrt{5}t\right) + b_2 \sin\left(\sqrt{5}t\right),$$
$$y(t) = c_1 \cos t + c_2 \sin t + d_1 \cos\left(\sqrt{5}t\right) + d_2 \sin\left(\sqrt{5}t\right).$$

Upon substitution in either differential equation we see that $c_1 = a_1$ and $c_2 = a_2$, and $d_1 = -b_1$ and $d_2 = -b_2$. This gives

$$x(t) = a_1 \cos t + a_2 \sin t + b_1 \cos\left(\sqrt{5}t\right) + b_2 \sin\left(\sqrt{5}t\right),$$
$$y(t) = a_1 \cos t + a_2 \sin t - b_1 \cos\left(\sqrt{5}t\right) - b_2 \sin\left(\sqrt{5}t\right).$$

In the natural mode with frequency $\omega_1 = 1$ the masses move in the same direction (figure **a**), while in the natural mode with frequency $\omega_2 = \sqrt{5}$ they move in opposite directions (figure **b**). In each case the amplitudes of motion of the two masses are equal.

Problem 43a

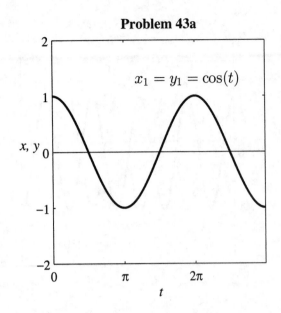

$x_1 = y_1 = \cos(t)$

Problem 43b

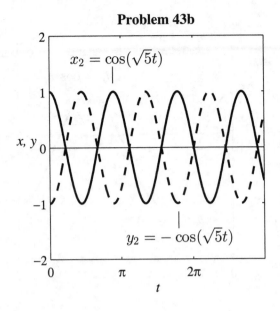

$x_2 = \cos(\sqrt{5}t)$

$y_2 = -\cos(\sqrt{5}t)$

45.   The system has operational determinant $2D^4 + 20D^2 + 32 = 2(D^2 + 2)(D^2 + 8)$. Hence the general form of the solution is

$$x(t) = a_1 \cos(\sqrt{2}t) + a_2 \sin(\sqrt{2}t) + b_1 \cos(\sqrt{8}t) + b_2 \sin(\sqrt{8}t),$$
$$y(t) = c_1 \cos(\sqrt{2}t) + c_2 \sin(\sqrt{2}t) + d_1 \cos(\sqrt{8}t) + d_2 \sin(\sqrt{8}t).$$

Upon substitution in either differential equation we see that $c_1 = a_1$ and $c_2 = a_2$, and $d_1 = -\frac{1}{2}b_1$ and $d_2 = -\frac{1}{2}b_2$. This gives

$$x(t) = a_1 \cos(\sqrt{2}t) + a_2 \sin(\sqrt{2}t) + b_1 \cos(\sqrt{8}t) + b_2 \sin(\sqrt{8}t),$$
$$y(t) = a_1 \cos(\sqrt{2}t) + a_2 \sin(\sqrt{2}t) - \frac{1}{2}b_1 \cos(\sqrt{8}t) - \frac{1}{2}b_2 \sin(\sqrt{8}t).$$

In the natural mode with frequency $\omega_1 = \sqrt{2}$ the two masses move in the same direction with equal amplitudes of oscillation (figure **a**). In the natural mode with frequency $\omega_2 = \sqrt{8} = 2\sqrt{2}$ the two masses move in opposite directions with the amplitude of m_2 being half that of m_1 (figure **b**).

Problem 45a

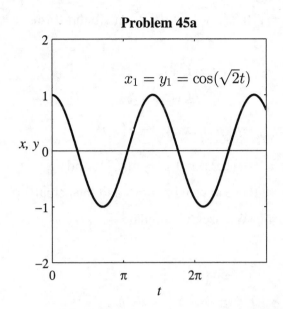

$$x_1 = y_1 = \cos(\sqrt{2}t)$$

Problem 45b

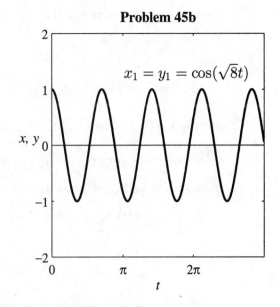

$$x_1 = y_1 = \cos(\sqrt{8}t)$$

47. **(a)** Looking at Fig. 4.2.7 in the text, we see that the first spring is stretched by x, the second spring is stretched by $y - x$, the third spring is stretched by $z - y$, and the fourth spring is compressed by z. Hence Newton's second law gives $mx'' = -k(x) + k(y - x)$, $my'' = -k(y - x) + k(z - y)$, and $mz'' = -k(z - y) - k(z)$.

(b) The operational determinant is

$$\left(D^2 + 2\right)\left[\left(D^2 + 2\right)^2 - 1\right] + \left[-\left(D^2 + 2\right)\right] = \left(D^2 + 2\right)\left[\left(D^2 + 2\right)^2 - 2\right],$$

and the characteristic equation $\left(r^2 + 2\right)\left[\left(r^2 + 2\right)^2 - 2\right] = 0$ has roots $\pm i\sqrt{2}$ and

$$\pm i\sqrt{2 \pm \sqrt{2}}.$$

SECTION 4.3

NUMERICAL METHODS FOR SYSTEMS

In Problems 1-8 we first write the given system in the form $x' = f(t, x, y)$ $y' = g(t, x, y)$. Then we use the template

$$h = 0.1; \qquad\qquad t_1 = t_0 + h$$
$$x_1 = x_0 + hf(t_0, x_0, y_0); \quad y_1 = y_0 + hg(t_0, x_0, y_0)$$
$$x_2 = x_1 + hf(t_1, x_1, y_1); \quad y_2 = y_1 + hg(t_1, x_1, y_1)$$

(with the given values of t_0, x_0, and y_0) to calculate the Euler approximations $x_1 \approx x(0.1)$ and $y_1 \approx y(0.1)$, and $x_2 \approx x(0.2)$ and $y_2 \approx y(0.2)$, in part (a). We give these approximations and the actual values $x_{act} = x(0.2)$, $y_{act} = y(0.2)$ in tabular form. We use the template

$$h = 0.2; \qquad\qquad\qquad t_1 = t_0 + h$$
$$u_1 = x_0 + hf(t_0, x_0, y_0); \qquad\qquad v_1 = y_0 + hg(t_0, x_0, y_0)$$
$$x_1 = x_0 + \frac{1}{2}h\left[f(t_0, x_0, y_0) + f(t_1, u_1, v_1) \right] \quad y_1 = y_0 + \frac{1}{2}h\left[g(t_0, x_0, y_0) + g(t_1, u_1, v_1) \right]$$

to calculate the improved Euler approximations $u_1 \approx x(0.2)$ and $u_1 \approx y(0.2)$, and $x_1 \approx x(0.2)$ and $y_1 \approx y(0.2)$, in part (b). We give these approximations and the actual values $x_{act} = x(0.2)$, $y_{act} = y(0.2)$ in tabular form. We use the template

$$h = 0.2;$$

$$F_1 = f(t_0, x_0, y_0); \qquad\qquad G_1 = g(t_0, x_0, y_0);$$
$$F_2 = f\left(t_0 + \frac{h}{2}, x_0 + \frac{h}{2}F_1, y_0 + \frac{1}{2}hG_1 \right); \quad G_2 = g\left(t_0 + \frac{h}{2}, x_0 + \frac{h}{2}F_1, y_0 + \frac{h}{2}G_1 \right);$$
$$F_3 = f\left(t_0 + \frac{h}{2}, x_0 + \frac{h}{2}F_2, y_0 + \frac{h}{2}G_2 \right); \quad G_3 = g\left(t_0 + \frac{h}{2}, x_0 + \frac{h}{2}F_2, y_0 + \frac{h}{2}G_2 \right);$$
$$F_4 = f(t_0 + h, x_0 + hF_3, y_0 + hG_3); \quad G_4 = g(t_0 + h, x_0 + hF_3, y_0 + hG_3);$$
$$x_1 = x_0 + \frac{h}{6}(F_1 + 2F_2 + 2F_3 + F_4); \qquad y_1 = y_0 + \frac{h}{6}(G_1 + 2G_2 + 2G_3 + G_4)$$

to calculate the intermediate slopes and Runge-Kutta approximations $x_1 \approx x(0.2)$ and $y_1 \approx y(0.2)$ for part (c). Again, we give the results in tabular form.

1. **(a)**

x_1	y_1	x_2	y_2	x_{act}	y_{act}
0.4	2.2	0.88	2.5	1.0034	2.6408

(b)

u_1	v_1	x_1	y_1	x_{act}	y_{act}
0.8	2.4	0.96	2.6	1.0034	2.6408

(c)

F_1	G_1	F_2	G_2	F_3	G_3	F_4	G_4
4	2	4.8	3	5.08	3.26	6.32	4.684

x_1	y_1	x_{act}	y_{act}				
1.0027	2.6401	1.0034	2.6408				

3.   **(a)**

x_1	y_1	x_2	y_2	x_{act}	y_{act}
1.7	1.5	2.81	2.31	3.6775	2.9628

(b)

u_1	v_1	x_1	y_1	x_{act}	y_{act}
2.4	2	3.22	2.62	3.6775	2.9628

(c)

F_1	G_1	F_2	G_2	F_3	G_3	F_4	G_4
7	5	11.1	8.1	13.57	9.95	23.102	17.122
x_1	y_1	x_{act}	y_{act}				
3.6481	2.9407	3.6775	2.9628				

5.   **(a)**

x_1	y_1	x_2	y_2	x_{act}	y_{act}
0.9	3.2	−0.52	2.92	-0.5793	2.4488

(b)

u_1	v_1	x_1	y_1	x_{act}	y_{act}
−0.2	3.4	−0.84	2.44	-0.5793	2.4488

(c)

F_1	G_1	F_2	G_2	F_3	G_3	F_4	G_4
−11	2	−14.2	−2.8	−12.44	−3.12	−12.856	−6.704

x_1	y_1	x_{act}	y_{act}				
−0.5712	2.4485	-0.5793	2.4488				

7. **(a)**

x_1	y_1	x_2	y_2	x_{act}	y_{act}
2.5	1.3	3.12	1.68	3.2820	1.7902

(b)

u_1	v_1	x_1	y_1	x_{act}	y_{act}
3	1.6	3.24	1.76	3.2820	1.7902

(c)

F_1	G_1	F_2	G_2	F_3	G_3	F_4	G_4
5	3	6.2	3.8	6.48	4	8.088	5.096

x_1	y_1	x_{act}	y_{act}				
3.2816	1.7899	3.2820	1.7902				

In Problems 9-11 we use the same Runge-Kutta template as in part (c) of Problems 1–8 above, and give both the Runge-Kutta approximate values with step sizes $h = 0.1$ and $h = 0.05$, and also the actual values.

9.

With $h = 0.1$:	$x(1) \approx 3.99261$	$y(1) \approx 6.21770$
With $h = 0.05$:	$x(1) \approx 3.99234$	$y(1) \approx 6.21768$
Actual values:	$x(1) \approx 3.99232$	$y(1) \approx 6.21768$

11.

With $h = 0.1$:	$x(1) \approx -0.05832$	$y(1) \approx 0.56664$
With $h = 0.05$:	$x(1) \approx -0.05832$	$y(1) \approx 0.56665$

Actual values:	$x(1) \approx -0.05832$	$y(1) \approx 0.56665$

13.   With $y = x'$ we want to solve numerically the initial value problem

$$x' = y, \qquad x(0) = 0,$$
$$y' = -32 - 0.04y, \quad y(0) = 288.$$

When we run Program RK2DIM with step size $h = 0.1$ we find that the change of sign in the velocity v occurs as follows:

t	x	v
7.6	1050.2	+2.8
7.7	1050.3	-0.4

Thus the bolt attains a maximum height of about 1050 feet in about 7.7 seconds.

15.   With $y = x'$ and with x in miles and in seconds, we want to solve numerically the initial value problem

$$x' = y, \qquad x(0) = 0,$$
$$y' = \frac{-95485.5}{x^2 + 7920x + 15681600}, \quad y(0) = 1.$$

We find (running RK2DIM with $h = 1$) that the projectile reaches a maximum height of about 83.83 miles in about $168 \sec = 2 \min 48 \sec$.

17.   The data in Problem 16 indicate that the range increases when the initial angle is decreased below $45°$. The further data

Angle	Range
41.0	352.1
40.5	352.6
40.0	352.9
39.5	352.8
39.0	352.7
35.0	350.8

indicate that a maximum range of about 353 ft is attained with $\alpha \approx 40°$.

19. First we run program **rkn** (with $h = 0.1$) with $v_0 = 250$ ft/sec and obtain the following results:

t	x	y
5.0	457.43	103.90
6.0	503.73	36.36

Interpolation gives $x = 494.4$ when $y = 50$. Then a run with $v_0 = 255$ ft/sec gives the following results:

t	x	y
5.5	486.75	77.46
6.0	508.86	41.62

Finally, a run with $v_0 = 253$ ft/sec gives these results:

t	x	y
5.5	484.77	75.44
6.0	506.82	39.53

Now $x \approx 500$ ft when $y = 50$ ft. Thus Babe Ruth's home run ball had an initial velocity of 253 ft/sec.

21. A run with $h = 0.1$ indicates that the projectile has a range of about $21,400$ ft ≈ 4.05 mi and a flight time of about 46 sec. It attains a maximum height of about 8970 ft in about 17.5 sec. At time $t \approx 23$ sec it has its minimum velocity of about 368 ft/sec. It hits the ground ($t \approx 23$ sec) at an angle of about $77°$ with a velocity of about 518 ft/sec.

CHAPTER 5

LINEAR SYSTEMS OF DIFFERENTIAL EQUATIONS

Along with Chapter 4, this chapter is designed to offer considerable flexibility in the treatment of linear systems, depending on the background in linear algebra that students are assumed to have. Sections 4.1 and 4.2 of the previous chapter can stand alone as a brief introduction to linear systems without the use of linear algebra and matrices. But this chapter employs the notation and terminology of elementary linear algebra. For ready reference and review, Section 5.1 includes a complete and self-contained account of the needed background of determinants, matrices, and vectors. The additional linear algebra that is needed in subsequent sections is introduced along the way.

SECTION 5.1

MATRICES AND LINEAR SYSTEMS

The first half-dozen pages of this section are devoted to a review of matrix notation and terminology. With students who've had some prior exposure to matrices and determinants, this review material can be skimmed rapidly. In this event serious study of the section can begin with the subsections on matrix-valued functions and first-order linear systems. About all that's actually needed for this purpose is some acquaintance with determinants, with matrix multiplication and inverse matrices, and with the fact that a square matrix is invertible if and only if its determinant is nonzero.

1. **(a)** $2\mathbf{A} + 3\mathbf{B} = \begin{bmatrix} 4 & -6 \\ 8 & 14 \end{bmatrix} + \begin{bmatrix} 9 & -12 \\ 15 & 3 \end{bmatrix} = \begin{bmatrix} 13 & -18 \\ 23 & 17 \end{bmatrix}$

   (b) $3\mathbf{A} - 2\mathbf{B} = \begin{bmatrix} 6 & -9 \\ 12 & 21 \end{bmatrix} - \begin{bmatrix} 6 & -8 \\ 10 & 2 \end{bmatrix} = \begin{bmatrix} 0 & -1 \\ 2 & 19 \end{bmatrix}$

   (c) $\mathbf{AB} = \begin{bmatrix} -9 & -11 \\ 47 & -9 \end{bmatrix}$ **(d)** $\mathbf{BA} = \begin{bmatrix} -10 & -37 \\ 14 & -8 \end{bmatrix}$

3. $\mathbf{AB} = \begin{bmatrix} -1 & 8 \\ 46 & -1 \end{bmatrix};$ $\mathbf{BA} = \begin{bmatrix} 11 & -12 & 14 \\ -14 & 0 & 7 \\ 0 & 8 & -13 \end{bmatrix}$

5.   **(a)** $7\mathbf{A} + 4\mathbf{B} = \begin{bmatrix} 21 & 14 & -7 \\ 0 & 28 & 21 \\ -35 & 14 & 49 \end{bmatrix} + \begin{bmatrix} 0 & -12 & 8 \\ 4 & 16 & -12 \\ 8 & 20 & -4 \end{bmatrix} = \begin{bmatrix} 21 & 2 & 1 \\ 4 & 44 & 9 \\ -27 & 34 & 45 \end{bmatrix}$

(b) $3\mathbf{A} - 5\mathbf{B} = \begin{bmatrix} 9 & 6 & -3 \\ 0 & 12 & 9 \\ -15 & 6 & 21 \end{bmatrix} - \begin{bmatrix} 0 & -15 & 10 \\ 5 & 20 & -15 \\ 10 & 25 & -5 \end{bmatrix} = \begin{bmatrix} 9 & 21 & -13 \\ -5 & -8 & 24 \\ -25 & -19 & 26 \end{bmatrix}$

(c) $\mathbf{AB} = \begin{bmatrix} 0 & -6 & 1 \\ 10 & 31 & -15 \\ 16 & 58 & -23 \end{bmatrix}$
   (d) $\mathbf{BA} = \begin{bmatrix} -10 & -8 & 5 \\ 18 & 12 & -10 \\ 11 & 22 & 6 \end{bmatrix}$

(e) $\mathbf{A} - t\mathbf{I} = \begin{bmatrix} 3 & 2 & -1 \\ 0 & 4 & 3 \\ -5 & 2 & 7 \end{bmatrix} - \begin{bmatrix} t & 0 & 0 \\ 0 & t & 0 \\ 0 & 0 & t \end{bmatrix} = \begin{bmatrix} 3-t & 2 & -1 \\ 0 & 4-t & 3 \\ -5 & 2 & 7-t \end{bmatrix}$

7. $\det(\mathbf{AB}) = 0 = 0 \cdot 0 = \det(\mathbf{A}) \cdot \det(\mathbf{B})$

9. $(\mathbf{AB})' = \begin{bmatrix} t - 4t^2 + 6t^3 & t + t^2 - 4t^3 + 8t^4 \\ 3t + t^3 - t^4 & 4t^2 + t^3 + t^4 \end{bmatrix}' = \begin{bmatrix} 1 - 8t + 18t^2 & 1 + 2t - 12t^2 + 32t^3 \\ 3 + 3t^2 - 4t^3 & 8t + 3t^2 + 4t^3 \end{bmatrix}$

$\mathbf{A}'\mathbf{B} + \mathbf{A}\mathbf{B}' = \begin{bmatrix} 1 & 2 \\ 3t^2 & -\dfrac{1}{t^2} \end{bmatrix} \begin{bmatrix} 1-t & 1+t \\ 3t^2 & 4t^3 \end{bmatrix} + \begin{bmatrix} t & 2t-1 \\ t^3 & \dfrac{1}{t} \end{bmatrix} \begin{bmatrix} -1 & 1 \\ 6t & 12t^2 \end{bmatrix}$

$= \begin{bmatrix} 1 - t + 6t^2 & 1 + t + 8t^3 \\ -3 + 3t^2 - 3t^3 & -4t + 3t^2 + 3t^3 \end{bmatrix} + \begin{bmatrix} -7t + 12t^2 & t - 12t^2 + 24t^3 \\ 6 - t^3 & 12t + t^3 \end{bmatrix}$

$= \begin{bmatrix} 1 - 8t + 18t^2 & 1 + 2t - 12t^2 + 32t^3 \\ 3 + 3t^2 - 4t^3 & 8t + 3t^2 + 4t^3 \end{bmatrix}$

11. $\mathbf{x} = \begin{bmatrix} x \\ y \end{bmatrix}; \; \mathbf{P}(t) = \begin{bmatrix} 0 & -3 \\ 3 & 0 \end{bmatrix}; \; \mathbf{f}(t) = \begin{bmatrix} 0 \\ 0 \end{bmatrix}$

13. $\mathbf{x} = \begin{bmatrix} x \\ y \end{bmatrix}; \; \mathbf{P}(t) = \begin{bmatrix} 2 & 4 \\ 5 & -1 \end{bmatrix}; \; \mathbf{f}(t) = \begin{bmatrix} 3e^t \\ -t^2 \end{bmatrix}$

15. $\mathbf{x} = \begin{bmatrix} x \\ y \\ z \end{bmatrix}$; $\mathbf{P}(t) = \begin{bmatrix} 0 & 1 & 1 \\ 1 & 0 & 1 \\ 1 & 1 & 0 \end{bmatrix}$; $\mathbf{f}(t) = \begin{bmatrix} 0 \\ 0 \\ 0 \end{bmatrix}$

17. $\mathbf{x} = \begin{bmatrix} x \\ y \\ z \end{bmatrix}$; $\mathbf{P}(t) = \begin{bmatrix} 3 & -4 & 1 \\ 1 & 0 & -3 \\ 0 & 6 & -7 \end{bmatrix}$; $\mathbf{f}(t) = \begin{bmatrix} t \\ t^2 \\ t^3 \end{bmatrix}$

19. $\mathbf{x} = \begin{bmatrix} x_1 \\ x_2 \\ x_3 \\ x_4 \end{bmatrix}$; $\mathbf{P}(t) = \begin{bmatrix} 0 & 1 & 0 & 0 \\ 0 & 0 & 2 & 0 \\ 0 & 0 & 0 & 3 \\ 4 & 0 & 0 & 0 \end{bmatrix}$; $\mathbf{f}(t) = \begin{bmatrix} 0 \\ 0 \\ 0 \\ 0 \end{bmatrix}$

21. $W(t) = \begin{vmatrix} 2e^t & e^{2t} \\ -3e^t & -e^{2t} \end{vmatrix} = e^{3t} \neq 0$;

$$\mathbf{x}_1' = \begin{bmatrix} 2e^t \\ -3e^t \end{bmatrix}' = \begin{bmatrix} 2e^t \\ -3e^t \end{bmatrix} = \begin{bmatrix} 4 & 2 \\ -3 & -1 \end{bmatrix} \begin{bmatrix} 2e^t \\ -3e^t \end{bmatrix} = \mathbf{A}\mathbf{x}_1$$

$$\mathbf{x}_2' = \begin{bmatrix} e^{2t} \\ -e^{2t} \end{bmatrix}' = \begin{bmatrix} 2e^{2t} \\ -2e^{2t} \end{bmatrix} = \begin{bmatrix} 4 & 2 \\ -3 & -1 \end{bmatrix} \begin{bmatrix} e^{2t} \\ -e^{2t} \end{bmatrix} = \mathbf{A}\mathbf{x}_2$$

$$\mathbf{x}(t) = c_1\mathbf{x}_1 + c_2\mathbf{x}_2 = c_1 \begin{bmatrix} 2e^t \\ -3e^t \end{bmatrix} + c_2 \begin{bmatrix} e^{2t} \\ -e^{2t} \end{bmatrix} = \begin{bmatrix} 2c_1e^t + c_2e^{2t} \\ -3c_1e^t - c_2e^{2t} \end{bmatrix}$$

In most of Problems 22–30, we omit the verifications of the given solutions. In each case, this is simply a matter of calculating both the derivative $\mathbf{x}_i'$ of the given solution vector and the product $\mathbf{A}\mathbf{x}_i$ (where $\mathbf{A}$ is the coefficient matrix in the given differential equation) to verify that $\mathbf{x}_i' = \mathbf{A}\mathbf{x}_i$ (just as in the verification of the solutions $\mathbf{x}_1$ and $\mathbf{x}_2$ in Problem 21 above).

23. $W(t) = \begin{vmatrix} e^{2t} & e^{-2t} \\ e^{2t} & 5e^{-2t} \end{vmatrix} = 4 \neq 0$

$$\mathbf{x}(t) = c_1\mathbf{x}_1 + c_2\mathbf{x}_2 = c_1 \begin{bmatrix} 1 \\ 1 \end{bmatrix} e^{2t} + c_2 \begin{bmatrix} 1 \\ 5 \end{bmatrix} e^{-2t} = \begin{bmatrix} c_1e^{2t} + c_2e^{-2t} \\ c_1e^{2t} + 5c_2e^{-2t} \end{bmatrix}$$

25. $W(t) = \begin{vmatrix} 3e^{2t} & e^{-5t} \\ 2e^{2t} & 3e^{-5t} \end{vmatrix} = 7e^{-3t} \neq 0$

$$\mathbf{x}(t) = c_1\mathbf{x}_1 + c_2\mathbf{x}_2 = c_1\begin{bmatrix} 3e^{2t} \\ 2e^{2t} \end{bmatrix} + c_2\begin{bmatrix} e^{-5t} \\ 3e^{-5t} \end{bmatrix} = \begin{bmatrix} 3c_1e^{2t} + c_2e^{-5t} \\ 2c_1e^{2t} + 3c_2e^{-5t} \end{bmatrix}$$

27.
$$W(t) = \begin{vmatrix} e^{2t} & e^{-t} & 0 \\ e^{2t} & 0 & e^{-t} \\ e^{2t} & -e^{-t} & -e^{-t} \end{vmatrix} = 3 \neq 0$$

$$\mathbf{x}(t) = c_1\mathbf{x}_1 + c_2\mathbf{x}_2 + c_3\mathbf{x}_3 = c_1\begin{bmatrix} 1 \\ 1 \\ 1 \end{bmatrix}e^{2t} + c_2\begin{bmatrix} 1 \\ 0 \\ -1 \end{bmatrix}e^{-t} + c_3\begin{bmatrix} 0 \\ 1 \\ -1 \end{bmatrix}e^{-t} = \begin{bmatrix} c_1e^{2t} + c_2e^{-t} \\ c_1e^{2t} + c_3e^{-t} \\ c_1e^{2t} - c_2e^{-t} - c_3e^{-t} \end{bmatrix}$$

$$\mathbf{x}_1' = \begin{bmatrix} 2 \\ 2 \\ 2 \end{bmatrix}e^t = \begin{bmatrix} 0 & 1 & 1 \\ 1 & 0 & 1 \\ 1 & 1 & 0 \end{bmatrix}\begin{bmatrix} 1 \\ 1 \\ 1 \end{bmatrix}e^t = \mathbf{A}\mathbf{x}_1; \quad \mathbf{x}_2' = \begin{bmatrix} -1 \\ 0 \\ 1 \end{bmatrix}e^{-t} = \begin{bmatrix} 0 & 1 & 1 \\ 1 & 0 & 1 \\ 1 & 1 & 0 \end{bmatrix}\begin{bmatrix} 1 \\ 0 \\ -1 \end{bmatrix}e^{-t} = \mathbf{A}\mathbf{x}_2;$$

$$\mathbf{x}_3' = \begin{bmatrix} 0 \\ -1 \\ 1 \end{bmatrix}e^t = \begin{bmatrix} 0 & 1 & 1 \\ 1 & 0 & 1 \\ 1 & 1 & 0 \end{bmatrix}\begin{bmatrix} 0 \\ 1 \\ -1 \end{bmatrix}e^{-t} = \mathbf{A}\mathbf{x}_3$$

29.
$$W(t) = \begin{vmatrix} 3e^{-2t} & e^t & e^{3t} \\ -2e^{-2t} & -e^t & -e^{3t} \\ 2e^{-2t} & e^t & 0 \end{vmatrix} = e^{2t} \neq 0$$

$$\mathbf{x}(t) = c_1\mathbf{x}_1 + c_2\mathbf{x}_2 + c_3\mathbf{x}_3 = c_1\begin{bmatrix} 3 \\ -2 \\ 2 \end{bmatrix}e^{-2t} + c_2\begin{bmatrix} 1 \\ -1 \\ 1 \end{bmatrix}e^t + c_3\begin{bmatrix} 1 \\ -1 \\ 0 \end{bmatrix}e^{3t} = \begin{bmatrix} 3c_1e^{-2t} + c_2e^t + c_3e^{3t} \\ -2c_1e^{-2t} - c_2e^t - c_3e^{3t} \\ 2c_1e^{-2t} + c_2e^t \end{bmatrix}$$

In Problems 31–34 (and similarly in Problems 35–40) we give first the scalar components $x_1(t)$ and $x_2(t)$ of a general solution, then the equations in the coefficients c_1 and c_2 that are obtained when the given initial conditions are imposed, and finally the resulting particular solution of the given system.

31. $x_1(t) = c_1e^{3t} + 2c_2e^{-2t}, \quad x_2(t) = 3c_1e^{3t} + c_2e^{-2t}$

$$c_1 + 2c_2 = 0, \qquad 3c_1 + c_2 = 5$$
$$x_1(t) = 2e^{3t} - 2e^{-2t}, \quad x_2(t) = 6e^{3t} - e^{-2t}$$

33. $x_1(t) = c_1e^{3t} + c_2e^{2t}, \quad x_2(t) = -c_1e^{3t} - 2c_2e^{2t}$

$$c_1 + c_2 = 11, \qquad -c_1 - 2c_2 = -7$$
$$x_1(t) = 15e^{3t} - 4e^{2t}, \quad x_2(t) = -15e^{3t} + 8e^{2t}$$

35. $x_1(t) = 2c_1 e^t - 2c_2 e^{3t} + 2c_3 e^{5t}, \quad x_2(t) = 2c_1 e^t - 2c_3 e^{5t}, \quad x_3(t) = c_1 e^t + c_2 e^{3t} + c_3 e^{5t}$

$$2c_1 - 2c_2 + 2c_3 = 0, \qquad 2c_1 - 2c_3 = 0, \qquad c_1 + c_2 + c_3 = 4$$
$$x_1(t) = 2e^t - 4e^{3t} + 2e^{5t}, \quad x_2(t) = 2e^t - 2e^{5t}, \quad x_3(t) = e^t + 2e^{3t} + e^{5t}$$

37. $x_1(t) = 3c_1 e^{-2t} + c_2 e^t + c_3 e^{3t}, \quad x_2(t) = -2c_1 e^{-2t} - c_2 e^t - c_3 e^{3t}, \quad x_3(t) = 2c_1 e^{-2t} + c_2 e^t$

$$3c_1 + c_2 + c_3 = 1, \qquad -2c_1 - c_2 - c_3 = 2, \qquad 2c_1 + c_2 = 3$$
$$x_1(t) = 9e^{-2t} - 3e^t - 5e^{3t}, \quad x_2(t) = -6e^{-2t} + 3e^t + 5e^{3t}, \quad x_3(t) = 6e^{-2t} - 3e^t$$

39. $x_1(t) = c_1 e^{-t} + c_4 e^t, \quad x_2(t) = c_3 e^t, \quad x_3(t) = c_2 e^{-t} + 3c_4 e^t, \quad x_4(t) = c_1 e^{-t} - 2c_3 e^t$

$$c_1 + c_4 = 1, \qquad c_3 = 1, \qquad c_2 + 3c_4 = 1, \qquad c_1 - 2c_3 = 1$$
$$x_1(t) = 3e^{-t} - 2e^t, \quad x_2(t) = e^t, \quad x_3(t) = 7e^{-t} - 6e^t, \quad x_4(t) = 3e^{-t} - 2e^t$$

41.   **(a)** $\mathbf{x}_2 = t\mathbf{x}_1$, so neither is a constant multiple of the other.

(b) $W(\mathbf{x}_1, \mathbf{x}_2) = 0$, whereas Theorem 2 would imply that $W \neq 0$ if $\mathbf{x}_1$ and $\mathbf{x}_2$ were independent solutions of a system of the indicated form.

43.   Suppose $W(a) = x_{11}(a)x_{22}(a) - x_{12}(a)x_{21}(a) = 0$. Then the coefficient determinant of the homogeneous linear system

$$c_1 x_{11}(a) + c_2 x_{12}(a) = 0, \quad c_1 x_{21}(a) + c_2 x_{22}(a) = 0$$

vanishes. The system therefore has a non-trivial solution $\{c_1, c_2\}$ such that $\mathbf{x}(a) = \mathbf{0}$. It therefore follows (by uniqueness of solutions) that $\mathbf{x}(t) \equiv \mathbf{0}$, that is, $c_1 \mathbf{x}_1(t) + c_2 \mathbf{x}_2(t) \equiv \mathbf{0}$ with c_1 and c_2 not both zero. Thus the solution vectors $\mathbf{x}_1$ and $\mathbf{x}_2$ are linearly dependent.

45.   Suppose that $c_1 \mathbf{x}_1(t) + c_2 \mathbf{x}_2(t) + \cdots + c_n \mathbf{x}_n(t) \equiv \mathbf{0}$. Then the ith scalar component of this vector equation is $c_1 x_{i1}(t) + c_2 x_{i2}(t) + \cdots + c_n x_{in}(t) \equiv 0$. Hence the fact that the scalar functions $x_{i1}(t)$, $x_{i2}(t)$, $x_{in}(t)$ are linearly independent implies that $c_1 = c_2 = \cdots = c_n = 0$. Consequently the vector functions $\mathbf{x}_1(t), \mathbf{x}_2(t), \cdots, \mathbf{x}_n(t)$ are linearly independent.

SECTION 5.2

THE EIGENVALUE METHOD FOR HOMOGENEOUS LINEAR SYSTEMS

In each of Problems 1–16 we give the characteristic equation, the eigenvalues λ_1 and λ_2 of the coefficient matrix of the given system, the corresponding equations determining the associated eigenvectors $\mathbf{v}_1 = \begin{bmatrix} a_1 & b_1 \end{bmatrix}^T$ and $\mathbf{v}_2 = \begin{bmatrix} a_2 & b_2 \end{bmatrix}^T$, these eigenvectors, and the resulting scalar components $x_1(t)$ and $x_2(t)$ of a general solution $\mathbf{x}(t) = c_1 \mathbf{v}_1 e^{\lambda_1 t} + c_2 \mathbf{v}_2 e^{\lambda_2 t}$ of the system. Finally, the figure for each Problem shows a direction field and some typical solution curves for the system.

1.   Characteristic equation $\lambda^2 - 2\lambda - 3 = 0$;

Eigenvalues $\lambda_1 = -1$ and $\lambda_2 = 3$;

Eigenvector equations $\begin{bmatrix} 2 & 2 \\ 2 & 2 \end{bmatrix}\begin{bmatrix} a_1 \\ b_1 \end{bmatrix} = \begin{bmatrix} 0 \\ 0 \end{bmatrix}$ and $\begin{bmatrix} -2 & 2 \\ 2 & -2 \end{bmatrix}\begin{bmatrix} a_2 \\ b_2 \end{bmatrix} = \begin{bmatrix} 0 \\ 0 \end{bmatrix}$;

Eigenvectors $\mathbf{v}_1 = \begin{bmatrix} 1 & -1 \end{bmatrix}^T$ and $\mathbf{v}_2 = \begin{bmatrix} 1 & 1 \end{bmatrix}^T$;

$$x_1(t) = c_1 e^{-t} + c_2 e^{3t}, \quad x_2(t) = -c_1 e^{-t} + c_2 e^{3t}$$

Problem 1	**Problem 3**
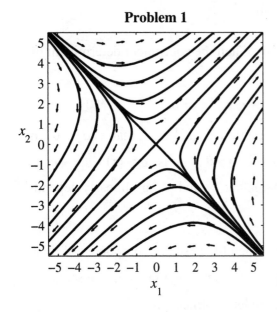	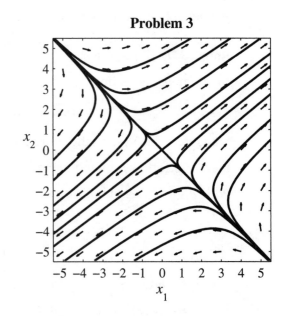

3.   Characteristic equation $\lambda^2 - 5\lambda - 6 = 0$;

Eigenvalues $\lambda_1 = -1$ and $\lambda_2 = 6$;

Eigenvector equations $\begin{bmatrix} 4 & 4 \\ 3 & 3 \end{bmatrix}\begin{bmatrix} a_1 \\ b_1 \end{bmatrix} = \begin{bmatrix} 0 \\ 0 \end{bmatrix}$ and $\begin{bmatrix} -3 & 4 \\ 3 & -4 \end{bmatrix}\begin{bmatrix} a_2 \\ b_2 \end{bmatrix} = \begin{bmatrix} 0 \\ 0 \end{bmatrix}$;

Eigenvectors $\mathbf{v}_1 = \begin{bmatrix} 1 & -1 \end{bmatrix}^T$ and $\mathbf{v}_1 = \begin{bmatrix} 4 & 3 \end{bmatrix}^T$;

$$x_1(t) = c_1 e^{-t} + 4c_2 e^{6t}, \quad x_2(t) = -c_1 e^{-t} + 3c_2 e^{6t};$$

The equations

$$x_1(0) = c_1 + 4c_2 = 1, \quad x_2(0) = -c_1 + 3c_2 = 1$$

yield $c_1 = -\dfrac{1}{7}$ and $c_2 = \dfrac{2}{7}$, so the desired particular solution is given by

$$x_1(t) = \frac{1}{7}\left(-e^{-t} + 8e^{6t}\right), \quad x_2(t) = \frac{1}{7}\left(e^{-t} + 6e^{6t}\right).$$

5.   Characteristic equation $\lambda^2 - 4\lambda - 5 = 0$;

Eigenvalues $\lambda_1 = -1$ and $\lambda_2 = 5$;

Eigenvector equations $\begin{bmatrix} 7 & -7 \\ 1 & -1 \end{bmatrix}\begin{bmatrix} a_1 \\ b_1 \end{bmatrix} = \begin{bmatrix} 0 \\ 0 \end{bmatrix}$ and $\begin{bmatrix} 1 & -7 \\ 1 & -7 \end{bmatrix}\begin{bmatrix} a_2 \\ b_2 \end{bmatrix} = \begin{bmatrix} 0 \\ 0 \end{bmatrix}$;

Eigenvectors $\mathbf{v}_1 = \begin{bmatrix} 1 & 1 \end{bmatrix}^T$ and $\mathbf{v}_2 = \begin{bmatrix} 7 & 1 \end{bmatrix}^T$;

$$x_1(t) = c_1 e^{-t} + 7c_2 e^{5t}, \quad x_2(t) = c_1 e^{-t} + c_2 e^{5t}$$

Problem 5	**Problem 7**
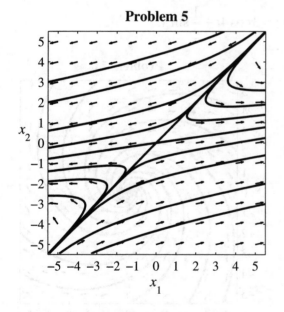	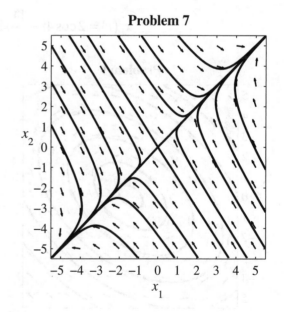

7.   Characteristic equation $\lambda^2 + 8\lambda - 9 = 0$;

Eigenvalues $\lambda_1 = 1$ and $\lambda_2 = -9$;

Eigenvector equations $\begin{bmatrix} -4 & 4 \\ 6 & -6 \end{bmatrix}\begin{bmatrix} a_1 \\ b_1 \end{bmatrix} = \begin{bmatrix} 0 \\ 0 \end{bmatrix}$ and $\begin{bmatrix} 6 & 4 \\ 6 & 4 \end{bmatrix}\begin{bmatrix} a_2 \\ b_2 \end{bmatrix} = \begin{bmatrix} 0 \\ 0 \end{bmatrix}$;

Eigenvectors $\mathbf{v}_1 = \begin{bmatrix} 1 & 1 \end{bmatrix}^T$ and $\mathbf{v}_2 = \begin{bmatrix} 2 & -3 \end{bmatrix}^T$;

$$x_1(t) = c_1 e^t + 2c_2 e^{-9t}, \quad x_2(t) = c_1 e^t - 3c_2 e^{-9t}$$

9.   Characteristic equation $\lambda^2 + 16 = 0$;

Eigenvalue $\lambda = 4i$;

Eigenvector equation $\begin{bmatrix} 2 - 4i & -5 \\ 4 & -2 - 4i \end{bmatrix} \begin{bmatrix} a \\ b \end{bmatrix} = \begin{bmatrix} 0 \\ 0 \end{bmatrix}$;

Eigenvector $\mathbf{v} = \begin{bmatrix} 5 & 2 - 4i \end{bmatrix}^T$;

The real and imaginary parts of

$$\mathbf{x}(t) = \mathbf{v}e^{4it} = \begin{bmatrix} 5\cos 4t + 5i\sin 4t \\ (2\cos 4t + 4\sin 4t) + i(2\sin 4t - 4\cos 4t) \end{bmatrix}$$

yield the general solution

$$x_1(t) = 5c_1 \cos 4t + 5c_2 \sin 4t,$$
$$x_2(t) = c_1(2\cos 4t + 4\sin 4t) + c_2(2\sin 4t - 4\cos 4t)$$

The initial conditions $x_1(0) = 2$ and $x_2(0) = 3$ give $c_1 = \dfrac{2}{5}$ and $c_2 = -\dfrac{11}{20}$, so the desired particular solution is

$$x_1(t) = 2\cos 4t - \frac{11}{4}\sin 4t, \quad x_2(t) = 3\cos 4t + \frac{1}{2}\sin 4t$$

Problem 9   **Problem 11**

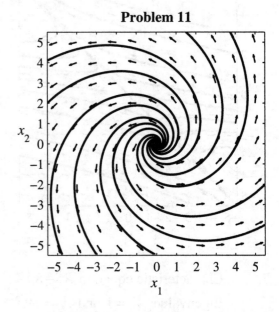

11.   Characteristic equation $\lambda^2 - 2\lambda + 5 = 0$;

Eigenvalue $\lambda = 1 - 2i$;

Eigenvector equation $\begin{bmatrix} 2i & -2 \\ 2 & 2i \end{bmatrix} \begin{bmatrix} a \\ b \end{bmatrix} = \begin{bmatrix} 0 \\ 0 \end{bmatrix}$;

Eigenvector $\mathbf{v} = \begin{bmatrix} 1 & i \end{bmatrix}^T$;

The real and imaginary parts of

$$\mathbf{x}(t) = \begin{bmatrix} 1 & i \end{bmatrix}^T e^t \left(\cos 2t - i \sin 2t \right) = e^t \begin{bmatrix} \cos 2t & \sin 2t \end{bmatrix}^T + ie^t \begin{bmatrix} -\sin 2t & \cos 2t \end{bmatrix}^T$$

yield the general solution

$$x_1(t) = e^t \left(c_1 \cos 2t - c_2 \sin 2t \right), \quad x_2(t) = e^t \left(c_1 \sin 2t + c_2 \cos 2t \right)$$

The particular solution with $x_1(0) = 0$ and $x_2(0) = 4$ is obtained with $c_1 = 0$ and $c_2 = 4$, so

$$x_1(t) = -4e^t \sin 2t, \quad x_2(t) = 4e^t \cos 2t.$$

13.   Characteristic equation $\lambda^2 - 4\lambda + 13 = 0$;

Eigenvalue $\lambda = 2 - 3i$;

Eigenvector equation $\begin{bmatrix} 3+3i & -9 \\ 2 & -3+3i \end{bmatrix} \begin{bmatrix} a \\ b \end{bmatrix} = \begin{bmatrix} 0 \\ 0 \end{bmatrix}$;

Eigenvector $\mathbf{v} = \begin{bmatrix} 3 & 1+i \end{bmatrix}^T$;

$$\mathbf{x}(t) = \mathbf{v}e^{(2-3i)t} = e^{2t} \begin{bmatrix} 3\cos 3t - 3i\sin 3t \\ (\cos 3t + \sin 3t) + i(\cos 3t - \sin 3t) \end{bmatrix}$$

$$x_1(t) = 3e^{2t} \left(c_1 \cos 3t - c_2 \sin 3t \right), \quad x_2(t) = e^{2t} \left[(c_1 + c_2) \cos 3t + (c_1 - c_2) \sin 3t \right]$$

Problem 13

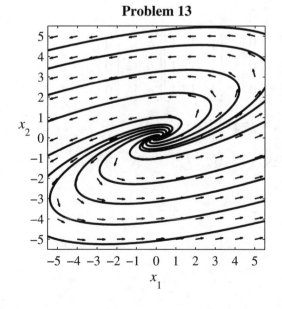

Problem 15

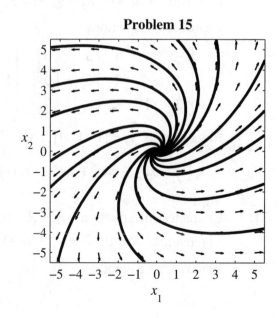

15. Characteristic equation $\lambda^2 - 10\lambda + 41 = 0$;

Eigenvalue $\lambda = 5 - 4i$;

Eigenvector equation $\begin{bmatrix} 2+4i & -5 \\ 4 & -2+4i \end{bmatrix} \begin{bmatrix} a \\ b \end{bmatrix} = \begin{bmatrix} 0 \\ 0 \end{bmatrix}$;

Eigenvector $\mathbf{v} = \begin{bmatrix} 5 & 2+4i \end{bmatrix}^{\mathrm{T}}$;

$$\mathbf{x}(t) = \mathbf{v}e^{(5-4i)t} = e^{5t} \begin{bmatrix} 5\cos 4t - 5i\sin 4t \\ (2\cos 4t + 4\sin 4t) + i(4\cos 4t - 2\sin 4t) \end{bmatrix};$$

$$x_1(t) = 5e^{5t}(c_1\cos 4t - c_2\sin 4t), \quad x_2(t) = e^{5t}\left[(2c_1 + 4c_2)\cos 4t + (4c_1 - 2c_2)\sin 4t\right]$$

17. Characteristic equation $-\lambda^3 + 15\lambda^2 - 54\lambda = 0$;

Eigenvalues $\lambda_1 = 9$, $\lambda_2 = 6$, $\lambda_3 = 0$;

Eigenvector equations

$$\begin{bmatrix} -5 & 1 & 4 \\ 1 & -2 & 1 \\ 4 & 1 & -5 \end{bmatrix} \begin{bmatrix} a_1 \\ b_1 \\ c_1 \end{bmatrix} = \begin{bmatrix} 0 \\ 0 \\ 0 \end{bmatrix}, \quad \begin{bmatrix} -2 & 1 & 4 \\ 1 & 1 & 1 \\ 4 & 1 & -2 \end{bmatrix} \begin{bmatrix} a_2 \\ b_2 \\ c_2 \end{bmatrix} = \begin{bmatrix} 0 \\ 0 \\ 0 \end{bmatrix}, \quad \begin{bmatrix} 4 & 1 & 4 \\ 1 & 7 & 1 \\ 4 & 1 & 4 \end{bmatrix} \begin{bmatrix} a_3 \\ b_3 \\ c_3 \end{bmatrix} = \begin{bmatrix} 0 \\ 0 \\ 0 \end{bmatrix};$$

Eigenvectors $\mathbf{v}_1 = \begin{bmatrix} 1 & 1 & 1 \end{bmatrix}^{\mathrm{T}}$, $\mathbf{v}_2 = \begin{bmatrix} 1 & -2 & 1 \end{bmatrix}^{\mathrm{T}}$, $\mathbf{v}_3 = \begin{bmatrix} 1 & 0 & -1 \end{bmatrix}^{\mathrm{T}}$;

$$x_1(t) = c_1 e^{9t} + c_2 e^{6t} + c_3, \quad x_2(t) = c_1 e^{9t} - 2c_2 e^{6t}, \quad x_3(t) = c_1 e^{9t} + c_2 e^{6t} - c_3$$

19. Characteristic equation $-\lambda^3 + 12\lambda^2 - 45\lambda + 54 = 0$;

Eigenvalues $\lambda_1 = 6$, $\lambda_2 = 3$, $\lambda_3 = 3$;

Eigenvector equations

$$\begin{bmatrix} -2 & 1 & 1 \\ 1 & -2 & 1 \\ 1 & 1 & -2 \end{bmatrix} \begin{bmatrix} a_1 \\ b_1 \\ c_1 \end{bmatrix} = \begin{bmatrix} 0 \\ 0 \\ 0 \end{bmatrix}, \quad \begin{bmatrix} 1 & 1 & 1 \\ 1 & 1 & 1 \\ 1 & 1 & 1 \end{bmatrix} \begin{bmatrix} a_2 \\ b_2 \\ c_2 \end{bmatrix} = \begin{bmatrix} 0 \\ 0 \\ 0 \end{bmatrix}, \quad \begin{bmatrix} 1 & 1 & 1 \\ 1 & 1 & 1 \\ 1 & 1 & 1 \end{bmatrix} \begin{bmatrix} a_3 \\ b_3 \\ c_3 \end{bmatrix} = \begin{bmatrix} 0 \\ 0 \\ 0 \end{bmatrix};$$

Eigenvectors $\mathbf{v}_1 = \begin{bmatrix} 1 & 1 & 1 \end{bmatrix}^{\mathrm{T}}$, $\mathbf{v}_2 = \begin{bmatrix} 1 & -2 & 1 \end{bmatrix}^{\mathrm{T}}$, $\mathbf{v}_3 = \begin{bmatrix} 1 & 0 & -1 \end{bmatrix}^{\mathrm{T}}$;

$$x_1(t) = c_1 e^{6t} + c_2 e^{3t} + c_3 e^{3t}, \quad x_2(t) = c_1 e^{6t} - 2c_2 e^{3t}, \quad x_3(t) = c_1 e^{6t} + c_2 e^{3t} - c_3 e^{3t}$$

21. Characteristic equation $-\lambda^3 + \lambda = 0$;

Eigenvalues $\lambda_1 = 0$, $\lambda_2 = 1$, $\lambda_3 = -1$;

Eigenvector equations

$$\begin{bmatrix} 5 & 0 & -6 \\ 2 & -1 & -2 \\ 4 & -2 & -4 \end{bmatrix}\begin{bmatrix} a_1 \\ b_1 \\ c_1 \end{bmatrix} = \begin{bmatrix} 0 \\ 0 \\ 0 \end{bmatrix}, \quad \begin{bmatrix} 4 & 0 & -6 \\ 2 & -2 & -2 \\ 4 & -2 & -5 \end{bmatrix}\begin{bmatrix} a_2 \\ b_2 \\ c_2 \end{bmatrix} = \begin{bmatrix} 0 \\ 0 \\ 0 \end{bmatrix}, \quad \begin{bmatrix} 6 & 0 & -6 \\ 2 & 0 & -2 \\ 4 & -2 & -3 \end{bmatrix}\begin{bmatrix} a_3 \\ b_3 \\ c_3 \end{bmatrix} = \begin{bmatrix} 0 \\ 0 \\ 0 \end{bmatrix};$$

Eigenvectors $\mathbf{v}_1 = \begin{bmatrix} 6 & 2 & 5 \end{bmatrix}^T$, $\mathbf{v}_2 = \begin{bmatrix} 3 & 1 & 2 \end{bmatrix}^T$, $\mathbf{v}_3 = \begin{bmatrix} 2 & 1 & 2 \end{bmatrix}^T$;

$$x_1(t) = 6c_1 + 3c_2 e^t + 2c_3 e^{-t}, \quad x_2(t) = 2c_1 + c_2 e^t + c_3 e^{-t}, \quad x_3(t) = 5c_1 + 2c_2 e^t + 2c_3 e^{-t}$$

23. Characteristic equation $-\lambda^3 + 3\lambda^2 + 4\lambda - 12 = 0$;

Eigenvalues $\lambda_1 = 2$, $\lambda_2 = -2$, $\lambda_3 = 3$;

Eigenvector equations

$$\begin{bmatrix} 1 & 1 & 1 \\ -5 & -5 & -1 \\ 5 & 5 & 1 \end{bmatrix}\begin{bmatrix} a_1 \\ b_1 \\ c_1 \end{bmatrix} = \begin{bmatrix} 0 \\ 0 \\ 0 \end{bmatrix}, \quad \begin{bmatrix} 5 & 1 & 1 \\ -5 & -1 & -1 \\ 5 & 5 & 5 \end{bmatrix}\begin{bmatrix} a_2 \\ b_2 \\ c_2 \end{bmatrix} = \begin{bmatrix} 0 \\ 0 \\ 0 \end{bmatrix}, \quad \begin{bmatrix} 0 & 1 & 1 \\ -5 & -6 & -1 \\ 5 & 5 & 0 \end{bmatrix}\begin{bmatrix} a_3 \\ b_3 \\ c_3 \end{bmatrix} = \begin{bmatrix} 0 \\ 0 \\ 0 \end{bmatrix};$$

Eigenvectors $\mathbf{v}_1 = \begin{bmatrix} 1 & -1 & 0 \end{bmatrix}^T$, $\mathbf{v}_2 = \begin{bmatrix} 0 & 1 & -1 \end{bmatrix}^T$, $\mathbf{v}_3 = \begin{bmatrix} 1 & -1 & 1 \end{bmatrix}^T$;

$$x_1(t) = c_1 e^{2t} + c_3 e^{3t}, \quad x_2(t) = -c_1 e^{2t} + c_2 e^{-2t} - c_3 e^{3t}, \quad x_3(t) = -c_2 e^{-2t} + c_3 e^{3t}$$

25. Characteristic equation $-\lambda^3 + 4\lambda^2 - 13\lambda = 0$;

Eigenvalues $\lambda = 0$ and $\lambda = 2 \pm 3i$;

With $\lambda = 1$ the eigenvector equation $\begin{bmatrix} 5 & 5 & 2 \\ -6 & -6 & -5 \\ 6 & 6 & 5 \end{bmatrix}\begin{bmatrix} a_1 \\ b_1 \\ c_1 \end{bmatrix} = \begin{bmatrix} 0 \\ 0 \\ 0 \end{bmatrix}$ gives the eigenvector

$\mathbf{v}_1 = \begin{bmatrix} 1 & -1 & 0 \end{bmatrix}^T$. With $\lambda = 2 + 3i$ we solve the eigenvector equation

$$\begin{bmatrix} 3-3i & 5 & 2 \\ -6 & -8-3i & -5 \\ 6 & 6 & 3-3i \end{bmatrix}\begin{bmatrix} a \\ b \\ c \end{bmatrix} = \begin{bmatrix} 0 \\ 0 \\ 0 \end{bmatrix}$$

to find the complex-valued eigenvector $\mathbf{v}_1 = \begin{bmatrix} 1+i & -2 & 2 \end{bmatrix}^T$. The corresponding complex-valued solution is

$$\mathbf{x}(t) = \mathbf{v}e^{(2+3i)t} = e^{2t}\begin{bmatrix} (\cos 3t - \sin 3t) + i(\cos 3t + \sin 3t) \\ -2\cos 3t - 2i\sin 3t \\ 2\cos 3t + 2i\sin 3t \end{bmatrix}.$$

The scalar components of the resulting general solution are

$$x_1(t) = c_1 + e^{2t}\left[(c_2 + c_3)\cos 3t + (-c_2 + c_3)\sin 3t\right],$$
$$x_2(t) = -c_1 + 2e^{2t}(-c_2\cos 3t - c_3\sin 3t),$$
$$x_3(t) = 2e^{2t}(c_2\cos 3t + c_3\sin 3t).$$

27. The coefficient matrix $\mathbf{A} = \begin{bmatrix} -0.2 & 0 \\ 0.2 & -0.4 \end{bmatrix}$ has characteristic equation

$\lambda^2 + 0.6\lambda + 0.08 = 0$ with eigenvalues $\lambda_1 = -0.2$ and $\lambda_2 = -0.4$. We find easily that the associated eigenvectors are $\mathbf{v}_1 = \begin{bmatrix} 1 & 1 \end{bmatrix}^T$ and $\mathbf{v}_2 = \begin{bmatrix} 0 & 1 \end{bmatrix}^T$, so we get the general solution

$$x_1(t) = c_1 e^{-0.2t}, \quad x_2(t) = c_1 e^{-0.2t} + c_2 e^{-0.4t}.$$

The initial conditions $x_1(0) = 15$, $x_2(0) = 0$ give $c_1 = 15$ and $c_2 = -15$, so we get

$$x_1(t) = 15e^{-0.2t}, \quad x_2(t) = 15e^{-0.2t} - 15e^{-0.4t}.$$

To find the maximum value of $x_2(t)$, we solve the equation $x_2'(t)$ for $t = 5\ln 2$, which gives the maximum value $x_2(5\ln 2) = 3.75$ lb. The figure shows the graphs of $x_1(t)$ and $x_2(t)$.

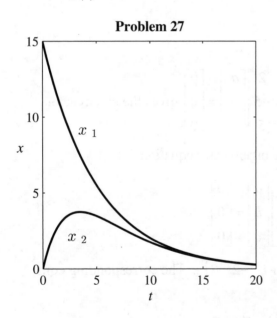

Problem 27

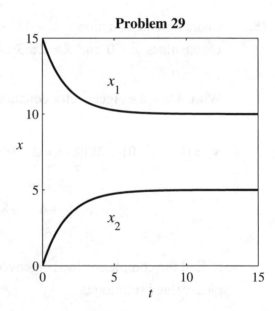

Problem 29

29. The coefficient matrix $\mathbf{A} = \begin{bmatrix} -0.2 & 0.4 \\ 0.2 & -0.4 \end{bmatrix}$ has eigenvalues $\lambda_1 = 0$ and $\lambda_2 = -0.6$, with eigenvectors $\mathbf{v}_1 = \begin{bmatrix} 2 & 1 \end{bmatrix}^T$ and $\mathbf{v}_2 = \begin{bmatrix} 1 & -1 \end{bmatrix}^T$ that yield the general solution

$$x_1(t) = 2c_1 + c_2 e^{-0.6t}, \quad x_2(t) = c_1 - c_2 e^{-0.6t}.$$

The initial conditions $x_1(0) = 15$, $x_2(0) = 0$ give $c_1 = c_2 = 5$, so we get

$$x_1(t) = 10 + 5e^{-0.6t}, \quad x_2(t) = 5 - 5e^{-0.6t}.$$

The figure shows the graphs of $x_1(t)$ and $x_2(t)$.

31. The coefficient matrix

$$\mathbf{A} = \begin{bmatrix} -1 & 0 & 0 \\ 1 & -2 & 0 \\ 0 & 2 & -3 \end{bmatrix}$$

has as eigenvalues its diagonal elements $\lambda_1 = -1$, $\lambda_2 = -2$, and $\lambda_3 = -3$. We find readily that the associated eigenvectors are $\mathbf{v}_1 = \begin{bmatrix} 1 & 1 & 1 \end{bmatrix}^T$, $\mathbf{v}_2 = \begin{bmatrix} 0 & 1 & 2 \end{bmatrix}^T$, and $\mathbf{v}_3 = \begin{bmatrix} 0 & 0 & 1 \end{bmatrix}^T$. The resulting general solution is given by

$$x_1(t) = c_1 e^{-t}, \quad x_2(t) = c_1 e^{-t} + c_2 e^{-2t}, \quad x_3(t) = c_1 e^{-t} + 2c_2 e^{-2t} + c_3 e^{-3t}.$$

The initial conditions $x_1(0) = 27$ and $x_2(0) = x_2(0) = 0$ give $c_1 = c_3 = 27$ and $c_2 = -27$, so we get

$$x_1(t) = 27e^{-t}, \quad x_2(t) = 27e^{-t} - 27e^{-2t}, \quad x_3(t) = 27e^{-t} - 54e^{-2t} + 27e^{-3t}.$$

The equation $x_3'(t) = 0$ simplifies to the equation

$$3e^{-2t} - 4e^{-t} + 1 = (3e^{-t} - 1)(e^{-t} - 1) = 0,$$

with positive solution $t_m = \ln 3$. Thus the maximum amount of salt ever in tank 3 is $x_3(\ln 3) = 4$ lb. The figure shows the graphs of $x_1(t)$, $x_2(t)$, and $x_3(t)$.

Problem 31

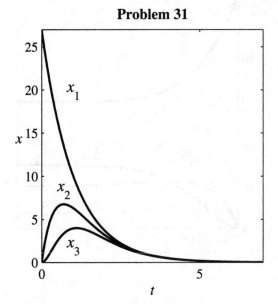

Problem 33

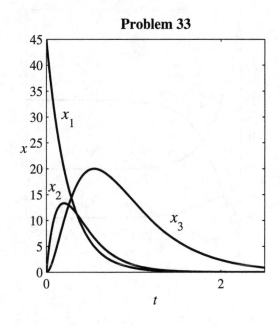

33.   The coefficient matrix

$$\mathbf{A} = \begin{bmatrix} -4 & 0 & 0 \\ 4 & -6 & 0 \\ 0 & 6 & -2 \end{bmatrix}$$

has as eigenvalues its diagonal elements $\lambda_1 = -4$, $\lambda_2 = -6$, and $\lambda_3 = -2$. We find readily that the associated eigenvectors are $\mathbf{v}_1 = \begin{bmatrix} -1 & -2 & 6 \end{bmatrix}^T$, $\mathbf{v}_2 = \begin{bmatrix} 0 & -2 & 3 \end{bmatrix}^T$, and $\mathbf{v}_3 = \begin{bmatrix} 0 & 0 & 1 \end{bmatrix}^T$. The resulting general solution is given by

$$x_1(t) = -c_1 e^{-4t}, \quad x_2(t) = -2c_1 e^{-4t} - 2c_2 e^{-6t}, \quad x_3(t) = 6c_1 e^{-4t} + 3c_2 e^{-6t} + c_3 e^{-2t}.$$

The initial conditions $x_1(0) = 45$ and $x_2(0) = x_2(0) = 0$ give $c_1 = -45$, $c_2 = 45$, and $c_3 = 135$, so we get

$$x_1(t) = 45e^{-4t}, \quad x_2(t) = 90e^{-4t} - 90e^{-6t}, \quad x_3(t) = -270e^{-4t} + 135e^{-6t} + 135e^{-2t}.$$

The equation $x_3'(t) = 0$ simplifies to the equation

$$3e^{-4t} - 4e^{-2t} + 1 = (3e^{-2t} - 1)(e^{-2t} - 1) = 0,$$

with positive solution $t_m = \frac{1}{2}\ln 3$. Thus the maximum amount of salt ever in tank 3 is

$x_3\left(\frac{1}{2}\ln 3\right) = 20$ lb. The figure shows the graphs of $x_1(t)$, $x_2(t)$, and $x_3(t)$.

Problem 35

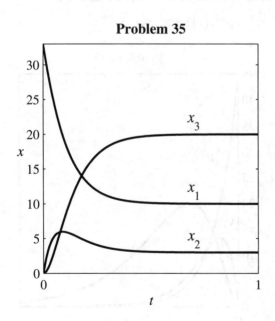

Problem 37

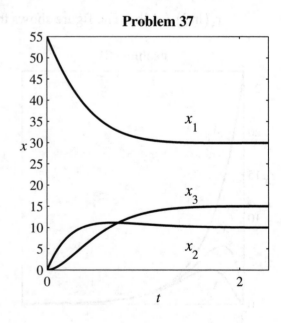

37.   The coefficient matrix

$$\mathbf{A} = \begin{bmatrix} -1 & 0 & 2 \\ 1 & -3 & 0 \\ 0 & 3 & -2 \end{bmatrix}$$

has characteristic equation $-\lambda^3 - 6\lambda^2 - 11\lambda = 0$, with eigenvalues $\lambda_0 = 0$, $\lambda_1 = -3 - i\sqrt{2}$, and $\lambda_1 = -3 + i\sqrt{2}$. The eigenvector equation

$$\begin{bmatrix} -1 & 0 & 2 \\ 1 & -3 & 0 \\ 0 & 3 & -2 \end{bmatrix} \begin{bmatrix} a \\ b \\ c \end{bmatrix} = \begin{bmatrix} 0 \\ 0 \\ 0 \end{bmatrix}$$

associated with the eigenvalue $\lambda_0 = 0$ yields the associated eigenvector $\mathbf{v}_0 = \begin{bmatrix} 6 & 2 & 3 \end{bmatrix}^T$, and consequently the constant solution $\mathbf{x}_0(t) \equiv \mathbf{v}_0$. Then the eigenvector equation

$$\begin{bmatrix} 2 + i\sqrt{2} & 0 & 2 \\ 1 & i\sqrt{2} & 0 \\ 0 & 3 & 1 + i\sqrt{2} \end{bmatrix} \begin{bmatrix} a \\ b \\ c \end{bmatrix} = \begin{bmatrix} 0 \\ 0 \\ 0 \end{bmatrix}$$

associated with $\lambda_1 = -3 - i\sqrt{2}$ yields the complex-valued eigenvector

$$\mathbf{v}_1 = \left[\frac{1}{3}\left(-2 + i\sqrt{2}\right) \quad \frac{1}{3}\left(-1 - i\sqrt{2}\right) \quad 1 \right]^T.$$

The corresponding complex-valued solution is

$$\mathbf{x}_1(t) = \mathbf{v}_1 e^{\left(-3 - i\sqrt{2}\right)t} = \frac{1}{3} e^{-3t} \begin{bmatrix} \left[-2\cos\left(\sqrt{2}t\right) + \sqrt{2}\sin\left(\sqrt{2}t\right) \right] + i\left[\sqrt{2}\cos\left(\sqrt{2}t\right) + 2\sin\left(\sqrt{2}t\right) \right] \\ \left[-\cos\left(\sqrt{2}t\right) - \sqrt{2}\sin\left(\sqrt{2}t\right) \right] + i\left[-\sqrt{2}\cos\left(\sqrt{2}t\right) + \sin\left(\sqrt{2}t\right) \right] \\ 3\cos\left(\sqrt{2}t\right) - 3i\sin\left(\sqrt{2}t\right) \end{bmatrix}.$$

The scalar components of resulting general solution $\mathbf{x} = c_0\mathbf{x}_0 + c_1\,\text{Re}[\mathbf{x}_1] + c_2\,\text{Im}[\mathbf{x}_1]$ are given by

$$x_1(t) = 6c_0 + \frac{1}{3} e^{-3t} \left[\left(-2c_1 + \sqrt{2}c_2\right)\cos\left(\sqrt{2}t\right) + \left(\sqrt{2}c_1 + 2c_2\right)\sin\left(\sqrt{2}t\right) \right]$$

$$x_2(t) = 2c_0 + \frac{1}{3} e^{-3t} \left[\left(-c_1 - \sqrt{2}c_2\right)\cos\left(\sqrt{2}t\right) + \left(-\sqrt{2}c_1 + c_2\right)\sin\left(\sqrt{2}t\right) \right]$$

$$x_3(t) = 3c_0 + e^{-3t} \left[c_1\cos\left(\sqrt{2}t\right) - c_2\sin\left(\sqrt{2}t\right) \right].$$

When we impose the initial conditions $x_1(0) = 55$ and $x_2(0) = x_2'(0) = 0$ we find that $c_0 = 5$, $c_1 = -15$, and $c_2 = \dfrac{45}{\sqrt{2}}$. This finally gives the particular solution

$$x_1(t) = 30 + e^{-3t}\left[25\cos\left(\sqrt{2}t\right) + 10\sqrt{2}\sin\left(\sqrt{2}t\right)\right]$$

$$x_2(t) = 10 - e^{-3t}\left[10\cos\left(\sqrt{2}t\right) - \frac{25}{2}\sqrt{2}\sin\left(\sqrt{2}t\right)\right]$$

$$x_3(t) = 15 - e^{-3t}\left[15\cos\left(\sqrt{2}t\right) + \frac{45}{2}\sqrt{2}\sin\left(\sqrt{2}t\right)\right].$$

Thus the limiting amounts of salt in tanks 1, 2, and 3 are 30 lb, 10 lb, and 15 lb. The figure shows the graphs of $x_1(t)$, $x_2(t)$, and $x_3(t)$.

In Problems 38-41 the Maple command `with(linalg):eigenvects(A)`, the *Mathematica* command `Eigensystem[A]`, or the MATLAB command `[V,D] = eig(A)` can be used to find the eigenvalues and associated eigenvectors of the given coefficient matrix **A**.

39.   Characteristic equation: $\left(\lambda^2 - 1\right)\left(\lambda^2 - 4\right) = 0$

Eigenvalues and associated eigenvectors:

λ	1	−1	2	−2
v	$[3 \ \ -2 \ \ 4 \ \ 1]^T$	$[0 \ \ 0 \ \ 1 \ \ 0]^T$	$[0 \ \ 1 \ \ 0 \ \ 0]^T$	$[1 \ \ -1 \ \ 0 \ \ 0]^T$

Scalar solution equations:

$$
\begin{aligned}
x_1(t) &= 3c_1 e^t & & + c_4 e^{-2t} \\
x_2(t) &= -2c_1 e^t & + c_3 e^{2t} &- c_4 e^{-2t} \\
x_3(t) &= 4c_1 e^t + c_2 e^{-t} \\
x_4(t) &= c_1 e^t
\end{aligned}
$$

41.   The eigenvalues and respective eigenvalues are given by the following table:

λ	−3	−6	10	15
v	$[1 \ \ 0 \ \ 0 \ \ -1]^T$	$[0 \ \ 1 \ \ -1 \ \ 0]^T$	$[-2 \ \ 1 \ \ 1 \ \ -2]^T$	$[1 \ \ 2 \ \ 2 \ \ 1]^T$

Hence the general solution has scalar component functions

$$
\begin{aligned}
x_1(t) &= c_1 e^{-3t} & & - 2c_3 e^{10t} &+ c_4 e^{15t} \\
x_2(t) &= & c_2 e^{-6t} &+ c_3 e^{10t} &+ 2c_4 e^{15t} \\
x_3(t) &= & - c_2 e^{-6t} &+ c_3 e^{10t} &+ 2c_4 e^{15t} \\
x_4(t) &= -c_1 e^{-3t} & & - 2c_3 e^{10t} &+ c_4 e^{15t}
\end{aligned}
$$

The given initial conditions are satisfied by choosing $c_1 = c_2 = 0$, $c_3 = -1$, and $c_4 = 1$, so the desired particular solution is given by

$$x_1(t) = 2e^{10t} + e^{15t} = x_4(t)$$
$$x_2(t) = -e^{10t} + 2e^{15t} = x_3(t)$$

In Problems 42–50 we give a general solution in the form $\mathbf{x}(t) = c_1\mathbf{v}_1 e^{\lambda_1 t} + c_2\mathbf{v}_2 e^{\lambda_2 t} + \cdots$ that exhibits explicitly the eigenvalues $\lambda_1, \lambda_2, \ldots$ and corresponding eigenvectors $\mathbf{v}_1, \mathbf{v}_2, \ldots$ of the given coefficient matrix $\mathbf{A}$.

43. $\mathbf{x}(t) = c_1 \begin{bmatrix} 3 \\ -1 \\ 5 \end{bmatrix} e^{-2t} + c_2 \begin{bmatrix} 1 \\ 1 \\ 1 \end{bmatrix} e^{4t} + c_3 \begin{bmatrix} 1 \\ -1 \\ 3 \end{bmatrix} e^{8t}$

45. $\mathbf{x}(t) = c_1 \begin{bmatrix} 1 \\ 1 \\ 1 \\ -1 \end{bmatrix} e^{-3t} + c_2 \begin{bmatrix} 1 \\ 2 \\ -1 \\ 1 \end{bmatrix} + c_3 \begin{bmatrix} 2 \\ 1 \\ 1 \\ 1 \end{bmatrix} e^{3t} + c_4 \begin{bmatrix} 1 \\ -1 \\ 2 \\ -1 \end{bmatrix} e^{6t}$

47. $\mathbf{x}(t) = c_1 \begin{bmatrix} 2 \\ 2 \\ 1 \\ -1 \end{bmatrix} e^{-3t} + c_2 \begin{bmatrix} 1 \\ 2 \\ -1 \\ 1 \end{bmatrix} e^{3t} + c_3 \begin{bmatrix} 2 \\ 1 \\ 1 \\ 1 \end{bmatrix} e^{6t} + c_4 \begin{bmatrix} 1 \\ -1 \\ 2 \\ -1 \end{bmatrix} e^{9t}$

49. $\mathbf{x}(t) = c_1 \begin{bmatrix} 1 \\ 0 \\ 3 \\ 1 \\ 1 \end{bmatrix} e^{-3t} + c_2 \begin{bmatrix} 0 \\ 3 \\ 0 \\ -1 \\ 1 \end{bmatrix} + c_3 \begin{bmatrix} 1 \\ 7 \\ 1 \\ 1 \\ 1 \end{bmatrix} e^{3t} + c_4 \begin{bmatrix} 0 \\ 1 \\ 0 \\ 1 \\ 1 \end{bmatrix} e^{6t} + c_5 \begin{bmatrix} 2 \\ 0 \\ 5 \\ 2 \\ 1 \end{bmatrix} e^{9t}$

SECTION 5.3

SOLUTION CURVES OF LINEAR SYSTEMS

This section emphasizes the connection between the algebraic properties of the matrix $\mathbf{A}$—specifically, its eigenvalues and eigenvectors—and the characteristic pattern of the phase diagram of the system $\mathbf{x}' = \mathbf{A}\mathbf{x}$.

In Problems 1-16 the eigenvalues, eigenvectors and phase portraits appear in the solutions to Section 5.2. Thus here we simply categorize each phase portrait according to the gallery in Fig. 5.3.16.

1.   Saddle point (real eigenvalues of opposite sign)

3.   Saddle point (real eigenvalues of opposite sign)

5.   Saddle point (real eigenvalues of opposite sign)

7.   Saddle point (real eigenvalues of opposite sign)

9.   Center (pure imaginary eigenvalues)

11.   Spiral source (complex conjugate eigenvalues with positive real part)

13.   Spiral source (complex conjugate eigenvalues with positive real part)

15.   Spiral source (complex conjugate eigenvalues with positive real part)

In Problems 17-28 we "pigeonhole" each phase portrait according to the gallery in Fig. 5.3.16 and give the nature of the eigenvalues of the matrix **A**. Where appropriate we further give approximate values of the corresponding eigenvectors.

17.   Center; pure imaginary eigenvalues

19.   Saddle point; real eigenvalues of opposite sign; $\mathbf{v}_1 \approx \begin{bmatrix} 0 & 1 \end{bmatrix}^T$ corresponds to the negative eigenvalue and $\mathbf{v}_2 \approx \begin{bmatrix} -1 & 1 \end{bmatrix}^T$ to the positive one.

21.   Proper nodal source; repeated positive real eigenvalue with linearly independent eigenvectors

23.   Spiral sink; complex conjugate eigenvalues with negative real part

25.   Saddle point; real eigenvalues of opposite sign; $\mathbf{v}_1 \approx \begin{bmatrix} 1 & 1 \end{bmatrix}^T$ corresponds to the positive eigenvalue and $\mathbf{v}_2 \approx \begin{bmatrix} 4 & -1 \end{bmatrix}^T$ to the negative one.

27.   Improper nodal source; distinct positive real eigenvalues; $\mathbf{v}_1 \approx \begin{bmatrix} 2 & 3 \end{bmatrix}^T$, $\mathbf{v}_2 \approx \begin{bmatrix} 2 & -1 \end{bmatrix}^T$

29.   **a)** If $v_0 = 0$, then $v(t) \equiv 0$ for all t, so that the point $(u, v) = \left(u_0 e^{-2t}, 0 \right)$ (in oblique coordinates) lies on the u-axis. The reverse argument applies if $u_0 = 0$.

b) If both u_0 and v_0 are nonzero, then solving $u = u_0 e^{-2t}$ for t gives $t = -\dfrac{1}{2}\ln\dfrac{u}{u_0}$, where-

as solving $v = v_0 e^{5t}$ for t gives $t = \dfrac{1}{5}\ln\dfrac{v}{v_0}$. We can thus eliminate t to conclude that

$-\dfrac{1}{2}\ln\dfrac{u}{u_0} = \dfrac{1}{5}\ln\dfrac{v}{v_0}$. Then solving for v gives $\ln\dfrac{v}{v_0} = -\dfrac{5}{2}\ln\dfrac{u}{u_0}$, or

$\dfrac{v}{v_0} = \exp\left(-\dfrac{5}{2}\ln\dfrac{u}{u_0}\right) = \left(\dfrac{u}{u_0}\right)^{-5/2}$, or finally $v = v_0\left(\dfrac{u}{u_0}\right)^{-5/2} = v_0 u_0^{5/2} u^{-5/2} = Cu^{-5/2}$, where

$C = v_0 u_0^{5/2}$.

31. If λ is an eigenvalue of $\mathbf{A}$ with associated eigenvector $\mathbf{v}$, then $(\mathbf{A} - \lambda\mathbf{I})\mathbf{v} = \mathbf{0}$. Taking the negative of both sides of this equation then gives $-(\mathbf{A} - \lambda\mathbf{I})\mathbf{v} = -\mathbf{0} = \mathbf{0}$. However, $-(\mathbf{A} - \lambda\mathbf{I})$ can be written as $(-\mathbf{A}) - (-\lambda)\mathbf{I}$, and so $\left[(-\mathbf{A}) - (-\lambda)\mathbf{I}\right]\mathbf{v} = \mathbf{0}$ as well. It follows that $-\lambda$ is an eigenvalue of the matrix $-\mathbf{A}$ with associated eigenvector $\mathbf{v}$. This means that if $\mathbf{A}$ has positive eigenvalues $0 < \lambda_2 < \lambda_1$ with associated eigenvectors $\mathbf{v}_1$ and $\mathbf{v}_2$, then $-\mathbf{A}$ has negative eigenvalues $-\lambda_1 < -\lambda_2 < 0$ associated to these same eigenvectors $\mathbf{v}_1$ and $\mathbf{v}_2$.

33. **a)** Let $\mathbf{v}_1$ and $\mathbf{v}_2$ denote the two linearly independent eigenvectors of $\mathbf{A}$ associated with the eigenvalue λ, so that $\mathbf{A}\mathbf{v}_1 = \lambda\mathbf{v}_1$ and $\mathbf{A}\mathbf{v}_2 = \lambda\mathbf{v}_2$. If $\mathbf{v}$ is any two-dimensional vector, then the fact that $\mathbf{v}_1$ and $\mathbf{v}_2$ are linearly independent implies that $\mathbf{v}$ can be written as a linear combination of $\mathbf{v}_1$ and $\mathbf{v}_2$, so that $\mathbf{v} = c_1\mathbf{v}_1 + c_2\mathbf{v}_2$ for some scalars c_1 and c_2. But then the linearity of matrix multiplication gives

$$\mathbf{A}\mathbf{v} = \mathbf{A}\left(c_1\mathbf{v}_1 + c_2\mathbf{v}_2\right) = c_1\mathbf{A}\mathbf{v}_1 + c_2\mathbf{A}\mathbf{v}_2 = c_1\lambda\mathbf{v}_1 + c_2\lambda\mathbf{v}_2 = \lambda\left(c_1\mathbf{v}_1 + c_2\mathbf{v}_2\right) = \lambda\mathbf{v},$$

proving that $\mathbf{v}$ is an eigenvector of A.

b) Let $\mathbf{A} = \begin{bmatrix} a & b \\ c & d \end{bmatrix}$. Part a) implies that $\mathbf{A}\mathbf{v} = \lambda\mathbf{v}$ for all vectors v, so in particular, if we take $\mathbf{v} = \begin{bmatrix} 1 & 0 \end{bmatrix}^{\mathrm{T}}$, then $\mathbf{A}\mathbf{v} = \begin{bmatrix} a & c \end{bmatrix}^{\mathrm{T}} = \lambda\mathbf{v} = \begin{bmatrix} \lambda & 0 \end{bmatrix}^{\mathrm{T}}$, proving that $a = \lambda$ and $c = 0$. Similarly, if we take $\mathbf{v} = \begin{bmatrix} 0 & 1 \end{bmatrix}^{\mathrm{T}}$, then $\mathbf{A}\mathbf{v} = \begin{bmatrix} b & d \end{bmatrix}^{\mathrm{T}} = \lambda\mathbf{v} = \begin{bmatrix} 0 & \lambda \end{bmatrix}^{\mathrm{T}}$, proving that $b = 0$ and $d = \lambda$. Thus A is given by Eq. (22).

35. Write the given equation as $M\left(x_1, x_2\right)dx_1 + N\left(x_1, x_2\right)dx_2 = 0$, where $M\left(x_1, x_2\right) = 6x_2 - 8x_1$ and $N\left(x_1, x_2\right) = 6x_1 - 17x_2$. The equation is exact because $\dfrac{\partial M}{\partial x_2} = 6 = \dfrac{\partial N}{\partial x_1}$. Its general solution is therefore given by $F\left(x_1, x_2\right) = k$, where $F\left(x_1, x_2\right)$

(as discussed in Section 1.6) satisfies the conditions $\dfrac{\partial F}{\partial x_1} = M$ and $\dfrac{\partial F}{\partial x_2} = N$ and k is a constant. The first condition implies that

$$F(x_1, x_2) = \int M\,dx_1 = 6x_1 x_2 - 4x_1^2 + h(x_2),$$

which specifies F up to the unknown function $h(x_2)$. Then the second condition gives

$$\frac{\partial}{\partial x_2}\left[6x_1 x_2 - 4x_1^2 + h(x_2)\right] = 6x_1 + h'(x_2) = 6x_1 - 17x_2,$$

or simply $h'(x_2) = -17x_2$, which means that up to a constant, $h(x_2) = -\dfrac{17}{2}x_2^2$. Altogether then, the general solution of the given equation is

$$F(x_1, x_2) = 6x_1 x_2 - 4x_1^2 - \frac{17}{2}x_2^2 = k.$$

37. With the same values of A, B, and C we find that $\dfrac{B}{A-C} = \dfrac{6}{-4+\dfrac{17}{2}} = \dfrac{6}{9/2} = \dfrac{4}{3}$. Thus it

suffices to confirm that $\theta = \arctan\dfrac{2}{4}$ satisfies $\tan 2\theta = \dfrac{4}{3}$. However, this is readily verified using the double-angle formula for the tangent function:

$$\tan 2\theta = \frac{2\tan\theta}{1-\tan^2\theta} = \frac{2\cdot\dfrac{2}{4}}{1-\left(\dfrac{2}{4}\right)^2} = \frac{4}{3}.$$

39.   a) The characteristic equation of $\mathbf{A}$ is given by $\det(\mathbf{A} - \lambda\mathbf{I}) = 0$, that is

$$\begin{vmatrix} a-\lambda & b \\ c & d-\lambda \end{vmatrix} = (a-\lambda)(d-\lambda) - bc = \lambda^2 - (a+d)\lambda + (ad-bc) = 0.$$

b) The quadratic formula shows that if the solutions to the characteristic equation are pure imaginary, then the coefficient $-(a+d)$ of λ must vanish. Hence the trace $T(\mathbf{A}) = a + d = 0$, which means that $d = -a$. For the same reason, the constant term $ad - bc$ must be positive. Substituting $d = -a$ then gives $ad - bc = -a^2 - bc > 0$, which is impossible if $c = 0$.

SECTION 5.4

SECOND-ORDER SYSTEMS AND MECHANICAL APPLICATIONS

This section uses the eigenvalue method to exhibit realistic applications of linear systems. If a computer system like Maple, *Mathematica*, MATLAB, or even a TI-85/86/89/92/Nspire calculator is available, then a system of more than three railway cars, or a multistory building with four or more floors (as in the project), can be investigated. However, the problems in the text are intended for manual solution.

Problems 1–7 involve the system

$$m_1 x_1'' = -(k_1 + k_2)x_1 + k_2 x_2$$
$$m_2 x_2'' = k_2 x_1 - (k_2 + k_3)x_2$$

with various values of m_1, m_2 and k_1, k_2, k_3. In each problem we divide the first equation by m_1 and the second one by m_2 to obtain a second-order linear system $\mathbf{x}'' = \mathbf{Ax}$ in the standard form of Theorem 1 in this section. If the eigenvalues λ_1 and λ_2 are both negative, then the natural (circular) frequencies of the system are $\omega_1 = \sqrt{-\lambda_1}$ and $\omega_2 = \sqrt{-\lambda_2}$, and—according to Eq. (11) in Theorem 1 of this section—the eigenvectors $\mathbf{v}_1$ and $\mathbf{v}_2$ associated with λ_1 and λ_2 determine the natural modes of oscillations at these frequencies.

1. The matrix $\mathbf{A} = \begin{bmatrix} -2 & 2 \\ 2 & -2 \end{bmatrix}$ has eigenvalues $\lambda_0 = 0$ and $\lambda_1 = -4$ with associated eigenvectors $\mathbf{v}_0 = \begin{bmatrix} 1 & 1 \end{bmatrix}^T$ and $\mathbf{v}_1 = \begin{bmatrix} 1 & -1 \end{bmatrix}^T$. Thus we have the special case described in Eq. (12) of Theorem 1, and a general solution is given by

$$x_1(t) = a_1 + a_2 t + b_1 \cos 2t + b_2 \sin 2t,$$
$$x_2(t) = a_1 + a_2 t - b_1 \cos 2t - b_2 \sin 2t.$$

The natural frequencies are $\omega_1 = 0$ and $\omega_2 = 2$. In the degenerate natural mode with "frequency" $\omega_1 = 0$ the two masses move by translation without oscillating. At frequency $\omega_2 = 2$ they oscillate in opposite directions with equal amplitudes.

3. The matrix $\mathbf{A} = \begin{bmatrix} -3 & 2 \\ 1 & -2 \end{bmatrix}$ has eigenvalues $\lambda_1 = -1$ and $\lambda_2 = -4$, with associated eigenvectors $\mathbf{v}_1 = \begin{bmatrix} 1 & 1 \end{bmatrix}^T$ and $\mathbf{v}_2 = \begin{bmatrix} 2 & -1 \end{bmatrix}^T$. Hence a general solution is given by

$$x_1(t) = a_1 \cos t + a_2 \sin t + 2b_1 \cos 2t + 2b_2 \sin 2t,$$
$$x_2(t) = a_1 \cos t + a_2 \sin t - b_1 \cos 2t - b_2 \sin 2t.$$

The natural frequencies are $\omega_1 = 1$ and $\omega_2 = 2$. In the natural mode with frequency ω_1, the two masses m_1 and m_2 move in the same direction with equal amplitudes of oscillation. In the natural mode with frequency ω_2 they move in opposite directions with the amplitude of oscillation of m_1 twice that of m_2.

5. The matrix $\mathbf{A} = \begin{bmatrix} -3 & 1 \\ 1 & -3 \end{bmatrix}$ has eigenvalues $\lambda_1 = -2$ and $\lambda_2 = -4$ with associated eigenvectors $\mathbf{v}_1 = \begin{bmatrix} 1 & 1 \end{bmatrix}^T$ and $\mathbf{v}_2 = \begin{bmatrix} 1 & -1 \end{bmatrix}^T$. Hence a general solution is given by

$$x_1(t) = a_1 \cos\left(\sqrt{2}t\right) + a_2 \sin\left(\sqrt{2}t\right) + b_1 \cos 2t + b_2 \sin 2t,$$
$$x_2(t) = a_1 \cos\left(\sqrt{2}t\right) + a_2 \sin\left(\sqrt{2}t\right) - b_1 \cos 2t - b_2 \sin 2t.$$

The natural frequencies are $\omega_1 = \sqrt{2}$ and $\omega_2 = 2$. In the natural mode with frequency ω_1, the two masses m_1 and m_2 move in the same direction with equal amplitudes of oscillation. At frequency ω_2 they move in opposite directions with equal amplitudes.

7. The matrix $\mathbf{A} = \begin{bmatrix} -10 & 6 \\ 6 & -10 \end{bmatrix}$ has eigenvalues $\lambda_1 = -4$ and $\lambda_2 = -16$ with associated eigenvectors $\mathbf{v}_1 = \begin{bmatrix} 1 & 1 \end{bmatrix}^T$ and $\mathbf{v}_2 = \begin{bmatrix} 1 & -1 \end{bmatrix}^T$. Hence a general solution is given by

$$x_1(t) = a_1 \cos 2t + a_2 \sin 2t + b_1 \cos 4t + b_2 \sin 4t,$$
$$x_2(t) = a_1 \cos 2t + a_2 \sin 2t - b_1 \cos 4t - b_2 \sin 4t.$$

The natural frequencies are $\omega_1 = 2$ and $\omega_2 = 4$. In the natural mode with frequency ω_1, the two masses m_1 and m_2 move in the same direction with equal amplitudes of oscillation. At frequency ω_2 they move in opposite directions with equal amplitudes.

9. Substitution of the trial solution $x_1 = c_1 \cos 3t$, $x_2 = c_2 \cos 3t$ in the system

$$x_1'' = -3x_1 + 2x_2 \quad 2x_2'' = 2x_1 - 4x_2 + 120\cos 3t$$

yields $c_1 = 3$ and $c_2 = -9$, so a general solution is given by

$$x_1(t) = a_1 \cos t + a_2 \sin t + 2b_1 \cos 2t + 2b_2 \sin 2t + 3\cos 3t,$$
$$x_2(t) = a_1 \cos t + a_2 \sin t - b_1 \cos 2t - b_2 \sin 2t - 9\cos 3t.$$

Imposition of the initial conditions $x_1(0) = x_2(0) = x_1'(0) = x_2'(0) = 0$ now yields $a_1 = 5$, $a_2 = 0$, $b_1 = -4$, and $b_2 = 0$. The resulting particular solution is

$$x_1(t) = 5\cos t - 8\cos 2t + 3\cos 3t,$$
$$x_2(t) = 5\cos t + 4\cos 2t - 9\cos 3t.$$

We have a superposition of three oscillations, in which the two masses move

- in the same direction with frequency $\omega_1 = 1$ and equal amplitudes;

- in opposite directions with frequency $\omega_2 = 2$ and with the amplitude of motion of m_1 being twice that of m_2;

- in opposite directions with frequency $\omega_3 = 3$ and with the amplitude of motion of m_2 being 3 times that of m_1.

11.   **(a)** The matrix $\mathbf{A} = \begin{bmatrix} -40 & 8 \\ 12 & -60 \end{bmatrix}$ has eigenvalues $\lambda_1 = -36$ and $\lambda_2 = -64$ with associated eigenvalues $\mathbf{v}_1 = \begin{bmatrix} 2 & 1 \end{bmatrix}^T$ and $\mathbf{v}_2 = \begin{bmatrix} 1 & -3 \end{bmatrix}^T$. Hence a general solution is given by

$$x(t) = 2a_1 \cos 6t + 2a_2 \sin 6t + b_1 \cos 8t + b_2 \sin 8t,$$
$$y(t) = a_1 \cos 6t + a_2 \sin 6t - 3b_1 \cos 8t - 3b_2 \sin 8t.$$

The natural frequencies are $\omega_1 = 6$ and $\omega_2 = 8$. In mode 1 the two masses oscillate in the same direction with frequency $\omega_1 = 6$ and with the amplitude of motion of m_1 being twice that of m_2. In mode 2 the two masses oscillate in opposite directions with frequency $\omega_2 = 8$ and with the amplitude of motion of m_2 being 3 times that of m_1.

(b)   Substitution of the trial solution $x = c_1 \cos 7t$, $y = c_2 \cos 7t$ in the system

$$x'' = -40x + 8y - 195\cos 7t, \quad y'' = 12x - 60y - 195\cos 7t$$

yields $c_1 = 19$ and $c_2 = 3$, so a general solution is given by

$$x(t) = 2a_1 \cos 6t + 2a_2 \sin 6t + b_1 \cos 8t + b_2 \sin 8t + 19\cos 7t,$$
$$y(t) = a_1 \cos 6t + a_2 \sin 6t - 3b_1 \cos 8t - 3b_2 \sin 8t + 3\cos 7t.$$

Imposition of the initial conditions $x(0) = 19$, $x'(0) = 12$, $y(0) = 3$, and $y'(0) = 6$ now yields $a_1 = 0$, $a_2 = 1$, $b_1 = 0$, and $b_2 = 0$. The resulting particular solution is

$$x(t) = 2\sin 6t + 19\cos 7t,$$
$$y(t) = \sin 6t + 3\cos 7t.$$

Thus the expected oscillation with frequency $\omega_2 = 8$ is missing, and we have a superposition of (only two) oscillations, in which the two masses move

- in the same direction with frequency $\omega_1 = 6$ and with the amplitude of motion of m_1 being twice that of m_2;

- in the same direction with frequency $\omega_3 = 7$ and with the amplitude of motion of m_1 being $\dfrac{19}{3}$ times that of m_2.

13. The coefficient matrix $\mathbf{A} = \begin{bmatrix} -4 & 2 & 0 \\ 2 & -4 & 2 \\ 0 & 2 & -4 \end{bmatrix}$ has characteristic polynomial

$$-\lambda^3 - 12\lambda^2 - 40\lambda - 32 = -(\lambda + 4)(\lambda^2 + 8\lambda + 8).$$

Its eigenvalues $\lambda_1 = -4$, $\lambda_2 = -4 - 2\sqrt{2}$, and $\lambda_3 = -4 + 2\sqrt{2}$ have associated eigenvectors $\mathbf{v}_1 = \begin{bmatrix} 1 & 0 & -1 \end{bmatrix}^T$, $\mathbf{v}_2 = \begin{bmatrix} 1 & -\sqrt{2} & 1 \end{bmatrix}^T$, and $\mathbf{v}_3 = \begin{bmatrix} 1 & \sqrt{2} & 1 \end{bmatrix}^T$. Hence the system's three natural modes of oscillation have

- Natural frequency $\omega_1 = 2$ with amplitude ratios $1 : 0 : -1$;

- Natural frequency $\omega_2 = \sqrt{4 + 2\sqrt{2}}$ with amplitude ratios $1 : -\sqrt{2} : 1$.

- Natural frequency $\omega_3 = \sqrt{4 - 2\sqrt{2}}$ with amplitude ratios $1 : \sqrt{2} : 1$.

15. First we need the general solution of the homogeneous system $\mathbf{x}'' = \mathbf{A}\mathbf{x}$ with

$$\mathbf{A} = \begin{bmatrix} -50 & 25/2 \\ 50 & -50 \end{bmatrix}.$$

The eigenvalues of A are $\lambda_1 = -25$ and $\lambda_2 = -75$, so the natural frequencies of the system are $\omega_1 = 5$ and $\omega_2 = 5\sqrt{3}$. The associated eigenvectors are $\mathbf{v}_1 = \begin{bmatrix} 1 & 2 \end{bmatrix}^T$ and $\mathbf{v}_2 = \begin{bmatrix} 1 & -2 \end{bmatrix}^T$, so the complementary solution $\mathbf{x}_c(t)$ is given by

$$x_1(t) = a_1 \cos 5t + a_2 \sin 5t + b_1 \cos\left(5\sqrt{3}t\right) + b_2 \sin\left(5\sqrt{3}t\right),$$

$$x_2(t) = 2a_1 \cos 5t + 2a_2 \sin 5t - 2b_1 \cos\left(5\sqrt{3}t\right) - 2b_2 \sin\left(5\sqrt{3}t\right).$$

When we substitute the trial solution $\mathbf{x}_p(t) = \begin{bmatrix} c_1 & c_2 \end{bmatrix}^T \cos 10t$ in the nonhomogeneous system, we find that $c_1 = \dfrac{4}{3}$ and $c_2 = -\dfrac{16}{3}$, so a particular solution $\mathbf{x}_p(t)$ is described by

$$x_1(t) = \frac{4}{3}\cos 10t, \quad x_2(t) = -\frac{16}{3}\cos 10t.$$

Finally, when we impose the zero initial conditions on the solution $\mathbf{x}(t) = \mathbf{x}_c(t) + \mathbf{x}_p(t)$ we find that $a_1 = \dfrac{2}{3}$, $a_2 = 0$, $b_1 = -2$, and $b_2 = 0$. Thus the solution we seek is described by

$$x_1(t) = \frac{2}{3}\cos 5t - 2\cos\left(5\sqrt{3}t\right) + \frac{4}{3}\cos 10t,$$

$$x_2(t) = \frac{4}{3}\cos 5t + 4\cos\left(5\sqrt{3}t\right) + \frac{16}{3}\cos 10t$$

We have a superposition of two oscillations with the natural frequencies $\omega_1 = 5$ and $\omega_2 = 5\sqrt{3}$ and a forced oscillation with frequency $\omega = 10$. In each of the two natural oscillations the amplitude of motion of m_2 is twice that of m_1, while in the forced oscillation the amplitude of motion of m_2 is four times that of m_1.

17.   With $c_1 = c_2 = 2$, it follows from Problem 16 that the natural frequencies and associated eigenvectors are $\omega_1 = 0$, $\mathbf{v}_1 = \begin{bmatrix} 1 & 1 \end{bmatrix}^T$ and $\omega_2 = 2$, $\mathbf{v}_2 = \begin{bmatrix} 1 & -1 \end{bmatrix}^T$. Hence Theorem 1 gives the general solution

$$x_1(t) = a_1 + b_1 t + a_2 \cos 2t + b_2 \sin 2t,$$
$$x_2(t) = a_1 + b_1 t - a_2 \cos 2t - b_2 \sin 2t.$$

The initial conditions $x_1'(0) = v_0$, $x_1(0) = x_2(0) = x_2'(0) = 0$ yield $a_1 = a_2 = 0$, $b_1 = \dfrac{v_0}{2}$, and $b_2 = \dfrac{v_0}{4}$, so

$$x_1(t) = \frac{v_0}{4}(2t + \sin 2t), \quad x_2(t) = \frac{v_0}{4}(2t - \sin 2t),$$

while $x_2 - x_1 = \dfrac{v_0}{4}(-2\sin 2t) < 0$; that is, until $t = \dfrac{\pi}{2}$. Finally, $x_1'\left(\dfrac{\pi}{2}\right) = 0$ and $x_2'\left(\dfrac{\pi}{2}\right) = v_0$.

19.   With $c_1 = 1$ and $c_2 = 3$, it follows from Problem 16 that the natural frequencies and associated eigenvectors are $\omega_1 = 0$, $\mathbf{v}_1 = \begin{bmatrix} 1 & 1 \end{bmatrix}^T$ and $\omega_2 = 2$, $\mathbf{v}_2 = \begin{bmatrix} 1 & -3 \end{bmatrix}^T$. Hence Theorem 1 gives the general solution

$$x_1(t) = a_1 + b_1 t + a_2 \cos 2t + b_2 \sin 2t,$$
$$x_2(t) = a_1 + b_1 t - 3a_2 \cos 2t - 3b_2 \sin 2t.$$

The initial conditions $x_1'(0) = v_0$, $x_1(0) = x_2(0) = x_2'(0) = 0$ yield $a_1 = a_2 = 0$, $b_1 = \dfrac{3v_0}{4}$, and $b_2 = \dfrac{v_0}{8}$, so

$$x_1(t) = \frac{v_0}{8}(6t + \sin 2t), \quad x_2(t) = \frac{v_0}{8}(6t - 3\sin 2t),$$

while $x_2 - x_1 = \dfrac{v_0}{8}(-4\sin 2t) < 0$; that is, until $t = \dfrac{\pi}{2}$. Finally, $x_1'\left(\dfrac{\pi}{2}\right) = \dfrac{v_0}{2}$ and $x_2'\left(\dfrac{\pi}{2}\right) = \dfrac{3v_0}{2}$.

The method of solution in each of Problems 20–23 is the same as that in Example 2 in this section. Thus, looking at the equations in (26), we need to solve the equations

$$\begin{array}{ccccccc} b_1 & + & 2b_2 & + & 4b_3 & = & x_1'(0) \\ b_1 & & & - & 12b_3 & = & x_2'(0) \\ b_1 & - & 2b_2 & + & 4b_3 & = & x_3'(0) \end{array}$$

for the coefficients b_1, b_2, b_3 after inserting given initial values $x_1'(0)$, $x_2'(0)$, $x_3'(0)$ of the three railway cars.

21. With $x_1'(0) = 2v_0$, $x_2'(0) = 0$, and $x_3'(0) = -v_0$, substitution of the resulting coefficient values b_1, b_2, b_3 in (25) gives the railway car displacement functions

$$x_1(t) = \frac{1}{32} v_0 (12t + 24\sin 2t + \sin 4t),$$

$$x_2(t) = \frac{1}{32} v_0 (12t \qquad\qquad - 3\sin 4t),$$

$$x_3(t) = \frac{1}{32} v_0 (12t - 24\sin 2t + \sin 4t).$$

We then see (substituting $\sin 4t = 2\sin 2t \cos 2t$) that

$$x_2(t) - x_1(t) = -\frac{1}{8} v_0 (\sin 4t + 6\sin 2t) = -\frac{1}{4} v_0 (\sin 2t)(\cos 2t + 3)$$

remains negative until $t = \frac{\pi}{2}$ (as does $x_3(t) - x_2(t)$, similarly) at which time the cars separate with velocities

$$x_1'\left(\frac{\pi}{2}\right) = -v_0, \quad x_2'\left(\frac{\pi}{2}\right) = 0, \quad x_3'\left(\frac{\pi}{2}\right) = 2v_0.$$

Thus the car in the center remains fixed thereafter, whereas the first and third cars rebound in opposite directions, having exchanged their original velocities.

23. With $x_1'(0) = 3v_0$, $x_2'(0) = 2v_0$, and $x_3'(0) = 2v_0$, substitution of the resulting coefficient values b_1, b_2, b_3 in (25) gives the railway car displacement functions

$$x_1(t) = \frac{1}{32} v_0 (76t + 8\sin 2t + \sin 4t),$$

$$x_2(t) = \frac{1}{32} v_0 (76t \qquad\qquad - 3\sin 4t),$$

$$x_3(t) = \frac{1}{32} v_0 (76t - 8\sin 2t + \sin 4t).$$

We then see (substituting $\sin 4t = 2\sin 2t \cos 2t$) that

$$x_2(t) - x_1(t) = -\frac{1}{8}v_0(2\sin 2t + \sin 4t) = -\frac{1}{4}v_0(\sin 2t)(1 + \cos 2t)$$

remains negative until $t = \frac{\pi}{2}$ (as does $x_3(t) - x_2(t)$, similarly) at which time the cars separate with velocities

$$x_1'\left(\frac{\pi}{2}\right) = 2v_0, \quad x_2'\left(\frac{\pi}{2}\right) = 2v_0, \quad x_3'\left(\frac{\pi}{2}\right) = 3v_0.$$

Thus the car in the center proceeds thereafter with the same velocity it had originally, whereas the first and third cars rebound in opposite directions, having exchanged their original velocities.

25.   **(a)** The matrix

$$\mathbf{A} = \begin{bmatrix} -160/3 & 320/3 \\ 8 & -116 \end{bmatrix}$$

has eigenvalues $\lambda_1 = -41.8285$ and $\lambda_2 = -127.5049$, so the natural frequencies are

$$\omega_1 \approx 6.4675\,\text{rad/sec} \approx 1.0293\,\text{Hz}, \quad \omega_2 \approx 11.2918\,\text{rad/sec} \approx 1.7971\,\text{Hz}.$$

(b) Resonance occurs at the two critical speeds

$$v_1 = \frac{20\omega_1}{\pi} \approx 41\,\text{ft/sec} \approx 28\,\text{mi/h}, \quad v_2 = \frac{20\omega_2}{\pi} \approx 72\,\text{ft/sec} \approx 49\,\text{mi/h}.$$

27. $100x'' = -4000x$, $800\theta'' = 100000\theta$.

Obviously the matrix $\mathbf{A} = \begin{bmatrix} -40 & 0 \\ 0 & -125 \end{bmatrix}$ has eigenvalues $\lambda_1 = -40$ and $\lambda_2 = -125$.

Up-and-down: $\omega_1 = \sqrt{40}$, $v_1 \approx 40.26\,\text{ft/sec} \approx 27\,\text{mi/h}$;

Angular: $\omega_2 = \sqrt{125}$, $v_2 \approx 71.18\,\text{ft/sec} \approx 49\,\text{mi/h}$.

29. $100x'' = -3000x - 5000\theta$, $800\theta'' = -5000x - 75000\theta$.

The matrix $\mathbf{A} = \begin{bmatrix} -30 & -50 \\ -25/4 & -375/4 \end{bmatrix}$ has eigenvalues $\lambda_1, \lambda_2 = \frac{5}{8}\left(-99 \pm \sqrt{3401}\right)$.

$\omega_1 \approx 5.0424$, $v_1 \approx 32.10\,\text{ft/sec} \approx 22\,\text{mi/h}$;

$\omega_2 \approx 9.9158$, $v_2 \approx 63.13\,\text{ft/sec} \approx 43\,\text{mi/h}$.

SECTION 5.5

MULTIPLE EIGENVALUE SOLUTIONS

In each of Problems 1–6 we give first the characteristic equation with repeated (multiplicity 2) eigenvalue λ. In each case we find that $(\mathbf{A} - \lambda\mathbf{I})^2 = \mathbf{0}$. Then $\mathbf{w} = \begin{bmatrix} 1 & 0 \end{bmatrix}^T$ is a generalized eigenvector and $\mathbf{v} = (\mathbf{A} - \lambda\mathbf{I})\mathbf{w} \neq \mathbf{0}$ is an ordinary eigenvector associated with λ. We give finally the scalar component functions $x_1(t)$, $x_2(t)$ of the general solution

$$\mathbf{x}(t) = c_1\mathbf{v}e^{\lambda t} + c_2(\mathbf{v}t + \mathbf{w})e^{\lambda t}$$

of the given system $\mathbf{x}' = \mathbf{A}\mathbf{x}$.

1. Characteristic equation $\lambda^2 + 6\lambda + 9 = 0$; repeated eigenvalue $\lambda = -3$; generalized eigenvector $\mathbf{w} = \begin{bmatrix} 1 & 0 \end{bmatrix}^T$;

$$\mathbf{v} = (\mathbf{A} - \lambda\mathbf{I})\mathbf{w} = \begin{bmatrix} 1 & 1 \\ -1 & -1 \end{bmatrix}\begin{bmatrix} 1 \\ 0 \end{bmatrix} = \begin{bmatrix} 1 \\ -1 \end{bmatrix};$$

$$x_1(t) = (c_1 + c_2 + c_2 t)e^{-3t}, \quad x_2(t) = (-c_1 - c_2 t)e^{-3t}.$$

Problem 1	**Problem 3**
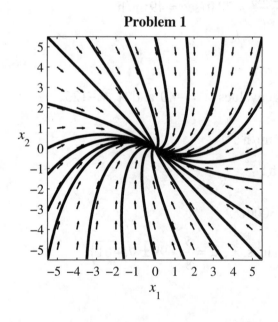	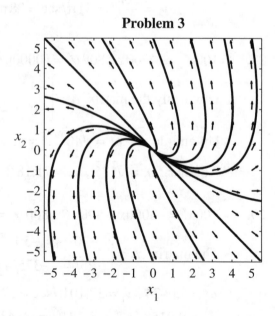

3. Characteristic equation $\lambda^2 - 6\lambda + 9 = 0$; repeated eigenvalue $\lambda = 3$; generalized eigenvector $\mathbf{w} = \begin{bmatrix} 1 & 0 \end{bmatrix}^T$;

$$\mathbf{v} = (\mathbf{A} - \lambda\mathbf{I})\mathbf{w} = \begin{bmatrix} -2 & -2 \\ 2 & 2 \end{bmatrix}\begin{bmatrix} 1 \\ 0 \end{bmatrix} = \begin{bmatrix} -2 \\ 2 \end{bmatrix};$$

$$x_1(t) = \left(-2c_1 + c_2 - 2c_2 t\right) e^{3t}, \quad x_2(t) = \left(2c_1 + 2c_2 t\right) e^{3t}.$$

5.   Characteristic equation $\lambda^2 - 10\lambda + 25 = 0$; repeated eigenvalue $\lambda = 5$; generalized eigenvector $\mathbf{w} = \begin{bmatrix} 1 & 0 \end{bmatrix}^T$;

$$\mathbf{v} = \left(\mathbf{A} - \lambda \mathbf{I}\right)\mathbf{w} = \begin{bmatrix} 2 & 1 \\ -4 & -2 \end{bmatrix}\begin{bmatrix} 1 \\ 0 \end{bmatrix} = \begin{bmatrix} 2 \\ -4 \end{bmatrix};$$

$$x_1(t) = \left(2c_1 + c_2 + 2c_2 t\right) e^{5t}, \quad x_2(t) = \left(-4c_1 - 4c_2 t\right) e^{5t}.$$

Problem 5

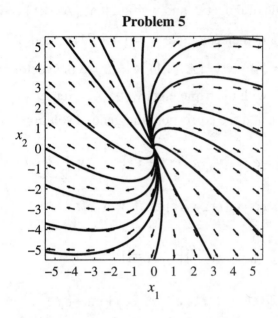

In each of Problems 7–10 the characteristic polynomial is easily calculated by expansion along the row or column of $\mathbf{A}$ that contains two zeros. The matrix $\mathbf{A}$ has only two distinct eigenvalues, so we write $\lambda_1, \lambda_2, \lambda_3$ with either $\lambda_1 = \lambda_2$ or $\lambda_2 = \lambda_3$. Nevertheless, we find that it has 3 linearly independent eigenvectors $\mathbf{v}_1$, $\mathbf{v}_2$, and $\mathbf{v}_3$. We list also the scalar components $x_1(t), x_2(t), x_3(t)$ of the general solution $\mathbf{x}(t) = c_1 \mathbf{v}_1 e^{\lambda_1 t} + c_2 \mathbf{v}_2 e^{\lambda_2 t} + c_3 \mathbf{v}_3 e^{\lambda_3 t}$ of the system.

7.   Characteristic equation $-\lambda^3 + 13\lambda^2 - 40\lambda + 36 = -\left(\lambda - 2\right)^2 \left(\lambda - 9\right)$;

Eigenvalues $\lambda = 2, 2, 9$;

Eigenvectors $\begin{bmatrix} 1 & 1 & 0 \end{bmatrix}^T, \begin{bmatrix} 1 & 0 & 1 \end{bmatrix}^T, \begin{bmatrix} 0 & 1 & 0 \end{bmatrix}^T$;

$$\begin{aligned}
x_1(t) &= c_1 e^{2t} + c_2 e^{2t} \\
x_2(t) &= c_1 e^{2t} \qquad\qquad + c_3 e^{9t} \\
x_3(t) &= \qquad\qquad c_2 e^{2t}
\end{aligned}$$

9.   Characteristic equation $-\lambda^3 + 19\lambda^2 - 115\lambda + 225 = -\left(\lambda - 5\right)^2 \left(\lambda - 9\right)$;

Eigenvalues $\lambda = 5, 5, 9$;

Eigenvectors $\begin{bmatrix} 1 & 2 & 0 \end{bmatrix}^T$, $\begin{bmatrix} 7 & 0 & 2 \end{bmatrix}^T$, $\begin{bmatrix} 3 & 0 & 1 \end{bmatrix}^T$;

$$
\begin{aligned}
x_1(t) &= c_1 e^{5t} + 7c_2 e^{5t} + 3c_3 e^{9t} \\
x_2(t) &= 2c_1 e^{5t} \\
x_3(t) &= \phantom{2c_1 e^{5t} +} 2c_2 e^{5t} + c_3 e^{9t}
\end{aligned}
$$

In each of Problems 11-14, the characteristic equation is $-\lambda^3 - 3\lambda^2 - 3\lambda - 1 = -(\lambda + 1)^3$. Hence $\lambda = -1$ is a triple eigenvalue of defect 2, and we find that $(\mathbf{A} - \lambda\mathbf{I})^3 = \mathbf{0}$. In each problem we start with $\mathbf{v}_3 = \begin{bmatrix} 1 & 0 & 0 \end{bmatrix}^T$ and then calculate $\mathbf{v}_2 = (\mathbf{A} - \lambda\mathbf{I})\mathbf{v}_3$ and $\mathbf{v}_1 = (\mathbf{A} - \lambda\mathbf{I})\mathbf{v}_2 \neq \mathbf{0}$. It follows that $(\mathbf{A} - \lambda\mathbf{I})\mathbf{v}_1 = (\mathbf{A} - \lambda\mathbf{I})^2\mathbf{v}_2 = (\mathbf{A} - \lambda\mathbf{I})^3\mathbf{v}_3 = \mathbf{0}$, so the vector $\mathbf{v}_1$ (if nonzero) is an ordinary eigenvector associated with the triple eigenvalue λ. Hence $\{\mathbf{v}_1, \mathbf{v}_2, \mathbf{v}_3\}$ is a length 3 chain of generalized eigenvectors, and the corresponding general solution is described by

$$
\mathbf{x}(t) = e^{-t}\left[c_1\mathbf{v}_1 + c_2\left(\mathbf{v}_1 t + \mathbf{v}_2\right) + c_3\left(\mathbf{v}_1 \frac{t^2}{2} + \mathbf{v}_2 t + \mathbf{v}_3\right) \right].
$$

We give the scalar components $x_1(t)$, $x_2(t)$, $x_3(t)$ of $\mathbf{x}(t)$.

11. $\mathbf{v}_1 = \begin{bmatrix} 0 & 1 & 0 \end{bmatrix}^T$, $\mathbf{v}_2 = \begin{bmatrix} -2 & -1 & 1 \end{bmatrix}^T$, $\mathbf{v}_3 = \begin{bmatrix} 1 & 0 & 0 \end{bmatrix}^T$;

$$
x_1(t) = e^{-t}\left(-2c_2 + c_3 - 2c_3 t\right), \quad x_2(t) = e^{-t}\left(c_1 - c_2 + c_2 t - c_3 t + c_3 \frac{t^2}{2}\right), \quad x_3(t) = e^{-t}\left(c_2 + c_3 t\right).
$$

13. Here we are stymied initially, because if $\mathbf{v}_3 = \begin{bmatrix} 1 & 0 & 0 \end{bmatrix}^T$, then $(\mathbf{A} - \lambda\mathbf{I})\mathbf{v}_3 = \mathbf{0}$ does not qualify as a (nonzero) generalized eigenvector. We therefore make a fresh start with $\mathbf{v}_3 = \begin{bmatrix} 0 & 1 & 0 \end{bmatrix}^T$, and now we get the desired nonzero generalized eigenvectors upon successive multiplication by $\mathbf{A} - \lambda\mathbf{I}$.

$$
\mathbf{v}_1 = \begin{bmatrix} 1 & 0 & 0 \end{bmatrix}^T, \quad \mathbf{v}_2 = \begin{bmatrix} 0 & 2 & 1 \end{bmatrix}^T, \quad \mathbf{v}_3 = \begin{bmatrix} 0 & 1 & 0 \end{bmatrix}^T;
$$

$$
x_1(t) = e^{-t}\left(c_1 + c_2 t + c_3 \frac{t^2}{2}\right), \quad x_2(t) = e^{-t}\left(2c_2 + c_3 + 2c_3 t\right), \quad x_3(t) = e^{-t}\left(c_2 + c_3 t\right).
$$

In each of Problems 15-18, the characteristic equation is $-\lambda^3 + 3\lambda^2 - 3\lambda + 1 = -(\lambda - 1)^3$. Hence $\lambda = 1$ is a triple eigenvalue of defect 1, and we find that $(\mathbf{A} - \lambda\mathbf{I})^2 = \mathbf{0}$. First we find the two linearly independent (ordinary) eigenvectors $\mathbf{u}_1$ and $\mathbf{u}_2$ associated with λ. Then we start with $\mathbf{v}_2 = \begin{bmatrix} 1 & 0 & 0 \end{bmatrix}^T$ and calculate $\mathbf{v}_1 = (\mathbf{A} - \lambda\mathbf{I})\mathbf{v}_2 \neq \mathbf{0}$. It follows that

$(\mathbf{A} - \lambda\mathbf{I})\mathbf{v}_1 = (\mathbf{A} - \lambda\mathbf{I})^2 \mathbf{v}_2 = \mathbf{0}$, so $\mathbf{v}_1$ is an ordinary eigenvector associated with λ. However, $\mathbf{v}_1$ is a linear combination of $\mathbf{u}_1$ and $\mathbf{u}_2$, so $\mathbf{v}_1 e^t$ is a linear combination of the independent solutions $\mathbf{u}_1 e^t$ and $\mathbf{u}_2 e^t$. But $\{\mathbf{v}_1, \mathbf{v}_2\}$ is a length 2 chain of generalized eigenvectors associated with λ, so $(\mathbf{v}_1 t + \mathbf{v}_2) e^t$ is the desired third independent solution. The corresponding general solution is described by

$$\mathbf{x}(t) = e^t \left[c_1\mathbf{u}_1 + c_2\mathbf{u}_2 + c_3(\mathbf{v}_1 t + \mathbf{v}_2) \right].$$

We give the scalar components $x_1(t)$, $x_2(t)$, $x_3(t)$ of $\mathbf{x}(t)$.

15. $\mathbf{u}_1 = \begin{bmatrix} 3 & -1 & 0 \end{bmatrix}^T$, $\mathbf{u}_2 = \begin{bmatrix} 0 & 0 & 1 \end{bmatrix}^T$;

$\mathbf{v}_1 = \begin{bmatrix} -3 & 1 & 1 \end{bmatrix}^T$, $\mathbf{v}_2 = \begin{bmatrix} 1 & 0 & 0 \end{bmatrix}^T$;

$x_1(t) = e^t(3c_1 + c_3 - 3c_3 t)$, $x_2(t) = e^t(-c_1 + c_3 t)$, $x_3(t) = e^t(c_2 + c_3 t)$.

17. $\mathbf{u}_1 = \begin{bmatrix} 2 & 0 & -9 \end{bmatrix}^T$, $\mathbf{u}_2 = \begin{bmatrix} 1 & -3 & 0 \end{bmatrix}^T$;

$\mathbf{v}_1 = \begin{bmatrix} 0 & 6 & -9 \end{bmatrix}^T$; $\mathbf{v}_2 = \begin{bmatrix} 0 & 1 & 0 \end{bmatrix}^T$;

(Either $\mathbf{v}_2 = \begin{bmatrix} 1 & 0 & 0 \end{bmatrix}^T$ or $\mathbf{v}_2 = \begin{bmatrix} 0 & 0 & 1 \end{bmatrix}^T$ can be used also, but they yield different forms of the solution than given in the book's answer section.)

$x_1(t) = e^t(2c_1 + c_2)$, $x_2(t) = e^t(-3c_2 + c_3 + 6c_3 t)$, $x_3(t) = e^t(-9c_1 - 9c_3 t)$.

19.   Characteristic equation $\lambda^4 - 2\lambda^2 + 1 = 0$;

Double eigenvalue $\lambda = -1$ with eigenvectors $\mathbf{v}_1 = \begin{bmatrix} 1 & 0 & 0 & 1 \end{bmatrix}^T$ and

$\mathbf{v}_2 = \begin{bmatrix} 0 & 0 & 1 & 0 \end{bmatrix}^T$;

Double eigenvalue $\lambda = +1$ with eigenvectors $\mathbf{v}_3 = \begin{bmatrix} 0 & 1 & 0 & -2 \end{bmatrix}^T$ and

$\mathbf{v}_4 = \begin{bmatrix} 1 & 0 & 3 & 0 \end{bmatrix}^T$;

General solution

$$\mathbf{x}(t) = e^{-t}(c_1\mathbf{v}_1 + c_2\mathbf{v}_2) + e^t(c_3\mathbf{v}_3 + c_4\mathbf{v}_4);$$

Scalar components

$x_1(t) = c_1 e^{-t} + c_4 e^t$, $x_2(t) = c_3 e^t$, $x_3(t) = c_2 e^{-t} + 3c_4 e^t$, $x_4(t) = c_1 e^{-t} - 2c_3 e^t$.

21.   Characteristic equation $\lambda^4 - 4\lambda^3 + 6\lambda^2 - 4\lambda + 1 = (\lambda - 1)^4 = 0$;

Eigenvalue $\lambda = 1$ with multiplicity 4 and defect 2.

We find that $(\mathbf{A} - \lambda\mathbf{I})^2 \neq \mathbf{0}$ but $(\mathbf{A} - \lambda\mathbf{I})^3 = \mathbf{0}$. We therefore start with $\mathbf{v}_3 = \begin{bmatrix} 1 & 0 & 0 & 0 \end{bmatrix}^T$ and define $\mathbf{v}_2 = (\mathbf{A} - \lambda\mathbf{I})\mathbf{v}_3$ and $\mathbf{v}_1 = (\mathbf{A} - \lambda\mathbf{I})\mathbf{v}_2 \neq \mathbf{0}$, thereby obtaining the length 3 chain $\{\mathbf{v}_1, \mathbf{v}_2, \mathbf{v}_3\}$ with

$$\mathbf{v}_1 = \begin{bmatrix} 0 & 0 & 0 & 1 \end{bmatrix}^T, \quad \mathbf{v}_2 = \begin{bmatrix} -2 & 1 & 1 & 0 \end{bmatrix}^T, \quad \mathbf{v}_3 = \begin{bmatrix} 1 & 0 & 0 & 0 \end{bmatrix}^T.$$

Then we find the second ordinary eigenvector $\mathbf{v}_4 = \begin{bmatrix} 0 & 0 & 1 & 0 \end{bmatrix}^T$. The corresponding general solution

$$\mathbf{x}(t) = e^t \left[c_1 \mathbf{v}_1 + c_2 (\mathbf{v}_1 t + \mathbf{v}_2) + c_3 \left(\mathbf{v}_1 \frac{t^2}{2} + \mathbf{v}_2 t + \mathbf{v}_3 \right) + c_4 \mathbf{v}_4 \right]$$

has scalar components

$$x_1(t) = e^t (-2c_2 + c_3 - 2c_3 t)$$
$$x_2(t) = e^t (c_2 + c_3 t)$$
$$x_3(t) = e^t (c_2 + c_4 + c_3 t)$$
$$x_4(t) = e^t \left(c_1 + c_2 t + c_3 \frac{t^2}{2} \right)$$

In Problems 23 and 24 there are only two distinct eigenvalues λ_1 and λ_2. However, the eigenvector equation $(\mathbf{A} - \lambda\mathbf{I})\mathbf{v} = \mathbf{0}$ yields the three linearly independent eigenvectors $\mathbf{v}_1$, $\mathbf{v}_2$, and $\mathbf{v}_3$ that are given. We list the scalar components of the corresponding general solution $\mathbf{x}(t) = c_1 \mathbf{v}_1 e^{\lambda_1 t} + c_2 \mathbf{v}_2 e^{\lambda_2 t} + c_3 \mathbf{v}_3 e^{\lambda_2 t}$.

23. $\lambda_1 = -1: \{\mathbf{v}_1\}$ with $\mathbf{v}_1 = \begin{bmatrix} 1 & -1 & 2 \end{bmatrix}^T$;

$\lambda_2 = 3: \{\mathbf{v}_2\}$ with $\mathbf{v}_2 = \begin{bmatrix} 4 & 0 & 9 \end{bmatrix}^T$ and $\{\mathbf{v}_3\}$ with $\mathbf{v}_3 = \begin{bmatrix} 0 & 2 & 1 \end{bmatrix}^T$.

Scalar components:

$$\begin{aligned} x_1(t) &= & c_1 e^{-t} &+& 4c_2 e^{3t} & & \\ x_2(t) &= & -c_1 e^{-t} & & &+& 2c_3 e^{3t} \\ x_3(t) &= & 2c_1 e^{-t} &+& 9c_2 e^{3t} &+& c_3 e^{3t} \end{aligned}$$

In Problems 25, 26, and 28 there is given a single eigenvalue λ of multiplicity 3. We find that $(\mathbf{A} - \lambda\mathbf{I})^2 \neq \mathbf{0}$ but $(\mathbf{A} - \lambda\mathbf{I})^3 = \mathbf{0}$. We therefore start with $\mathbf{v}_3 = \begin{bmatrix} 1 & 0 & 0 \end{bmatrix}^T$ and define $\mathbf{v}_2 = (\mathbf{A} - \lambda\mathbf{I})\mathbf{v}_3$ and $\mathbf{v}_1 = (\mathbf{A} - \lambda\mathbf{I})\mathbf{v}_2 \neq \mathbf{0}$, thereby obtaining the length 3 chain $\{\mathbf{v}_1, \mathbf{v}_2, \mathbf{v}_3\}$ of generalized eigenvectors based on the ordinary eigenvector $\mathbf{v}_1$. We list the scalar components of the corresponding general solution

$$\mathbf{x}(t) = c_1 \mathbf{v}_1 e^{\lambda t} + c_2 \left(\mathbf{v}_1 t + \mathbf{v}_2\right) e^{\lambda t} + c_3 \left(\mathbf{v}_1 \frac{t^2}{2} + \mathbf{v}_2 t + \mathbf{v}_3\right) e^{\lambda t} .$$

25. $\{\mathbf{v}_1, \mathbf{v}_2, \mathbf{v}_3\}$ with

$$\mathbf{v}_1 = \begin{bmatrix} -1 & 0 & -1 \end{bmatrix}^{\mathrm{T}}, \quad \mathbf{v}_2 = \begin{bmatrix} -4 & -1 & 0 \end{bmatrix}^{\mathrm{T}}, \quad \mathbf{v}_3 = \begin{bmatrix} 1 & 0 & 0 \end{bmatrix}^{\mathrm{T}}$$

Scalar components:

$$x_1(t) = e^{2t}\left(-c_1 - 4c_2 + c_3 - c_2 t - 4c_3 t - c_3 \frac{t^2}{2}\right)$$

$$x_2(t) = e^{2t}\left(-c_2 - c_3 t\right)$$

$$x_3(t) = e^{2t}\left(-c_1 - c_2 t - c_3 \frac{t^2}{2}\right)$$

27. We find that the triple eigenvalue $\lambda = 2$ has the two linearly independent eigenvectors $\begin{bmatrix} 1 & 1 & 0 \end{bmatrix}^{\mathrm{T}}$ and $\begin{bmatrix} -1 & 0 & 1 \end{bmatrix}^{\mathrm{T}}$. Next we find that $(\mathbf{A} - \lambda \mathbf{I}) \neq \mathbf{0}$ but $(\mathbf{A} - \lambda \mathbf{I})^2 = \mathbf{0}$. We therefore start with $\mathbf{v}_2 = \begin{bmatrix} 1 & 0 & 0 \end{bmatrix}^{\mathrm{T}}$ and define

$$\mathbf{v}_1 = (\mathbf{A} - \lambda \mathbf{I}) \mathbf{v}_2 = \begin{bmatrix} 5 & -3 & 8 \end{bmatrix}^{\mathrm{T}} \neq \mathbf{0},$$

thereby obtaining the length 2 chain $\{\mathbf{v}_1, \mathbf{v}_2\}$ of generalized eigenvectors based on the ordinary eigenvector $\mathbf{v}_1$. If we take $\mathbf{v}_3 = \begin{bmatrix} 1 & 1 & 0 \end{bmatrix}^{\mathrm{T}}$, then the general solution

$$\mathbf{x}(t) = e^{2t}\left[c_1 \mathbf{v}_1 + c_2 \left(\mathbf{v}_1 t + \mathbf{v}_2\right) + c_3 \mathbf{v}_3 \right]$$

has scalar components

$$x_1(t) = e^{2t}\left(-5c_1 + c_2 + c_3 - 5c_2 t\right)$$

$$x_2(t) = e^{2t}\left(3c_1 + 3c_2 t\right)$$

$$x_3(t) = e^{2t}\left(8c_1 + 8c_2 t\right)$$

In Problems 29 and 30 the matrix $\mathbf{A}$ has two distinct eigenvalues λ_1 and λ_2 each having multiplicity 2 and defect 1. First, we select $\mathbf{v}_2$ so that $\mathbf{v}_1 = (\mathbf{A} - \lambda_1 \mathbf{I}) \mathbf{v}_2 \neq \mathbf{0}$ but $(\mathbf{A} - \lambda_1 \mathbf{I}) \mathbf{v}_1 = \mathbf{0}$, so $\{\mathbf{v}_1, \mathbf{v}_2\}$ is a length 2 chain based on $\mathbf{v}_1$. Next, we select $\mathbf{u}_2$ so that $\mathbf{u}_1 = (\mathbf{A} - \lambda_1 \mathbf{I}) \mathbf{u}_2 \neq \mathbf{0}$ but $(\mathbf{A} - \lambda_1 \mathbf{I}) \mathbf{u}_1 = \mathbf{0}$, so $\{\mathbf{u}_1, \mathbf{u}_2\}$ is a length 2 chain based on $\mathbf{u}_1$. We give the scalar components of the corresponding general solution

$$\mathbf{x}(t) = e^{\lambda_1 t}\left[c_1 \mathbf{v}_1 + c_2 \left(\mathbf{v}_1 t + \mathbf{v}_2\right) \right] + e^{\lambda_2 t}\left[c_3 \mathbf{u}_1 + c_4 \left(\mathbf{u}_1 t + \mathbf{u}_2\right) \right].$$

29. $\lambda = -1$: $\{\mathbf{v}_1, \mathbf{v}_2\}$ with $\mathbf{v}_1 = \begin{bmatrix} 1 & -3 & -1 & -2 \end{bmatrix}^{\mathrm{T}}$ and $\mathbf{v}_2 = \begin{bmatrix} 0 & 1 & 0 & 0 \end{bmatrix}^{\mathrm{T}}$;

$\lambda = 2 : \{\mathbf{u}_1, \mathbf{u}_2\}$ with $\mathbf{u}_1 = \begin{bmatrix} 0 & -1 & 1 & 0 \end{bmatrix}^T$ and $\mathbf{u}_2 = \begin{bmatrix} 0 & 0 & 2 & 1 \end{bmatrix}^T$;

Scalar components

$$x_1(t) = e^{-t}(c_1 + c_2 t)$$
$$x_2(t) = e^{-t}(-3c_1 + c_2 - 3c_2 t) + e^{2t}(-c_3 - c_4 t)$$
$$x_3(t) = e^{-t}(-c_1 - c_2 t) + e^{2t}(c_3 + 2c_4 + c_4 t)$$
$$x_4(t) = e^{-t}(-2c_1 - 2c_2 t) + e^{2t}(c_4)$$

31.   We have the single eigenvalue $\lambda = 1$ of multiplicity 4. Starting with $\mathbf{v}_3 = \begin{bmatrix} 1 & 0 & 0 & 0 \end{bmatrix}^T$, we calculate $\mathbf{v}_2 = (\mathbf{A} - \lambda\mathbf{I})\mathbf{v}_3$ and $\mathbf{v}_1 = (\mathbf{A} - \lambda\mathbf{I})\mathbf{v}_2 \neq \mathbf{0}$, and find that $(\mathbf{A} - \lambda\mathbf{I})\mathbf{v}_1 = \mathbf{0}$. Therefore $\{\mathbf{v}_1, \mathbf{v}_2, \mathbf{v}_3\}$ is a length 3 chain based on the ordinary eigenvector $\mathbf{v}_1$. Next, the eigenvector equation $(\mathbf{A} - \lambda\mathbf{I})\mathbf{v} = \mathbf{0}$ yields the second linearly independent eigenvector $\mathbf{v}_4 = \begin{bmatrix} 0 & 1 & 3 & 0 \end{bmatrix}^T$. With

$$\mathbf{v}_1 = \begin{bmatrix} 42 & 7 & -21 & -42 \end{bmatrix}^T, \quad \mathbf{v}_2 = \begin{bmatrix} 34 & 22 & -10 & -27 \end{bmatrix}^T,$$
$$\mathbf{v}_3 = \begin{bmatrix} 1 & 0 & 0 & 0 \end{bmatrix}^T, \quad\quad\quad \mathbf{v}_4 = \begin{bmatrix} 0 & 1 & 3 & 0 \end{bmatrix}^T$$

the general solution

$$\mathbf{x}(t) = e^t \left[c_1\mathbf{v}_1 + c_2(\mathbf{v}_1 t + \mathbf{v}_2) + c_3\left(\mathbf{v}_1 \frac{t^2}{2} + \mathbf{v}_2 t + \mathbf{v}_3\right) + c_4\mathbf{v}_4 \right]$$

has scalar components

$$x_1(t) = e^t\left(42c_1 + 34c_2 + c_3 + 42c_2 t + 34c_3 t + 21c_3 t^2\right)$$
$$x_2(t) = e^t\left(7c_1 + 22c_2 + c_4 + 7c_2 t + 22c_3 t + 7c_3 \frac{t^2}{2}\right)$$
$$x_3(t) = e^t\left(-21c_1 - 10c_2 + 3c_4 - 21c_2 t - 10c_3 t - 21c_3 \frac{t^2}{2}\right)$$
$$x_4(t) = e^t\left(-42c_1 - 27c_2 - 42c_2 t - 27c_3 t - 21c_3 t^2\right)$$

33.   The chain $\{\mathbf{v}_1, \mathbf{v}_2\}$ was found using the matrices

$$\mathbf{A} - \lambda\mathbf{I} = \begin{bmatrix} 4i & -4 & 1 & 0 \\ 4 & 4i & 0 & 1 \\ 0 & 0 & 4i & -4 \\ 0 & 0 & 4 & 4i \end{bmatrix} \rightarrow \begin{bmatrix} 1 & i & 0 & 0 \\ 0 & 0 & 1 & 0 \\ 0 & 0 & 0 & 1 \\ 0 & 0 & 0 & 0 \end{bmatrix}$$

and

$$(\mathbf{A}-\lambda\mathbf{I})^2 \;=\; \begin{bmatrix} -32 & -32i & 8i & -8 \\ 32i & -32 & 8 & 8i \\ 0 & 0 & -32 & -32i \\ 0 & 0 & 32i & -32 \end{bmatrix} \;\rightarrow\; \begin{bmatrix} 1 & i & 0 & 0 \\ 0 & 0 & 1 & i \\ 0 & 0 & 0 & 0 \\ 0 & 0 & 0 & 0 \end{bmatrix},$$

where $\rightarrow$ signifies reduction to row-echelon form. The resulting real-valued solution vectors are

$$\mathbf{x}_1(t) = e^{3t}\begin{bmatrix} \cos 4t & \sin 4t & 0 & 0 \end{bmatrix}^{\mathrm{T}}$$

$$\mathbf{x}_2(t) = e^{3t}\begin{bmatrix} -\sin 4t & \cos 4t & 0 & 0 \end{bmatrix}^{\mathrm{T}}$$

$$\mathbf{x}_3(t) = e^{3t}\begin{bmatrix} t\cos 4t & t\sin 4t & \cos 4t & \sin 4t \end{bmatrix}^{\mathrm{T}}$$

$$\mathbf{x}_4(t) = e^{3t}\begin{bmatrix} -t\sin 4t & t\cos 4t & -\sin 4t & \cos 4t \end{bmatrix}^{\mathrm{T}}$$

35. The coefficient matrix

$$\mathbf{A} = \begin{bmatrix} 0 & 0 & 1 & 0 \\ 0 & 0 & 0 & 1 \\ -1 & 1 & -2 & 1 \\ 1 & -1 & 1 & -2 \end{bmatrix}$$

has the following eigenvalues and corresponding eigenvectors:

λ	Corresponding eigenvector(s)
0	$\mathbf{v}_1 = \begin{bmatrix} 1 & 1 & 0 & 0 \end{bmatrix}^{\mathrm{T}}$
-1	$\mathbf{v}_2 = \begin{bmatrix} 1 & 0 & -1 & 0 \end{bmatrix}^{\mathrm{T}}$ and $\mathbf{v}_3 = \begin{bmatrix} 0 & 1 & 0 & -1 \end{bmatrix}^{\mathrm{T}}$
-2	$\mathbf{v}_4 = \begin{bmatrix} 1 & -1 & -2 & 2 \end{bmatrix}^{\mathrm{T}}$

When we impose the given initial conditions on the general solution

$$\mathbf{x}(t) = c_1\mathbf{v}_1 + c_2\mathbf{v}_2 e^{-t} + c_3\mathbf{v}_3 e^{-t} + c_4\mathbf{v}_4 e^{-2t}$$

we find that $c_1 = v_0$, $c_2 = c_3 = -v_0$, $c_4 = 0$. Hence the position functions of the two masses are given by

$$x_1(t) = x_2(t) = v_0\left(1 - e^{-t}\right).$$

Each mass travels a distance v_0 before stopping.

SECTION 5.6

MATRIX EXPONENTIALS AND LINEAR SYSTEMS

In Problems 1–8 we first use the eigenvalues and eigenvectors of the coefficient matrix **A** to find first a fundamental matrix $\Phi(t)$ for the homogeneous system $\mathbf{x}' = \mathbf{A}\mathbf{x}$. Then we apply the formula $\mathbf{x}(t) = \Phi(t)\Phi(0)^{-1}\mathbf{x}_0$ to find the solution vector $\mathbf{x}(t)$ that satisfies the initial condition $\mathbf{x}(0) = \mathbf{x}_0$. Formulas (11) and (12) in the text provide inverses of 2×2 and 3×3 matrices.

1. Eigensystem: $\lambda_1 = 1$, $\mathbf{v}_1 = \begin{bmatrix} 1 & -1 \end{bmatrix}^T$; $\lambda_2 = 3$, $\mathbf{v}_1 = \begin{bmatrix} 1 & 1 \end{bmatrix}^T$

$$\Phi(t) = \begin{bmatrix} e^{\lambda_1 t}\mathbf{v}_1 & e^{\lambda_2 t}\mathbf{v}_2 \end{bmatrix} = \begin{bmatrix} e^t & e^{3t} \\ -e^t & e^{3t} \end{bmatrix}$$

$$\mathbf{x}(t) = \begin{bmatrix} e^t & e^{3t} \\ -e^t & e^{3t} \end{bmatrix} \cdot \frac{1}{2}\begin{bmatrix} 1 & -1 \\ 1 & 1 \end{bmatrix}\begin{bmatrix} 3 \\ -2 \end{bmatrix} = \frac{1}{2}\begin{bmatrix} 5e^t + e^{3t} \\ -5e^t + e^{3t} \end{bmatrix}$$

3. Eigensystem: $\lambda = 4i$, $\mathbf{v} = \begin{bmatrix} 1+2i & 2 \end{bmatrix}^T$;

$$\Phi(t) = \begin{bmatrix} \mathrm{Re}(\mathbf{v}e^{\lambda t}) & \mathrm{Im}(\mathbf{v}e^{\lambda t}) \end{bmatrix} = \begin{bmatrix} \cos 4t - 2\sin 4t & 2\cos 4t + \sin 4t \\ 2\cos 4t & 2\sin 4t \end{bmatrix}$$

$$\mathbf{x}(t) = \begin{bmatrix} \cos 4t - 2\sin 4t & 2\cos 4t + \sin 4t \\ 2\cos 4t & 2\sin 4t \end{bmatrix} \cdot \frac{1}{4}\begin{bmatrix} 0 & 2 \\ 2 & -1 \end{bmatrix} \cdot \begin{bmatrix} 0 \\ 1 \end{bmatrix} = \frac{1}{4}\begin{bmatrix} -5\sin 4t \\ 4\cos 4t - 2\sin 4t \end{bmatrix}$$

5. Eigensystem: $\lambda = 3i$, $\mathbf{v} = \begin{bmatrix} -1+i & 3 \end{bmatrix}^T$;

$$\Phi(t) = \begin{bmatrix} \mathrm{Re}(\mathbf{v}e^{\lambda t}) & \mathrm{Im}(\mathbf{v}e^{\lambda t}) \end{bmatrix} = \begin{bmatrix} -\cos 3t - \sin 3t & \cos 3t - \sin 3t \\ 3\cos 3t & 3\sin 3t \end{bmatrix}$$

$$\mathbf{x}(t) = \begin{bmatrix} -\cos 3t - \sin 3t & \cos 3t - \sin 3t \\ 3\cos 3t & 3\sin 3t \end{bmatrix} \cdot \frac{1}{3}\begin{bmatrix} 0 & 1 \\ 3 & 1 \end{bmatrix} \cdot \begin{bmatrix} 1 \\ -1 \end{bmatrix} = \frac{1}{3}\begin{bmatrix} 3\cos 3t - \sin 3t \\ -3\cos 3t + 6\sin 3t \end{bmatrix}$$

7. Eigensystem:

$\lambda_1 = 0$, $\mathbf{v}_1 = \begin{bmatrix} 6 & 2 & 5 \end{bmatrix}^T$; $\lambda_2 = 1$, $\mathbf{v}_2 = \begin{bmatrix} 3 & 1 & 2 \end{bmatrix}^T$; $\lambda_3 = -1$, $\mathbf{v}_3 = \begin{bmatrix} 2 & 1 & 2 \end{bmatrix}^T$;

$$\Phi(t) = \begin{bmatrix} e^{\lambda_1 t}\mathbf{v}_1 & e^{\lambda_2 t}\mathbf{v}_2 & e^{\lambda_3 t}\mathbf{v}_3 \end{bmatrix} = \begin{bmatrix} 6 & 3e^t & 2e^{-t} \\ 2 & e^t & e^{-t} \\ 5 & 2e^t & 2e^{-t} \end{bmatrix}$$

$$\mathbf{x}(t) = \begin{bmatrix} 6 & 3e^t & 2e^{-t} \\ 2 & e^t & e^{-t} \\ 5 & 2e^t & 2e^{-t} \end{bmatrix} \cdot \begin{bmatrix} 0 & -2 & 1 \\ 1 & 2 & -2 \\ -1 & 3 & 0 \end{bmatrix} \cdot \begin{bmatrix} 2 \\ 1 \\ 0 \end{bmatrix} = \begin{bmatrix} -12 + 12e^t + 2e^{-t} \\ -4 + 4e^t + e^{-t} \\ -10 + 8e^t + 2e^{-t} \end{bmatrix}$$

In each of Problems 9-20 we first solve the given linear system to find two linearly independent solutions $\mathbf{x}_1$ and $\mathbf{x}_2$, then set up the fundamental matrix $\mathbf{\Phi}(t) = \begin{bmatrix} \mathbf{x}_1(t) & \mathbf{x}_2(t) \end{bmatrix}$, and finally calculate the matrix exponential $e^{\mathbf{A}t} = \mathbf{\Phi}(t)\mathbf{\Phi}(0)^{-1}$.

9. Eigensystem: $\lambda_1 = 1$, $\mathbf{v}_1 = \begin{bmatrix} 1 & 1 \end{bmatrix}^T$; $\lambda_2 = 3$, $\mathbf{v}_2 = \begin{bmatrix} 2 & 1 \end{bmatrix}^T$;

$$\mathbf{\Phi}(t) = \begin{bmatrix} e^{\lambda_1 t}\mathbf{v}_1 & e^{\lambda_2 t}\mathbf{v}_2 \end{bmatrix} = \begin{bmatrix} e^t & 2e^{3t} \\ e^t & e^{3t} \end{bmatrix}$$

$$e^{\mathbf{A}t} = \begin{bmatrix} e^t & 2e^{3t} \\ e^t & e^{3t} \end{bmatrix} \begin{bmatrix} -1 & 2 \\ 1 & -1 \end{bmatrix} = \begin{bmatrix} -e^t + 2e^{3t} & 2e^t - 2e^{3t} \\ -e^t + e^{3t} & 2e^t - e^{3t} \end{bmatrix}$$

11. Eigensystem: $\lambda_1 = 2$, $\mathbf{v}_1 = \begin{bmatrix} 1 & 1 \end{bmatrix}^T$; $\lambda_2 = 3$, $\mathbf{v}_2 = \begin{bmatrix} 3 & 2 \end{bmatrix}^T$;

$$\mathbf{\Phi}(t) = \begin{bmatrix} e^{\lambda_1 t}\mathbf{v}_1 & e^{\lambda_2 t}\mathbf{v}_2 \end{bmatrix} = \begin{bmatrix} e^{2t} & 3e^{3t} \\ e^{2t} & 2e^{3t} \end{bmatrix}$$

$$e^{\mathbf{A}t} = \begin{bmatrix} e^{2t} & 3e^{3t} \\ e^{2t} & 2e^{3t} \end{bmatrix} \begin{bmatrix} -2 & 3 \\ 1 & -1 \end{bmatrix} = \begin{bmatrix} -2e^{2t} + 3e^{3t} & 3e^{2t} - 3e^{3t} \\ -2e^{2t} + 2e^{3t} & 3e^{2t} - 2e^{3t} \end{bmatrix}$$

13. Eigensystem: $\lambda_1 = 1$, $\mathbf{v}_1 = \begin{bmatrix} 1 & 1 \end{bmatrix}^T$; $\lambda_2 = 3$, $\mathbf{v}_2 = \begin{bmatrix} 4 & 3 \end{bmatrix}^T$

$$\mathbf{\Phi}(t) = \begin{bmatrix} e^{\lambda_1 t}\mathbf{v}_1 & e^{\lambda_2 t}\mathbf{v}_2 \end{bmatrix} = \begin{bmatrix} e^t & 4e^{3t} \\ e^t & 3e^{3t} \end{bmatrix}$$

$$e^{\mathbf{A}t} = \begin{bmatrix} e^t & 4e^{3t} \\ e^t & 3e^{3t} \end{bmatrix} \begin{bmatrix} -3 & 4 \\ 1 & -1 \end{bmatrix} = \begin{bmatrix} -3e^t + 4e^{3t} & 4e^t - 4e^{3t} \\ -3e^t + 3e^{3t} & 4e^t - 3e^{3t} \end{bmatrix}$$

15. Eigensystem: $\lambda_1 = 1$, $\mathbf{v}_1 = \begin{bmatrix} 2 & 1 \end{bmatrix}^T$; $\lambda_2 = 2$, $\mathbf{v}_2 = \begin{bmatrix} 5 & 2 \end{bmatrix}^T$;

$$\mathbf{\Phi}(t) = \begin{bmatrix} e^{\lambda_1 t}\mathbf{v}_1 & e^{\lambda_2 t}\mathbf{v}_2 \end{bmatrix} = \begin{bmatrix} 2e^t & 5e^{2t} \\ e^t & 2e^{2t} \end{bmatrix}$$

$$e^{At} = \begin{bmatrix} 2e^t & 5e^{2t} \\ e^t & 2e^{2t} \end{bmatrix} \begin{bmatrix} -2 & 5 \\ 1 & -2 \end{bmatrix} = \begin{bmatrix} -4e^t + 5e^{2t} & 10e^t - 10e^{2t} \\ -2e^t + 2e^{2t} & 5e^t - 4e^{2t} \end{bmatrix}$$

17. Eigensystem: $\lambda_1 = 2, \quad \mathbf{v}_1 = \begin{bmatrix} 1 & -1 \end{bmatrix}^T; \quad \lambda_2 = 4, \quad \mathbf{v}_2 = \begin{bmatrix} 1 & 1 \end{bmatrix}^T;$

$$\Phi(t) = \begin{bmatrix} e^{\lambda_1 t}\mathbf{v}_1 & e^{\lambda_2 t}\mathbf{v}_2 \end{bmatrix} = \begin{bmatrix} e^{2t} & e^{4t} \\ -e^{2t} & e^{4t} \end{bmatrix}$$

$$e^{At} = \begin{bmatrix} e^{2t} & e^{4t} \\ -e^{2t} & e^{4t} \end{bmatrix} \cdot \frac{1}{2}\begin{bmatrix} 1 & -1 \\ 1 & 1 \end{bmatrix} = \frac{1}{2}\begin{bmatrix} e^{2t} + e^{4t} & -e^{2t} + e^{4t} \\ -e^{2t} + e^{4t} & e^{2t} + e^{4t} \end{bmatrix}$$

19. Eigensystem: $\lambda_1 = 5, \quad \mathbf{v}_1 = \begin{bmatrix} 1 & -2 \end{bmatrix}^T; \quad \lambda_2 = 10, \quad \mathbf{v}_2 = \begin{bmatrix} 2 & 1 \end{bmatrix}^T;$

$$\Phi(t) = \begin{bmatrix} e^{\lambda_1 t}\mathbf{v}_1 & e^{\lambda_2 t}\mathbf{v}_2 \end{bmatrix} = \begin{bmatrix} e^{5t} & 2e^{10t} \\ -2e^{5t} & e^{10t} \end{bmatrix}$$

$$e^{At} = \begin{bmatrix} e^{5t} & 2e^{10t} \\ -2e^{5t} & e^{10t} \end{bmatrix} \cdot \frac{1}{5}\begin{bmatrix} 1 & -2 \\ 2 & 1 \end{bmatrix} = \frac{1}{5}\begin{bmatrix} e^{5t} + 4e^{10t} & -2e^{5t} + 2e^{10t} \\ -2e^{5t} + 2e^{10t} & 4e^{5t} + e^{10t} \end{bmatrix}$$

21. $\mathbf{A}^2 = \mathbf{0}$, so $e^{At} = \mathbf{I} + \mathbf{A}t = \begin{bmatrix} 1+t & -t \\ t & 1-t \end{bmatrix}$.

23. $\mathbf{A}^3 = \mathbf{0}$, so $e^{At} = \mathbf{I} + \mathbf{A}t + \frac{1}{2}\mathbf{A}^2 t^2 = \begin{bmatrix} 1+t & -t & -t-t^2 \\ t & 1-t & t-t^2 \\ 0 & 0 & 1 \end{bmatrix}$.

25. $\mathbf{A} = 2\mathbf{I} + \mathbf{B}$, where $\mathbf{B}^2 = \mathbf{0}$, so $e^{At} = e^{2It}e^{Bt} = (e^{2t}\mathbf{I})(\mathbf{I} + \mathbf{B}t)$. Hence

$$e^{At} = \begin{bmatrix} e^{2t} & 5te^{2t} \\ 0 & e^{2t} \end{bmatrix}, \quad \mathbf{x}(t) = e^{At}\begin{bmatrix} 4 \\ 7 \end{bmatrix} = e^{2t}\begin{bmatrix} 4+35t \\ 7 \end{bmatrix}.$$

27. $\mathbf{A} = \mathbf{I} + \mathbf{B}$, where $\mathbf{B}^3 = \mathbf{0}$, so $e^{At} = e^{It}e^{Bt} = (e^t\mathbf{I})\left(\mathbf{I} + t + \frac{1}{2}\mathbf{B}^2 t^2\right)$. Hence

$$e^{At} = \begin{bmatrix} e^t & 2te^t & (3t+2t^2)e^t \\ 0 & e^t & 2te^t \\ 0 & 0 & e^t \end{bmatrix}, \quad \mathbf{x}(t) = e^{At}\begin{bmatrix} 4 \\ 5 \\ 6 \end{bmatrix} = e^t\begin{bmatrix} 4+28t+12t^2 \\ 5+12t \\ 6 \end{bmatrix}.$$

29. $\mathbf{A} = \mathbf{I} + \mathbf{B}$, where $\mathbf{B}^4 = \mathbf{0}$, so $e^{\mathbf{A}t} = e^{\mathbf{I}t}e^{\mathbf{B}t} = \left(e^t\mathbf{I}\right)\left(\mathbf{I} + \mathbf{B}t + \frac{1}{2}\mathbf{B}^2t^2 + \frac{1}{6}\mathbf{B}^3t^3\right)$. Hence

$$e^{\mathbf{A}t} = e^t\begin{bmatrix} 1 & 2t & 3t+6t^2 & 4t+6t^2+4t^3 \\ 0 & 1 & 6t & 3t+6t^2 \\ 0 & 0 & 1 & 2t \\ 0 & 0 & 0 & 1 \end{bmatrix}, \quad \mathbf{x}(t) = e^{\mathbf{A}t}\begin{bmatrix} 1 \\ 1 \\ 1 \\ 1 \end{bmatrix} = e^t\begin{bmatrix} 1+9t+12t^2+4t^3 \\ 1+9t+6t^2 \\ 1+2t \\ 1 \end{bmatrix}.$$

33. $e^{\mathbf{A}t} = \mathbf{I}\cosh t + \mathbf{A}\sinh t = \begin{bmatrix} \cosh t & \sinh t \\ \sinh t & \cosh t \end{bmatrix}$, so the general solution of $\mathbf{x}' = \mathbf{A}\mathbf{x}$ is

$$\mathbf{x}(t) = e^{\mathbf{A}t}\mathbf{c} = \begin{bmatrix} c_1\cosh t + c_2\sinh t \\ c_1\sinh t + c_2\cosh t \end{bmatrix}.$$

In Problems 35–40 we give first the linearly independent generalized eigenvectors $\mathbf{u}_1, \mathbf{u}_2, \ldots, \mathbf{u}_n$ of the matrix $\mathbf{A}$ and the corresponding solution vectors $\mathbf{x}_1(t), \mathbf{x}_2(t), \ldots, \mathbf{x}_n(t)$ defined by Eq. (34) in the text, then the fundamental matrix $\Phi(t) = \begin{bmatrix} \mathbf{x}_1(t) & \mathbf{x}_2(t) & \cdots & \mathbf{x}_n(t) \end{bmatrix}$. Finally we calculate the exponential matrix $e^{\mathbf{A}t} = \Phi(t)\Phi(0)^{-1}$.

35. $\lambda = 3: \mathbf{u}_1 = \begin{bmatrix} 4 & 0 \end{bmatrix}^T, \mathbf{u}_2 = \begin{bmatrix} 0 & 1 \end{bmatrix}^T$;

$\{\mathbf{u}_1, \mathbf{u}_2\}$ is a length 2 chain based on the ordinary (rank 1) eigenvector $\mathbf{u}_1$, so $\mathbf{u}_2$ is a generalized eigenvector of rank 2.

$$\mathbf{x}_1(t) = e^{\lambda t}\mathbf{u}_1, \quad \mathbf{x}_2(t) = e^{\lambda t}\left[\mathbf{u}_2 + (\mathbf{A} - \lambda\mathbf{I})\mathbf{u}_2 t\right]$$

$$\Phi(t) = \begin{bmatrix} \mathbf{x}_1(t) & \mathbf{x}_2(t) \end{bmatrix} = e^{3t}\begin{bmatrix} 4 & 4t \\ 0 & 1 \end{bmatrix}$$

$$e^{\mathbf{A}t} = e^{3t}\begin{bmatrix} 4 & 4t \\ 0 & 1 \end{bmatrix}\cdot\frac{1}{4}\begin{bmatrix} 1 & 0 \\ 0 & 4 \end{bmatrix} = e^{3t}\begin{bmatrix} 1 & 4t \\ 0 & 1 \end{bmatrix}$$

37. $\lambda_1 = 2: \mathbf{u}_1 = \begin{bmatrix} 1 & 0 & 0 \end{bmatrix}^T, \mathbf{x}_1(t) = e^{\lambda_1 t}\mathbf{u}_1$;

$\lambda_2 = 1: \mathbf{u}_2 = \begin{bmatrix} 9 & -3 & 0 \end{bmatrix}^T, \mathbf{u}_3 = \begin{bmatrix} 10 & 1 & -1 \end{bmatrix}^T$;

$\{\mathbf{u}_2, \mathbf{u}_3\}$ is a length 2 chain based on the ordinary (rank 1) eigenvector $\mathbf{u}_2$, so $\mathbf{u}_3$ is a generalized eigenvector of rank 2.

$$\mathbf{x}_2(t) = e^{\lambda_2 t}\mathbf{u}_2, \quad \mathbf{x}_3(t) = e^{\lambda_2 t}\left[\mathbf{u}_3 + (\mathbf{A} - \lambda_2\mathbf{I})\mathbf{u}_3 t\right]$$

$$\Phi(t) = \begin{bmatrix} \mathbf{x}_1(t) & \mathbf{x}_2(t) & \mathbf{x}_3(t) \end{bmatrix} = \begin{bmatrix} e^{2t} & 9e^t & (10+9t)e^t \\ 0 & -3e^t & (1-3t)e^t \\ 0 & 0 & -e^t \end{bmatrix}$$

$$e^{\mathbf{A}t} = \begin{bmatrix} e^{2t} & 9e^t & (10+9t)e^t \\ 0 & -3e^t & (1-3t)e^t \\ 0 & 0 & -e^t \end{bmatrix} \cdot \frac{1}{3}\begin{bmatrix} 3 & 9 & 39 \\ 0 & -1 & -1 \\ 0 & 0 & -3 \end{bmatrix}$$

$$= \begin{bmatrix} e^{2t} & -3e^t + 3e^{2t} & (-13-9t)e^t + 13e^{2t} \\ 0 & e^t & 3te^t \\ 0 & 0 & e^t \end{bmatrix}$$

$$= \begin{bmatrix} e^{2t} & -3e^t + 3e^{2t} & (-13-9t)e^t + 13e^{2t} \\ 0 & e^t & 3te^t \\ 0 & 0 & e^t \end{bmatrix}$$

39. $\quad \lambda_2 = 1: \mathbf{u}_1 = \begin{bmatrix} 3 & 0 & 0 & 0 \end{bmatrix}^T, \mathbf{u}_2 = \begin{bmatrix} 0 & 1 & 0 & 0 \end{bmatrix}^T;$

$\{\mathbf{u}_1, \mathbf{u}_2\}$ is a length 2 chain based on the ordinary (rank 1) eigenvector $\mathbf{u}_1$, so $\mathbf{u}_2$ is a generalized eigenvector of rank 2.

$$\mathbf{x}_1(t) = e^{\lambda_1 t}\mathbf{u}_1, \quad \mathbf{x}_2(t) = e^{\lambda_1 t}\left[\mathbf{u}_2 + (\mathbf{A} - \lambda_1\mathbf{I})\mathbf{u}_2 t\right]$$

$\lambda_2 = 2: \mathbf{u}_3 = \begin{bmatrix} 144 & 36 & 12 & 0 \end{bmatrix}^T, \mathbf{u}_4 = \begin{bmatrix} 0 & 27 & 17 & 4 \end{bmatrix}^T;$

$\{\mathbf{u}_3, \mathbf{u}_4\}$ is a length 2 chain based on the ordinary (rank 1) eigenvector $\mathbf{u}_3$, so $\mathbf{u}_4$ is a generalized eigenvector of rank 2.

$$\mathbf{x}_3(t) = e^{\lambda_2 t}\mathbf{u}_3, \quad \mathbf{x}_4(t) = e^{\lambda_2 t}\left[\mathbf{u}_4 + (\mathbf{A} - \lambda_2\mathbf{I})\mathbf{u}_4 t\right]$$

$$\Phi(t) = \begin{bmatrix} \mathbf{x}_1(t) & \mathbf{x}_2(t) & \mathbf{x}_3(t) & \mathbf{x}_4(t) \end{bmatrix} = \begin{bmatrix} 3e^t & 3te^t & 144e^{2t} & 144te^{2t} \\ 0 & e^t & 36e^{2t} & (27+36t)e^{2t} \\ 0 & 0 & 12e^{2t} & (17+12t)e^{2t} \\ 0 & 0 & 0 & 4e^{2t} \end{bmatrix}$$

$$e^{\mathbf{A}t} = \begin{bmatrix} 3e^t & 3te^t & 144e^{2t} & 144te^{2t} \\ 0 & e^t & 36e^{2t} & (27+36t)e^{2t} \\ 0 & 0 & 12e^{2t} & (17+12t)e^{2t} \\ 0 & 0 & 0 & 4e^{2t} \end{bmatrix} \cdot \frac{1}{48}\begin{bmatrix} 16 & 0 & -192 & 816 \\ 0 & 48 & -144 & 288 \\ 0 & 0 & 4 & -17 \\ 0 & 0 & 0 & 12 \end{bmatrix}$$

$$= \begin{bmatrix} e^t & 3te^t & (-12-9t)e^t + 12te^{2t} & (51+18t)e^t + (-51+36t)e^{2t} \\ 0 & e^t & -3e^t + 3e^{2t} & 6e^t + (-6+9t)e^{2t} \\ 0 & 0 & e^{2t} & 3te^{2t} \\ 0 & 0 & 0 & e^{2t} \end{bmatrix}$$

SECTION 5.7

NONHOMOGENEOUS LINEAR SYSTEMS

1. Substitution of the trial solution $x_p(t) = a$, $y_p(t) = b$ yields the equations

$$a + 2b + 3 = 0, \quad 2a + b - 2 = 0$$

with solution $a = \dfrac{7}{3}$, $b = -\dfrac{8}{3}$. Thus we obtain the particular solution

$$x(t) = \frac{7}{3}, \quad y(t) = -\frac{8}{3}.$$

3. When we substitute the trial solution

$$x_p = a_1 + b_1 t + c_1 t^2, \quad y_p = a_2 + b_2 t + c_2 t^2$$

and collect coefficients, we get the equations

$$3a_1 + 4a_2 = b_1 \quad 3b_1 + 4b_2 = 2c_1 \quad 3c_1 + 4c_2 = 0$$
$$3a_1 + 2a_2 = b_2 \quad 3b_1 + 2b_2 = 2c_2 \quad 3c_1 + 2c_2 + 1 = 0$$

Working backwards, we solve first for $c_1 = -\dfrac{2}{3}$ and $c_2 = \dfrac{1}{2}$, then for $b_1 = \dfrac{10}{9}$ and

$b_2 = -\dfrac{7}{6}$, and finally for $a_1 = -\dfrac{31}{27}$ and $a_2 = \dfrac{41}{36}$. This determines the particular solution $x_p(t)$, $y_p(t)$. Next, the coefficient matrix of the associated homogeneous system has eigenvalues $\lambda_1 = -1$ and $\lambda_2 = 6$, with eigenvectors $\mathbf{v}_1 = \begin{bmatrix} 1 & -1 \end{bmatrix}^T$ and $\mathbf{v}_2 = \begin{bmatrix} 4 & 3 \end{bmatrix}^T$, respectively, so the complementary solution is given by

$$x_c(t) = c_1 e^{-t} + 4c_2 e^{6t}, \quad y_c(t) = -c_1 e^{-t} + 3c_2 e^{6t}.$$

When we impose the initial conditions $x(0) = 0$, $y(0) = 0$ on the general solution

$x(t) = x_c(t) + x_p(t)$, $y(t) = y_c(t) + y_p(t)$ we find that $c_1 = \dfrac{8}{7}$ and $c_2 = \dfrac{1}{756}$. This finally gives the desired particular solution

$$x(t) = \frac{1}{756}\left(864e^{-t} + 4e^{6t} - 868 + 840t - 504t^2\right),$$

$$y(t) = \frac{1}{756}\left(-864e^{-t} + 3e^{6t} + 861 - 882t + 378t^2\right).$$

5. The coefficient matrix of the associated homogeneous system has eigenvalues $\lambda_1 = -1$ and $\lambda_2 = 5$, so the nonhomogeneous term e^{-t} duplicates part of the complementary solution. We therefore try the particular solution

$$x_p(t) = a_1 + b_1 e^{-t} + c_1 t e^{-t}, \quad y_p(t) = a_2 + b_2 e^{-t} + c_2 t e^{-t}.$$

Upon solving the six linear equations we get by collecting coefficients after substitution of this trial solution into the given nonhomogeneous system, we obtain the particular solution

$$x(t) = \frac{1}{3}\left(-12 - e^{-t} - 7te^{-t}\right), \quad y(t) = \frac{1}{3}\left(-6 - 7te^{-t}\right).$$

7. First we try the particular solution

$$x_p(t) = a_1 \sin t + b_1 \cos t, \quad y_p(t) = a_2 \sin t + b_2 \cos t.$$

Upon solving the four linear equations we get by collecting coefficients after substitution of this trial solution into the given nonhomogeneous system, we find that $a_1 = -\frac{21}{82}$, $b_1 = -\frac{25}{82}$, $a_2 = -\frac{15}{41}$, and $b_2 = -\frac{12}{41}$. The coefficient matrix of the associated homogeneous system has eigenvalues $\lambda_1 = 1$ and $\lambda_2 = -9$, with eigenvectors $\mathbf{v}_1 = \begin{bmatrix} 1 & 1 \end{bmatrix}^T$ and $\mathbf{v}_2 = \begin{bmatrix} 2 & -3 \end{bmatrix}^T$, respectively, so the complementary solution is given by

$$x_c(t) = c_1 e^t + 2c_2 e^{-9t}, \quad y_c(t) = c_1 e^t - 3c_2 e^{-9t}.$$

When we impose the initial conditions $x(0) = 1$, $y(0) = 0$ we find that $c_1 = \frac{9}{10}$ and $c_2 = \frac{83}{410}$. It follows that the desired particular solution $x = x_c + x_p$, $y = y_c + y_p$ is given by

$$x(t) = \frac{1}{410}\left(369e^t + 166e^{-9t} - 125\cos t - 105\sin t\right),$$

$$y(t) = \frac{1}{410}\left(369e^t - 249e^{-9t} - 120\cos t - 150\sin t\right).$$

9. Here the associated homogeneous system is the same as in Problem 8, so the nonhomogeneous term $\cos 2t$ duplicates the complementary function. We therefore substitute the trial solution

$$x_p(t) = a_1 \sin 2t + b_1 \cos 2t + c_1 t \sin 2t + d_1 t \cos 2t$$

$$y_p(t) = a_2 \sin 2t + b_2 \cos 2t + c_2 t \sin 2t + d_2 t \cos 2t$$

and use a computer algebra system to solve the system of 8 linear equations that results when we collect coefficients in the usual way. This gives the particular solution

$$x(t) = \frac{1}{4}(\sin 2t + 2t \cos 2t + t \sin 2t), \quad y(t) = \frac{1}{4} t \sin 2t.$$

11.   The coefficient matrix of the associated homogeneous system has eigenvalues $\lambda_1 = 0$ and $\lambda_2 = 4$, so there is duplication of constant terms.   We therefore substitute the particular solution

$$x_p(t) = a_1 + b_1 t, \quad y_p(t) = a_2 + b_2 t$$

and solve the resulting equations for $a_1 = -2$, $a_2 = 0$, $b_1 = -2$, and $b_2 = 1$. The eigenvectors of the coefficient matrix associated with the eigenvalues $\lambda_1 = 0$ and $\lambda_2 = 4$ are $\mathbf{v}_1 = \begin{bmatrix} 2 & -1 \end{bmatrix}^T$ and $\mathbf{v}_2 = \begin{bmatrix} 2 & 1 \end{bmatrix}^T$, respectively, so the general solution of the given nonhomogeneous system is given by

$$x(t) = 2c_1 + 2c_2 e^{4t} - 2 - 2t, \quad y(t) = -c_1 + c_2 e^{4t} + t.$$

When we impose the initial conditions $x(0) = 1$, $y(0) = -1$ we find readily that $c_1 = \frac{5}{4}$, $c_2 = \frac{1}{4}$. This gives the desired particular solution

$$x(t) = \frac{1}{2}(1 - 4t + e^{4t}), \quad y(t) = \frac{1}{4}(-5 + 4t + e^{4t}).$$

13.   The coefficient matrix of the associated homogeneous system has eigenvalues $\lambda_1 = 1$ and $\lambda_2 = 3$, so there is duplication of e^t terms. We therefore substitute the trial solution

$$x_p(t) = (a_1 + b_1 t)e^t, \quad y_p(t) = (a_2 + b_2 t)e^t$$

This leads readily to the particular solution

$$x(t) = \frac{1}{2}(1 + 5t)e^t, \quad y(t) = -\frac{5}{2} te^t.$$

In Problems 15 and 16 the amounts $x_1(t)$ and $x_2(t)$ in the two tanks satisfy the equations

$$x_1' = rc_0 - k_1 x_1, \quad x_2' = k_1 x_1 - k_2 x_2,$$

where $k_i = \dfrac{r}{V_i}$, in terms of the flow rate r, the inflowing concentration c_0, and the volumes V_1 and V_2 of the two tanks.

15.   **(a)** We solve the initial value problem

$$x_1' = 20 - \frac{x_1}{10}, \quad x_1(0) = 0$$

$$x_2' = \frac{x_1}{10} - \frac{x_2}{20}, \quad x_2(0) = 0$$

for $x_1(t) = 200\left(1 - e^{-t/10}\right)$, $x_2(t) = 400\left(1 + e^{-t/10} - 2e^{-t/20}\right)$.

(b) Evidently $x_1(t) \to 200\,\text{gal}$ and $x_2(t) \to 400\,\text{gal}$ as $t \to \infty$.

(c) It takes about 6 min 56 sec for tank 1 to reach a salt concentration of 1 lb/gal, and about 24 min 34 sec for tank 2 to reach this concentration.

In Problems 17–34 we apply the variation of parameters formula in Eq. (28) of Section 5.6. The answers shown below were actually calculated using the *Mathematica* code listed in the computing project for Section 5.7. For instance, for Problem 17 we first enter the coefficient matrix

```
A = {{6, -7}, {1, -2}};
```

the initial vector

```
x0 = {{0}, {0}};
```

and the vector

```
f[t_] := {{60}, {90}};
```

of nonhomogeneous terms. It simplifies the notation to rename *Mathematica*'s exponential matrix function by defining

```
exp[A_] := MatrixExp[A]
```

Then the integral in the variation of parameters formula is given by

```
integral = Integrate[exp[-A*s] . f[s], {s, 0, t}] // Simplify
```

and yields the output

$$\begin{bmatrix} -102 + 7e^{-5t} + 95e^t \\ -96 + e^{-5t} + 95e^t \end{bmatrix}.$$

Finally the desired particular solution is given by

```
solution = exp[A*t] . (x0 + integral) // Simplify
```

which yields

$$\begin{bmatrix} 102 - 7e^{-5t} - 95e^t \\ 96 - e^{-5t} - 95e^t \end{bmatrix}.$$

(Maple and MATLAB versions of this computation are provided in the applications manual that accompanies the textbook.)

In each succeeding problem, we need only substitute the given coefficient matrix **A**, initial vector **x0**, and the vector **f** of nonhomogeneous terms in the above commands, and then re-execute them in turn. We give below only the component functions of the final results.

17. $x_1(t) = 102 - 95e^{-t} - 7e^{5t}$, $x_2(t) = 96 - 95e^{-t} - e^{5t}$

19. $x_1(t) = -70 - 60t + 16e^{-3t} + 54e^{2t}$, $x_2(t) = 5 - 60t - 32e^{-3t} + 27e^{2t}$

21. $x_1(t) = -e^{-t} - 14e^{2t} + 15e^{3t}$, $x_2(t) = -5e^{-t} - 10e^{2t} + 15e^{3t}$

23. $x_1(t) = 3 + 11t + 8t^2$, $x_2(t) = 5 + 17t + 24t^2$

25. $x_1(t) = -1 + 8t + \cos t - 8\sin t$, $x_2(t) = -2 + 4t + 2\cos t - 3\sin t$

27. $x_1(t) = 8t^3 + 6t^4$, $x_2(t) = 3t^2 - 2t^3 + 3t^4$

29. $x_1(t) = t\cos t - \ln(\cos t)\sin t$, $x_2(t) = t\sin t - \ln(\cos t)\cos t$

31. $x_1(t) = (9t^2 + 4t^3)e^t$, $x_2(t) = 6t^2 e^t$, $x_3(t) = 6te^t$

33. $x_1(t) = 15t^2 + 60t^3 + 95t^4 + 12t^5$, $x_2(t) = 15t^2 + 55t^3 + 15t^4$, $x_3(t) = 15t^2 + 20t^3$, $x_4(t) = 15t^2$

CHAPTER 6

NONLINEAR SYSTEMS AND PHENOMENA

SECTION 6.1

STABILITY AND THE PHASE PLANE

1.   The only solution of the homogeneous system $2x - y = 0$, $x - 3y = 0$ is the origin $(0,0)$. The only figure among Figs. 6.1.11 through 6.1.18 showing a single critical point at the origin is Fig. 6.1.13. Thus the only critical point of the given autonomous system is the saddle point $(0,0)$ shown in Figure 6.1.13 in the text.

3.   The only solution of the system $x - 2y + 3 = 0$, $x - y + 2 = 0$ is the point $(-1,1)$. The only figure among Figs. 6.1.11 through 6.1.18 showing a single critical point at $(-1,1)$ is Fig. 6.1.18. Thus the only critical point of the given autonomous system is the stable center $(-1,1)$ shown in Figure 6.1.18 in the text.

5.   The first equation $1 - y^2 = 0$ gives $y = 1$ or $y = -1$ at a critical point. Then the second equation $x + 2y = 0$ gives $x = -2$ or $x = 2$, respectively. The only figure among Figs. 6.1.11 through 6.1.18 showing two critical points at $(-2,1)$ and $(2,-1)$ is Fig. 6.1.11. Thus the critical points of the given autonomous system are the spiral point $(-2,1)$ and the saddle point $(2,1)$ shown in Figure 6.1.11 in the text.

7.   The first equation $4x - x^3 = 0$ gives $x = -2$, $x = 0$, or $x = 2$ at a critical point. Then the second equation $x - 2y = 0$ gives $y = -1$, $y = 0$, or $y = 1$, respectively. The only figure among Figs. 6.1.11 through 6.1.18 showing three critical points at $(-2,-1)$, $(0,0)$, and $(2,1)$ is Fig. 6.1.14. Thus the critical points of the given autonomous system are the spiral point $(0,0)$ and the saddle points $(-2,-1)$ and $(2,1)$ shown in Figure 6.1.14 in the text.

In each of Problems 9-12 we need only set $x' = x'' = 0$ and solve the resulting equation for x.

9.   The equation $4x - x^3 = x(4 - x^2) = 0$ has the three solutions $x = 0, \pm 2$. This gives the three equilibrium solutions $x(t) \equiv 0$, $x(t) \equiv 2$, and $x(t) \equiv -2$ of the given 2nd-order differential equation. A phase plane portrait for the equivalent 1st-order system $x' = y$,

$y' = -4x + x^3$ is shown in the figure. We observe that the critical point $(0,0)$ in the phase plane appears to be a center, whereas the points $(\pm 2, 0)$ appear to be saddle points.

Problem 9

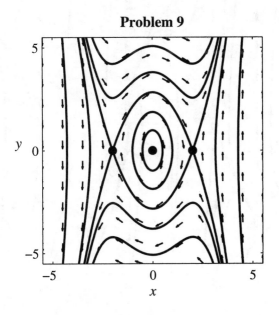

11. The equation $4\sin x = 0$ is satisfied by $x = n\pi$ for any integer n. Thus the given 2nd-order equation has infinitely many equilibrium solutions: $x(t) = n\pi$ for any integer n. A phase portrait for the equivalent 1st-order system $x' = y$, $y' = -3y - 4\sin x$ is shown. We observe that the critical point $(n\pi, 0)$ in the phase plane looks like a spiral sink if n is even, but a saddle point if n is odd.

Problem 11

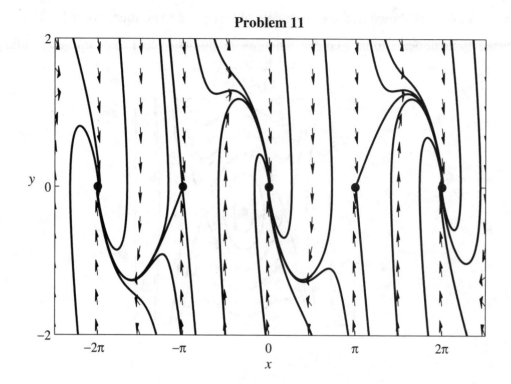

In Problems 13–16, the given x- and y-equations are independent exponential differential equations that we can solve immediately by inspection.

13. Solution: $x(t) = x_0 e^{-2t}$, $y(t) = y_0 e^{-2t}$.

Then $y = \left(\dfrac{y_0}{x_0}\right) x = kx$, so the trajectories are straight lines through the origin. Clearly

$x(t), y(t) \to 0$ as $t \to +\infty$, so the origin is a stable proper node like the one shown.

Problem 13

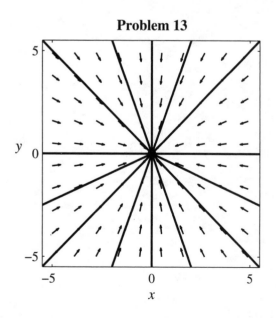

Problem 15

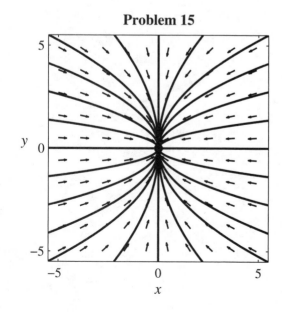

15.   Solution: $x(t) = x_0 e^{-2t}$, $y(t) = y_0 e^{-t}$.

Then $x = \left(\dfrac{x_0}{y_0^2}\right)(y_0 e^{-t})^2 = ky^2$, so the trajectories are parabolas of the form $x = ky^2$, and

clearly $x(t), y(t) \to 0$ as $t \to +\infty$. Thus the origin is a stable improper node like the one shown.

17.   Differentiation of the first equation and substitution using the second one gives $x'' = y' = -x$, so $x'' + x = 0$. Solving this differential equation for $x(t)$ and using the fact that $y = x'$ lead to the general solution

$$x(t) = A\cos t + B\sin t, \quad y(t) = B\cos t - A\sin t.$$

Then

$$\begin{aligned}
x^2 + y^2 &= (A\cos t + B\sin t)^2 + (B\cos t - A\sin t)^2 \\
&= (A^2 + B^2)\cos^2 t + (A^2 + B^2)\sin^2 t \\
&= A^2 + B^2.
\end{aligned}$$

Therefore the trajectories are clockwise-oriented circles centered at the origin, and the origin is a stable center as shown.

Problem 17

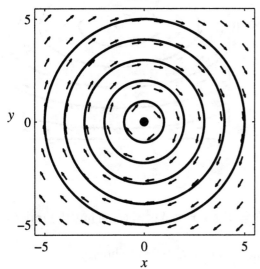

Problem 19

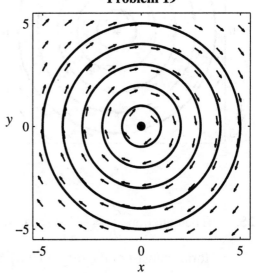

19.   Elimination of y as in Problem 17 gives $x'' + 4x = 0$, and then using $y = \dfrac{1}{2}x'$ we get the general solution

$$x(t) = A\cos 2t + B\sin 2t, \quad y(t) = B\cos 2t - A\sin 2t.$$

Then $x^2 + y^2 = A^2 + B^2$, so the origin is a stable center, and the trajectories are clockwise-oriented circles centered at $(0,0)$, as shown.

21.   We want to solve the system

$$-ky + x\left(1 - x^2 - y^2\right) = 0,$$
$$kx + y\left(1 - x^2 - y^2\right) = 0.$$

If we multiply the first equation by $-y$ and the second one by x, then add the two results, we get $k\left(x^2 + y^2\right) = 0$. It therefore follows that $x = y = 0$.

23.   The equation $\dfrac{dy}{dx} = -\dfrac{x}{y}$ separates to $x\,dx + y\,dy = 0$, so $x^2 + y^2 = C$. Thus the trajectories consist of the origin $(0,0)$ and the circles $x^2 + y^2 = C > 0$, as shown.

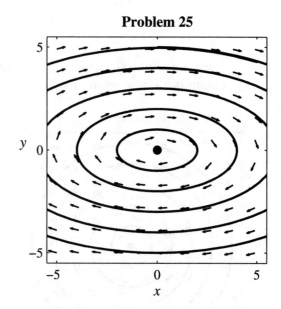

Problem 23

Problem 25

25.   The equation $\dfrac{dy}{dx} = -\dfrac{x}{4y}$ separates to $x\,dx + 4y\,dy = 0$, so $x^2 + 4y^2 = C$. Thus the trajectories consist of the origin $(0,0)$ and the ellipses $x^2 + 4y^2 = C > 0$, as shown.

27.   If $\phi(t) = x(t+\gamma)$ and $\psi(t) = y(t+\gamma)$, then

$$\phi'(t) = x'(t+\gamma) = y(t+\gamma) = \psi(t),$$

but

$$\psi'(t) = y'(t+\gamma) = (t+\gamma)x(t+\gamma) = t\phi(t) + \gamma\phi(t) \neq t\phi(t).$$

SECTION 6.2

LINEAR AND ALMOST LINEAR SYSTEMS

In Problems 1–10 we first find the roots λ_1 and λ_2 of the characteristic equation of the coefficient matrix of the given linear system. We can then read the type and stability of the critical point $(0,0)$ from the table of Figure 6.2.4 in the text.

1.   The roots $\lambda_1 = -1$ and $\lambda_2 = -3$ of the characteristic equation $\lambda^2 + 4\lambda + 3 = 0$ are both negative, so $(0,0)$ is an asymptotically stable node.

Problem 1	Problem 3
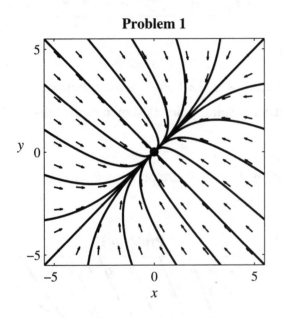	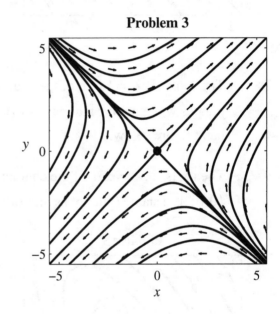

3.   The roots $\lambda_1 = -1$ and $\lambda_2 = 3$ of the characteristic equation $\lambda^2 - 2\lambda - 3 = 0$ have different signs, so $(0,0)$ is an unstable saddle point.

5.   The roots $\lambda_1 = \lambda_2 = -1$ of the characteristic equation $\lambda^2 + 2\lambda + 1 = 0$ are negative and equal, so $(0,0)$ is an asymptotically stable node.

Problem 5

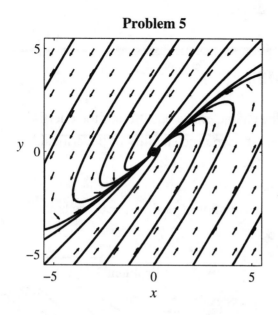

Problem 7

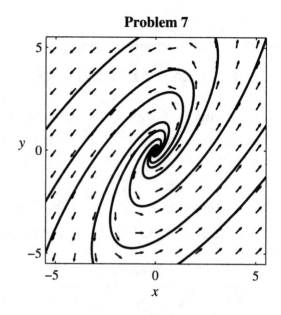

7.   The roots $\lambda_1, \lambda_2 = 1 \pm 2i$ of the characteristic equation $\lambda^2 - 2\lambda + 5 = 0$ are complex conjugates with positive real part, so $(0,0)$ is an unstable spiral point as shown in the left-hand figure below.

9.   The roots $\lambda_1, \lambda_2 = \pm 2i$ of the characteristic equation $\lambda^2 + 4 = 0$ are pure imaginary, so $(0,0)$ is a stable (but not asymptotically stable) center.

Problem 9

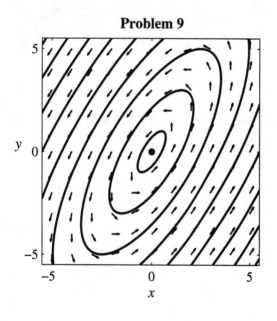

Problem 11

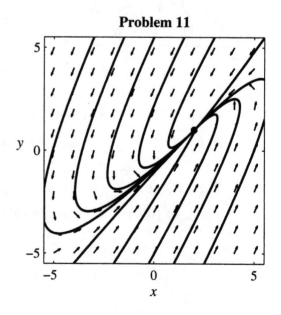

11. The Jacobian matrix $\mathbf{J} = \begin{bmatrix} 1 & -2 \\ 3 & -4 \end{bmatrix}$ has characteristic equation $\lambda^2 + 3\lambda + 2 = 0$ and eigenvalues $\lambda_1 = -1$, $\lambda_2 = -2$ that are both negative. Hence the critical point $(2,1)$ is an asymptotically stable node.

13. The Jacobian matrix $\mathbf{J} = \begin{bmatrix} 2 & -1 \\ 3 & -2 \end{bmatrix}$ has characteristic equation $\lambda^2 - 1 = 0$ and eigenvalues $\lambda_1 = -1$, $\lambda_2 = +1$ having different signs. Hence the critical point $(2,2)$ is an unstable saddle point.

Problem 13	**Problem 15**
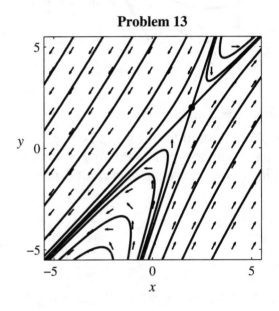	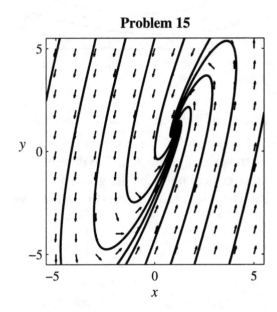

15. The Jacobian matrix $\mathbf{J} = \begin{bmatrix} 1 & -1 \\ 5 & -3 \end{bmatrix}$ has characteristic equation $\lambda^2 + 2\lambda + 2 = 0$ and eigenvalues $\lambda_1, \lambda_2 = -1 \pm i$ that are complex conjugates with negative real part. Hence the critical point $(1,1)$ is an asymptotically stable spiral point.

17. The Jacobian matrix $\mathbf{J} = \begin{bmatrix} 1 & -5 \\ 1 & -1 \end{bmatrix}$ has characteristic equation $\lambda^2 + 4 = 0$ and pure imaginary eigenvalues $\lambda_1, \lambda_2 = \pm 2i$. Hence $\left(\dfrac{5}{2}, -\dfrac{1}{2} \right)$ is a stable (but not asymptotically stable) center.

Problem 17

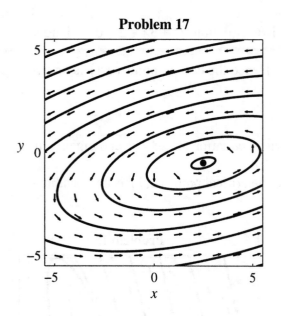

Problem 19

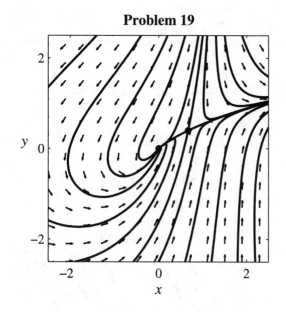

In each of Problems 19-28 we first calculate the Jacobian matrix **J** and its eigenvalues at $(0,0)$ and at each of the other critical points we observe in our phase portrait for the given system. Then we apply Theorem 2 to determine as much as we can about the type and stability of each of these critical points of the given almost linear system. Finally we draw a phase portrait that shows typical solution curves for the given system.

19. $\mathbf{J} = \begin{bmatrix} 1+2y & -3+2x \\ 4-y & -6-x \end{bmatrix}$

At $(0,0)$: The Jacobian matrix $\mathbf{J} = \begin{bmatrix} 1 & -3 \\ 4 & -6 \end{bmatrix}$ has characteristic equation $\lambda^2 + 5\lambda + 6 = 0$ and eigenvalues $\lambda_1 = -3$, $\lambda_2 = -2$ that are both negative. Hence $(0,0)$ is an asymptotically stable node of the given almost linear system.

At $(2/3, 2/5)$: The Jacobian matrix $\mathbf{J} = \begin{bmatrix} 9/5 & -5/3 \\ 18/5 & -20/3 \end{bmatrix}$ has characteristic equation $\lambda^2 + \dfrac{73}{15}\lambda - 6 = 0$ and approximate eigenvalues $\lambda_1 \approx -5.89$, $\lambda_2 \approx 1.02$ with different signs. Hence $(2/3, 2/5)$ is a saddle point.

21. $\mathbf{J} = \begin{bmatrix} 1+2x & 2+2y \\ 2-3y & -2-3x \end{bmatrix}$

At $(0,0)$: The Jacobian matrix $\mathbf{J} = \begin{bmatrix} 1 & 2 \\ 2 & -2 \end{bmatrix}$ has characteristic equation $\lambda^2 + \lambda - 6 = 0$ and eigenvalues $\lambda_1 = -3$, $\lambda_2 = 2$ with different signs. Hence $(0,0)$ is a saddle point of the given almost linear system.

At $(-0.51, -2.12)$: The Jacobian matrix $\mathbf{J} \approx \begin{bmatrix} -0.014 & -2.236 \\ 8.354 & -0.479 \end{bmatrix}$ has complex conjugate eigenvalues $\lambda_1, \lambda_2 \approx -0.25 \pm 4.32i$ with negative real parts. Hence $(-0.51, -2.12)$ is a spiral sink.

Problem 21	**Problem 23**
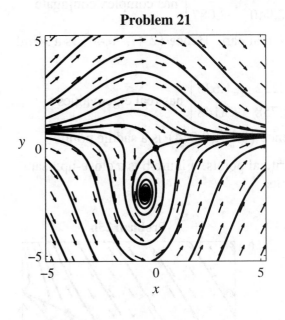	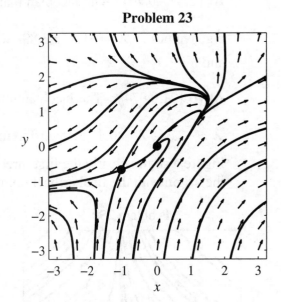

23. $\mathbf{J} = \begin{bmatrix} 2 + 3x^2 & -5 \\ 4 & -6 + 4y^3 \end{bmatrix}$

At $(0,0)$: The Jacobian matrix $\mathbf{J} = \begin{bmatrix} 2 & -5 \\ 4 & -6 \end{bmatrix}$ has characteristic equation $\lambda^2 + 4\lambda + 8 = 0$ and complex conjugate eigenvalues $\lambda_1, \lambda_2 = -2 \pm 2i$, with negative real part. Hence $(0,0)$ is a spiral sink of the given almost linear system.

At $(-1.08, -0.68)$: The Jacobian matrix $\mathbf{J} \approx \begin{bmatrix} 5.495 & -5 \\ 4 & -7.276 \end{bmatrix}$ has eigenvalues $\lambda_1 \approx -5.45$, $\lambda_2 \approx 3.67$ with different signs. Hence $(-1.08, -0.68)$ is a saddle point.

25. $\mathbf{J} = \begin{bmatrix} 1 + 3y & -2 + 3x \\ 2 - 2x & -3 - 2y \end{bmatrix}$

At $(0,0)$: The Jacobian matrix $\mathbf{J} = \begin{bmatrix} 1 & -2 \\ 2 & -3 \end{bmatrix}$ has characteristic equation $\lambda^2 + 2\lambda + 1 = 0$ and equal negative eigenvalues $\lambda_1 = -1$, $\lambda_1 = -1$. Hence $(0,0)$ is either a nodal sink or a spiral sink of the given almost linear system.

At $(0.74, -3.28)$: The Jacobian matrix $\mathbf{J} \approx \begin{bmatrix} -8.853 & 0.226 \\ 0.516 & 3.568 \end{bmatrix}$ has real eigenvalues $\lambda_1 \approx -8.86$, $\lambda_2 \approx 3.58$ with different signs. Hence $(0.74, -3.28)$ is a saddle point.

At $(2.47, -0.46)$: The Jacobian matrix $\mathbf{J} \approx \begin{bmatrix} -0.370 & 5.410 \\ -2.940 & -2.087 \end{bmatrix}$ has complex conjugate eigenvalues $\lambda_1, \lambda_2 \approx -1.23 \pm 3.89i$ with negative real part. Hence $(2.47, -0.46)$ is a spiral sink.

At $(0.121, 0.074)$: The Jacobian matrix $\mathbf{J} \approx \begin{bmatrix} 1.222 & -1.636 \\ 1.758 & -3.148 \end{bmatrix}$ has real eigenvalues $\lambda_1 \approx -2.34$, $\lambda_2 \approx 0.42$ with different signs. Hence $(0.121, 0.074)$ is a saddle point.

Figure a shows clearly the first three of these critical points. Figure b is a close-up near the origin with the final critical point now visible.

Problem 25a

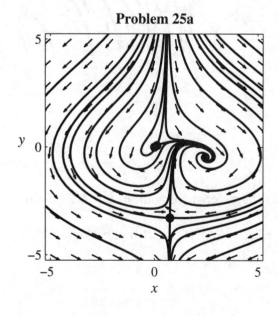

Problem 25b

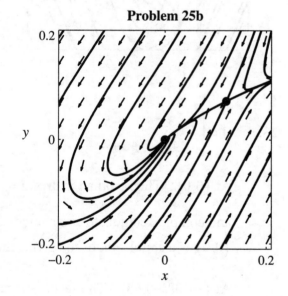

27. $\mathbf{J} = \begin{bmatrix} 1+4x^3 & -1-2y \\ 2-2x & -1+4y^3 \end{bmatrix}$

At $(0,0)$: The Jacobian matrix $\mathbf{J} = \begin{bmatrix} 1 & -1 \\ 2 & -1 \end{bmatrix}$ has characteristic equation $\lambda^2 + 1 = 0$ and equal positive eigenvalues $\lambda_1, \lambda_2 = \pm i$. Hence $(0,0)$ is either a center or a spiral point, but its stability is not determined by Theorem 2.

At $(-0.254, -0.507)$: The Jacobian matrix $\mathbf{J} \approx \begin{bmatrix} 0.934 & 0.014 \\ 2.508 & -1.521 \end{bmatrix}$ has real eigenvalues

$\lambda_1 \approx -1.53$, $\lambda_2 \approx 0.95$ with different signs. Hence $(-0.254, -0.507)$ is a saddle point.

At $(-1.557, 1.637)$: The Jacobian matrix $\mathbf{J} \approx \begin{bmatrix} -14.087 & -4.273 \\ 5.113 & 16.532 \end{bmatrix}$ has real eigenvalues

$\lambda_1 \approx -13.36$, $\lambda_2 \approx 15.80$ with different signs. Hence $(-1.557, 1.637)$ is a saddle point.

At $(-1.070, -1.202)$: The Jacobian matrix $\mathbf{J} \approx \begin{bmatrix} -3.905 & 1.403 \\ 4.141 & -7.940 \end{bmatrix}$ has unequal negative

eigenvalues $\lambda_1 \approx -9.07$, $\lambda_2 \approx -2.78$. Hence $(-1.070, -1.202)$ is a nodal sink.

Figure a shows these four critical points. Figure b, a close-up, suggests that the origin *may* be a stable center; note both the trajectory in grey that seems to spiral inward, and the oval-shaped trajectories that lie closer to the origin.

Problem 27a

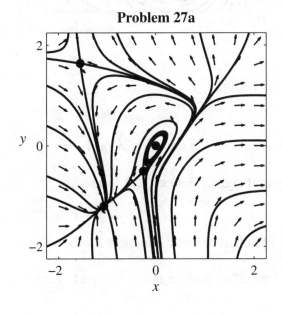

Problem 27b

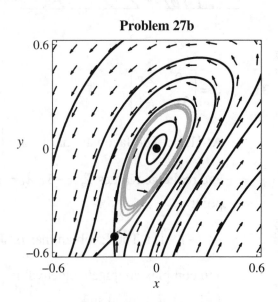

29. $\mathbf{J} = \begin{bmatrix} 1 & -1 \\ 2x & -1 \end{bmatrix}$

At $(0,0)$: The Jacobian matrix $\mathbf{J} = \begin{bmatrix} 1 & -1 \\ 0 & -1 \end{bmatrix}$ has characteristic equation $\lambda^2 - 1 = 0$ and

real eigenvalues $\lambda_1 = -1$, $\lambda_2 = +1$ with different signs. Hence $(0,0)$ is a saddle point.

At $(1,1)$: The Jacobian matrix $\mathbf{J} = \begin{bmatrix} 1 & -1 \\ 2 & -1 \end{bmatrix}$ has characteristic equation $\lambda^2 + 1 = 0$ and

pure imaginary eigenvalues $\lambda_1, \lambda_2 = \pm i$. Hence $(1,1)$ is either a center or a spiral point, but its stability is not determined by Theorem 2. However the figure suggests that $(1,1)$ is a stable center.

Problem 29

Problem 31

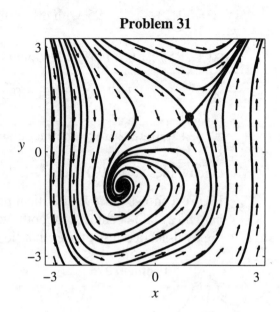

31. $\mathbf{J} = \begin{bmatrix} 0 & 2y \\ 3x^2 & -1 \end{bmatrix}$

At $(1,1)$: The Jacobian matrix $\mathbf{J} = \begin{bmatrix} 0 & 2 \\ 3 & -1 \end{bmatrix}$ has characteristic equation $\lambda^2 + \lambda - 6 = 0$ and real eigenvalues $\lambda_1 = -3$, $\lambda_2 = +2$ with different signs. Hence $(1,1)$ is a saddle point.

At $(-1,-1)$: The Jacobian matrix $\mathbf{J} = \begin{bmatrix} 0 & -2 \\ 3 & -1 \end{bmatrix}$ has characteristic equation $\lambda^2 + \lambda + 6 = 0$ and complex conjugate eigenvalues $\lambda_1, \lambda_2 \approx -0.5 \pm 2.398i$ with negative real part. Hence $(-1,-1)$ is a spiral sink.

33.   The characteristic equation of the given linear system is $(\lambda - \varepsilon)^2 + 1 = 0$, with characteristic roots $\lambda_1, \lambda_2 = \varepsilon \pm i$.

(a) So if $\varepsilon < 0$, then λ_1, λ_2 are complex conjugates with negative real part, and hence $(0,0)$ is an asymptotically stable spiral point.

(b) If $\varepsilon = 0$, then $\lambda_1, \lambda_2 = \pm i$ (pure imaginary), so $(0,0)$ is a stable center.

(c) If $\varepsilon > 0$, the situation is the same as in (a) except that the real part is positive, so $(0,0)$ is an unstable spiral point.

35. **(a)** If $h = 0$, then we have the familiar system $x' = y$, $y' = -x$ with circular trajectories about the origin, which is therefore a center.

(b) The change to polar coordinates as in Example 6 of Section 6.1 is routine, yielding $r' = hr^3$ and $\theta' = -1$.

(c) If $h = -1$, then $r' = -r^3$ integrates to give $2r^2 = \dfrac{1}{t+C}$, where C is a positive constant, so clearly $r \to 0$ as $t \to +\infty$, and thus the origin is a stable spiral point.

(d) If $h = +1$, then $r' = r^3$ integrates to give $2r^2 = -\dfrac{1}{t+C}$, where $C = -B$ is a positive constant. It follows that $2r^2 = \dfrac{1}{B-t}$, so now r increases as t starts at 0 and increases.

37. The substitution $y = vx$ in the homogeneous first-order equation

$$\frac{dy}{dx} = \frac{y\left(2x^3 - y^3\right)}{x\left(x^3 - 2y^3\right)}$$

yields

$$x\frac{dv}{dx} = -\frac{v^4 + v}{2v^3 - 1}.$$

Separating the variables and integrating by partial fractions, we get

$$\int\left(-\frac{1}{v} + \frac{1}{v+1} + \frac{2v-1}{v^2 - v+1}\right) dv = -\int \frac{dx}{x},$$

or

$$\ln\left[(v+1)\left(v^2 - v+1\right)\right] = \ln v - \ln x + \ln C,$$

or

$$(v+1)\left(v^2 - v+1\right) = \frac{Cv}{x},$$

or

$$v^3 + 1 = \frac{Cv}{x}.$$

Finally, the replacement $v = \dfrac{y}{x}$ yields $x^3 + y^3 = Cxy$.

SECTION 6.3

ECOLOGICAL APPLICATIONS: PREDATORS AND COMPETITORS

1. $\mathbf{J} = \begin{bmatrix} 200 - 4y & -4x \\ 2y & -150 + 2x \end{bmatrix}$

At $(0,0)$: The Jacobian matrix $\mathbf{J} = \begin{bmatrix} 200 & 0 \\ 0 & -150 \end{bmatrix}$ has characteristic equation

$(200 - \lambda)(-150 - \lambda) = 0$ and real eigenvalues $\lambda_1 = -150$, $\lambda_2 = 200$ with different signs.
Hence $(0,0)$ is a saddle point of the linearized system $x' = 200x$, $y' = -150y$ (Figure a).

At $(75,50)$: The Jacobian matrix $\mathbf{J} = \begin{bmatrix} 0 & -300 \\ 100 & 0 \end{bmatrix}$ has characteristic equation

$\lambda^2 + 30000 = 0$ and pure imaginary eigenvalues $\lambda_1, \lambda_2 = \pm 100i\sqrt{3}$. Hence $(75,50)$ is a
stable center of the linearization $u' = -300v$, $v' = 100u$ (Figure b).

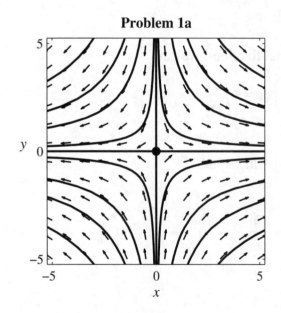

Problem 1a

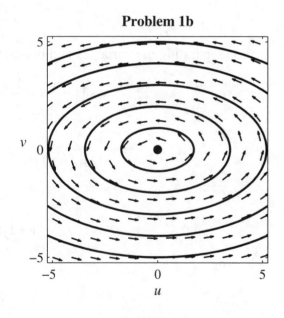

Problem 1b

3. The effect of using the insecticide is to replace b by $b + f$ and a by $a - f$ in the predator-prey equations, while leaving p and q unchanged. Hence the new harmful population
is

$$\frac{b + f}{q} > \frac{b}{q} = x_E$$

and the new benign population is

$$\frac{a-f}{p} < \frac{a}{p} = y_E.$$

Problems 4–7 deal with the competition system

$$x' = 60x - 4x^2 - 3xy, \quad y' = 42y - 2y^2 - 3xy,$$

that has Jacobian matrix $\mathbf{J} = \begin{bmatrix} 60 - 8x - 3y & -3x \\ -3y & 42 - 4y - 3x \end{bmatrix}.$

5. At $(0, 21)$ the Jacobian matrix $\mathbf{J} = \begin{bmatrix} -3 & 0 \\ -63 & -42 \end{bmatrix}$ has characteristic equation

$(-3 - \lambda)(-42 - \lambda) = 0$ and negative real eigenvalues $\lambda_1 = -42$, $\lambda_2 = -3$. Hence $(0, 21)$ is a nodal sink of the linearized system $u' = -3u$, $v' = -63u - 42v$.

7. At $(6, 12)$ the Jacobian matrix $\mathbf{J} = \begin{bmatrix} -24 & -18 \\ -36 & -24 \end{bmatrix}$ has characteristic equation

$(-24 - \lambda)^2 - (-36)(-18) = 0$ and real eigenvalues $\lambda_1 = -24 + 18\sqrt{2} > 0$, $\lambda_2 = -24 - 18\sqrt{2} < 0$ with different signs. Hence $(6, 12)$ is a saddle point of the linearized system $u' = -24u - 18v$, $v' = -36u - 24v$. Figure **a** illustrates this saddle point, whereas Figure **b** shows all four critical points of the system.

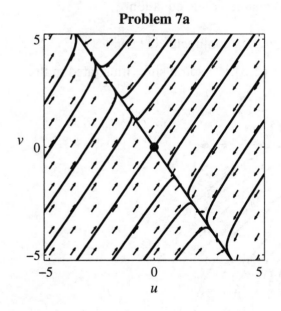

Problem 7a

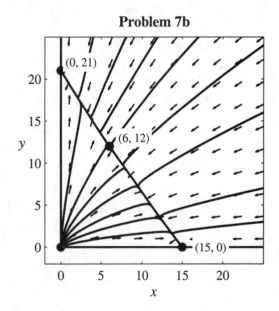

Problem 7b

Problems 8–10 deal with the competition system

$$x' = 60x - 3x^2 - 4xy, \quad y' = 42y - 3y^2 - 2xy$$

that has Jacobian matrix $\mathbf{J} = \begin{bmatrix} 60 - 6x - 4y & -4x \\ -2y & 42 - 6y - 2x \end{bmatrix}$.

9. At $(20,0)$ the Jacobian matrix $\mathbf{J} = \begin{bmatrix} -60 & -80 \\ 0 & 2 \end{bmatrix}$ has characteristic equation

$(-60 - \lambda)(2 - \lambda) = 0$ and real eigenvalues $\lambda_1 = -60$, $\lambda_2 = 2$ with different signs. Hence $(20,0)$ is a saddle point of the linearized system $u' = -60u - 80v$, $v' = 2v$.

Problems 11–13 deal with the predator-prey system

$$x' = 5x - x^2 - xy, \quad y' = -2y + xy$$

that has Jacobian matrix $\mathbf{J} = \begin{bmatrix} 5 - 2x - y & -x \\ y & -2 + x \end{bmatrix}$.

11. At $(0,0)$ the Jacobian matrix $\mathbf{J} = \begin{bmatrix} 5 & 0 \\ 0 & -2 \end{bmatrix}$ has characteristic equation

$(5 - \lambda)(-2 - \lambda) = 0$ and real eigenvalues $\lambda_1 = -2$, $\lambda_2 = 5$ with different signs. Hence $(0,0)$ is a saddle point of the linearized system $x' = 5x$, $y' = -2y$.

13. At $(2,3)$ the Jacobian matrix $\mathbf{J} = \begin{bmatrix} -2 & -2 \\ 3 & 0 \end{bmatrix}$ has characteristic equation

$(-2 - \lambda)(-\lambda) - 3 \cdot (-2) = \lambda^2 + 2\lambda + 6 = 0$ and complex conjugate eigenvalues $\lambda_1, \lambda_2 = -1 \pm i\sqrt{5}$ with negative real part. Hence $(2,3)$ is a spiral sink (illustrated in the figure) of the linearized system $u' = -2u - 2v$, $v' = 3u$.

Problem 13

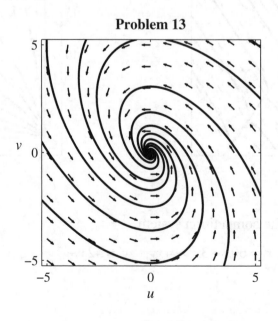

Problems 14–17 deal with the predator-prey system

$$x' = x^2 - 2x - xy, \quad y' = y^2 - 4y + xy$$

that has Jacobian matrix $\mathbf{J} = \begin{bmatrix} 2x - 2 - y & -x \\ y & 2y - 4 + x \end{bmatrix}.$

15.   At $(0,4)$ the Jacobian matrix $\mathbf{J} = \begin{bmatrix} -6 & 0 \\ 4 & 4 \end{bmatrix}$ has characteristic equation

$(-6 - \lambda)(4 - \lambda) = 0$ and real eigenvalues $\lambda_1 = -6$, $\lambda_2 = 4$ with different signs. Hence $(0,4)$ is a saddle point of the linearized system $u' = -6u$, $v' = 4u + 4v$.

17.   At $(3,1)$ the Jacobian matrix $\mathbf{J} = \begin{bmatrix} 3 & -3 \\ 1 & 1 \end{bmatrix}$ has characteristic equation

$(3 - \lambda)(1 - \lambda) - 1 \cdot (-3) = \lambda^2 - 4\lambda + 6 = 0$ and complex conjugate eigenvalues $\lambda_1, \lambda_2 = 2 \pm i\sqrt{2}$ with positive real part. Hence $(3,1)$ is a spiral source of the linearized system $u' = 3u - 3v$, $v' = u + v$ (illustrated in the figure).

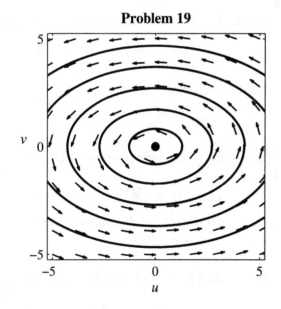

Problem 17   **Problem 19**

Problems 18 and 19 deal with the predator-prey system

$$x' = 2x - xy, \quad y' = -5y + xy$$

that has Jacobian matrix $\mathbf{J} = \begin{bmatrix} 2 - y & -x \\ y & -5 + x \end{bmatrix}.$

19.   At $(5,2)$ the Jacobian matrix $\mathbf{J} = \begin{bmatrix} 0 & -5 \\ 2 & 0 \end{bmatrix}$ has characteristic equation

$(-\lambda)(-\lambda) - 2 \cdot (-5) = \lambda^2 + 10 = 0$ and pure imaginary roots $\lambda = \pm i\sqrt{10}$, so the origin is a stable center for the linearized system $u' = -5v$, $v' = 2u$. This is the indeterminate case, but the figure suggests that $(5,2)$ is also a stable center for the original system.

Problems 20–22 deal with the predator-prey system

$$x' = -3x + x^2 - xy, \quad y' = -5y + xy$$

that has Jacobian matrix $\mathbf{J} = \begin{bmatrix} -3 + 2x - y & -x \\ y & -5 + x \end{bmatrix}$.

21.   At $(3,0)$ the Jacobian matrix $\mathbf{J} = \begin{bmatrix} 3 & -3 \\ 0 & -2 \end{bmatrix}$ has characteristic equation

$(3 - \lambda)(-2 - \lambda) = 0$ and real eigenvalues $\lambda_1 = -2$, $\lambda_2 = 3$ with different signs. Hence $(3,0)$ is a saddle point of the linearized system $u' = 3u - 3v$, $v' = -2v$.

Problems 23–25 deal with the predator-prey system

$$x' = 7x - x^2 - xy, \quad y' = -5y + xy$$

that has Jacobian matrix $\mathbf{J} = \begin{bmatrix} 7 - 2x - y & -x \\ y & -5 + x \end{bmatrix}$.

23.   At $(0,0)$ the Jacobian matrix $\mathbf{J} = \begin{bmatrix} 7 & 0 \\ 0 & -5 \end{bmatrix}$ has characteristic equation

$(7 - \lambda)(-5 - \lambda) = 0$ and real eigenvalues $\lambda_1 = -5$, $\lambda_2 = 7$ with different signs. Hence $(0,0)$ is a saddle point of the linearized system $x' = 7x$, $y' = -5y$.

25.   At $(5,2)$ the Jacobian matrix $\mathbf{J} = \begin{bmatrix} -5 & -5 \\ 2 & 0 \end{bmatrix}$ has characteristic equation

$(-5 - \lambda)(-\lambda) - 2 \cdot (-5) = \lambda^2 + 5\lambda + 10 = 0$ and complex conjugate eigenvalues

$\lambda_1, \lambda_2 = \frac{1}{2}\left(-5 \pm i\sqrt{15}\right)$ with negative real part. Hence $(5,2)$ is a spiral sink of the linearized system $u' = -5u - 5v$, $v' = 2u$ (illustrated in the figure).

Problem 25

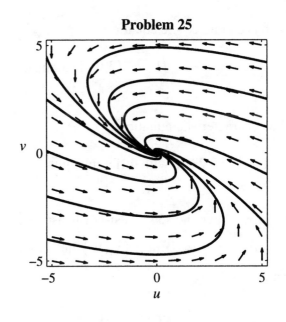

Problem 27

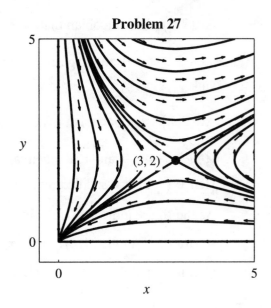

27. $\mathbf{J} = \begin{bmatrix} 2y-4 & 2x \\ y & x-3 \end{bmatrix}$

At $(0,0)$: The Jacobian matrix $\mathbf{J} = \begin{bmatrix} -4 & 0 \\ 0 & -3 \end{bmatrix}$ has characteristic equation

$\lambda^2 + 7\lambda + 12 = 0$ and negative real eigenvalues $\lambda_1 = -4$, $\lambda_2 = -3$. Hence $(0,0)$ is a nodal sink.

At $(3,2)$: The Jacobian matrix $\mathbf{J} = \begin{bmatrix} 0 & 6 \\ 2 & 0 \end{bmatrix}$ has characteristic equation $\lambda^2 - 12 = 0$ and

real eigenvalues $\lambda_1, \lambda_2 = \pm 2\sqrt{3}$ with different signs. Hence $(3,2)$ is a saddle point.

As the figure indicates, if the initial point (x_0, y_0) lies below the northwest-southeast

separatrix through $(3,2)$, then $(x(t), y(t)) \to (0,0)$ as $t \to \infty$. But if (x_0, y_0) lies above

this separatrix, then $(x(t), y(t)) \to (\infty, \infty)$ as $t \to \infty$.

29. $\mathbf{J} = \begin{bmatrix} -2x - \dfrac{1}{2}y + 3 & -\dfrac{1}{2}x \\ -2y & 4 - 2x \end{bmatrix}$

At $(0,0)$: The Jacobian matrix $\mathbf{J} = \begin{bmatrix} 3 & 0 \\ 0 & 4 \end{bmatrix}$ has characteristic equation $\lambda^2 - 7\lambda + 12 = 0$

and positive real eigenvalues $\lambda_1 = 3$, $\lambda_2 = 4$. Hence $(0,0)$ is a nodal source.

At $(3,0)$: The Jacobian matrix $\mathbf{J} = \begin{bmatrix} -3 & -3/2 \\ 0 & -2 \end{bmatrix}$ has characteristic equation

$\lambda^2 + 5\lambda + 6 = 0$ and negative real eigenvalues $\lambda_1 = -3$, $\lambda_2 = -2$. Hence $(3,0)$ is a nodal sink.

At $(2,2)$: The Jacobian matrix $\mathbf{J} = \begin{bmatrix} -2 & -1 \\ -4 & 0 \end{bmatrix}$ has characteristic equation

$\lambda^2 + 2\lambda - 4 = 0$ and real eigenvalues $\lambda_1 \approx -3.2361$, $\lambda_2 = 1.2361$ with different signs. Hence $(2,2)$ is a saddle point.

As the figure indicates, if the initial point (x_0, y_0) lies above the southwest-northeast separatrix through $(2,2)$, then $(x(t), y(t)) \to (0, \infty)$ as $t \to \infty$. But if (x_0, y_0) lies below this separatrix, then $(x(t), y(t)) \to (3,0)$ as $t \to \infty$.

Problem 29

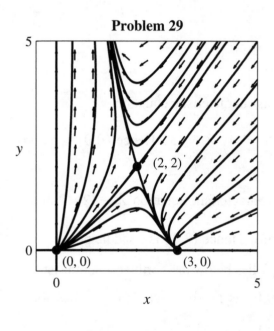

Problem 31

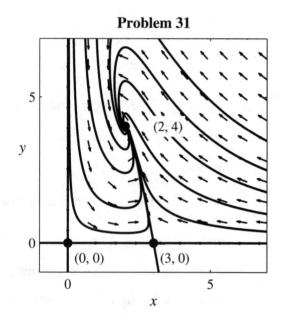

31. $\mathbf{J} = \begin{bmatrix} -2x - \dfrac{1}{4}y + 3 & -\dfrac{1}{4}x \\ y & x - 2 \end{bmatrix}$

At $(0,0)$: The Jacobian matrix $\mathbf{J} = \begin{bmatrix} 3 & 0 \\ 0 & -2 \end{bmatrix}$ has characteristic equation $\lambda^2 - \lambda - 6 = 0$ and real eigenvalues $\lambda_1 = -2$, $\lambda_2 = 3$ of opposite sign. Hence $(0,0)$ is a saddle point.

At $(3,0)$: The Jacobian matrix $\mathbf{J} = \begin{bmatrix} -3 & -3/4 \\ 0 & 1 \end{bmatrix}$ has characteristic equation

$\lambda^2 + 2\lambda - 3 = 0$ and real eigenvalues $\lambda_1 = -3$, $\lambda_2 = 1$ of opposite sign. Hence $(3,0)$ is a saddle point.

At $(2,4)$: The Jacobian matrix $\mathbf{J} = \begin{bmatrix} -2 & -1/2 \\ 4 & 0 \end{bmatrix}$ has characteristic equation

$\lambda^2 + 2\lambda + 2 = 0$ and complex conjugate eigenvalues $\lambda_1, \lambda_2 = -1 \pm i$ with negative real part. Hence $(2,4)$ is a spiral sink.

As $t \to \infty$, each solution point $(x(t), y(t))$ with nonzero initial conditions approaches the spiral sink $(2,4)$, as indicated by the direction arrows in the figure.

33. $\mathbf{J} = \begin{bmatrix} -6x + y + 30 & x \\ 4y & 4x - 6y + 60 \end{bmatrix}$

At $(0,0)$: The Jacobian matrix $\mathbf{J} = \begin{bmatrix} 30 & 0 \\ 0 & 80 \end{bmatrix}$ has characteristic equation

$\lambda^2 - 110\lambda + 2400 = 0$ and positive real eigenvalues $\lambda_1 = 30$, $\lambda_2 = 80$. Hence $(0,0)$ is a nodal source.

At $(0,20)$: The Jacobian matrix $\mathbf{J} = \begin{bmatrix} 10 & 0 \\ 40 & -80 \end{bmatrix}$ has characteristic equation

$\lambda^2 + 70\lambda - 800 = 0$ and real eigenvalues $\lambda_1 = -80$, $\lambda_2 = 10$ of opposite sign. Hence $(0,20)$ is a saddle point.

At $(15,0)$: The Jacobian matrix $\mathbf{J} = \begin{bmatrix} -30 & 15 \\ 0 & 110 \end{bmatrix}$ has characteristic equation

$\lambda^2 - 80\lambda - 3300 = 0$ and real eigenvalues $\lambda_1 = -30$, $\lambda_2 = 110$ of opposite sign. Hence $(15,0)$ is a saddle point.

At $(4,22)$: The Jacobian matrix $\mathbf{J} = \begin{bmatrix} -8 & -4 \\ 44 & -88 \end{bmatrix}$ has characteristic equation

$\lambda^2 + 96\lambda + 880 = 0$ and negative real eigenvalues $\lambda_1 \approx -85.736$, $\lambda_2 = -10.264$. Hence $(4,22)$ is a nodal sink.

As the figure shows, as $t \to \infty$, each solution point $(x(t), y(t))$ with nonzero initial conditions approaches the nodal sink $(4,22)$.

Problem 33

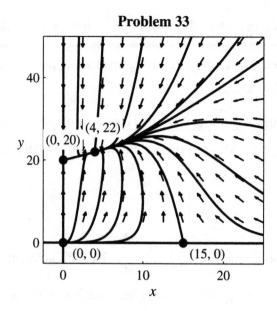

SECTION 6.4

NONLINEAR MECHANICAL SYSTEMS

In each of Problems 1–4 we need only substitute the familiar power series for the exponential, sine, and cosine functions, and then discard all higher-order terms. For each problem we give the corresponding linear system, the eigenvalues λ_1 and λ_2, and the type of this critical point.

1.

$$x' = 1 - \left(1 + x + \frac{1}{2}x^2 + \cdots\right) + 2y \approx -x + 2y$$

$$y' = -x - 4\left(y - \frac{1}{6}y^3 + \cdots\right) \approx -x - 4y$$

The coefficient matrix $\mathbf{A} = \begin{bmatrix} -1 & 2 \\ -1 & -4 \end{bmatrix}$ has negative eigenvalues $\lambda_1 = -2$ and $\lambda_2 = -3$ indicating a stable nodal sink as illustrated in the figure. Alternatively, we can calculate the Jacobian matrix

$$\mathbf{J}(x, y) = \begin{bmatrix} -e^x & 2 \\ -1 & -4\cos x \end{bmatrix}, \quad \text{so} \quad \mathbf{J}(0, 0) = \begin{bmatrix} -1 & 2 \\ -1 & -4 \end{bmatrix}.$$

Problem 1

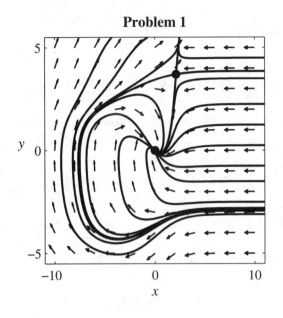

Problem 3

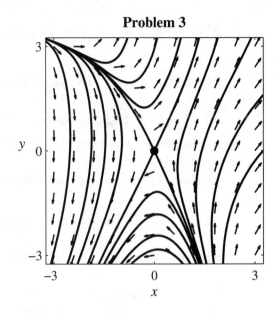

3. $x' = \left(1 + x + \dfrac{1}{2}x^2 + \cdots\right) + 2y - 1 \approx x + 2y$

$y' = 8x + \left(1 + y + \dfrac{1}{2}y^2 + \cdots\right) - 1 \approx 8x + y$

The coefficient matrix $\mathbf{A} = \begin{bmatrix} 1 & 2 \\ 8 & 1 \end{bmatrix}$ has real eigenvalues $\lambda_1 = -3$ and $\lambda_2 = 5$ of opposite sign, indicating an unstable saddle point as illustrated in the figure. Alternatively, we can calculate the Jacobian matrix

$$\mathbf{J}(x,y) = \begin{bmatrix} e^x & 2 \\ 8 & e^y \end{bmatrix}, \quad \text{so} \quad \mathbf{J}(0,0) = \begin{bmatrix} 1 & 2 \\ 8 & 1 \end{bmatrix}.$$

5.   The critical points are of the form $(0, n\pi)$, where n is an integer, so we substitute $x = u$, $y = v + n\pi$. Then

$$u' = x' = -u + \sin(v + n\pi) = -u + (\cos n\pi)v = -u + (-1)^n v.$$

Hence the linearized system at $(0, n\pi)$ is

$$u' = -u \pm v, \quad v' = 2u,$$

where we take the plus sign if n is even, the minus sign if n is odd. If n is even, then the eigenvalues are $\lambda_1 = 1$ and $\lambda_2 = -2$, so $(0, n\pi)$ is an unstable saddle point. If n is odd, then the eigenvalues are $\lambda_1, \lambda_2 = \dfrac{1}{2}\left(-1 \pm i\sqrt{7}\right)$, so $(0, n\pi)$ is a stable spiral point.

Alternatively, we can start by calculating the Jacobian matrix $\mathbf{J}(x,y) = \begin{bmatrix} -1 & \cos y \\ 2 & 0 \end{bmatrix}$.

At $(0,n\pi)$, n even: The Jacobian matrix $\mathbf{J} = \begin{bmatrix} -1 & 1 \\ 2 & 0 \end{bmatrix}$ has characteristic equation

$\lambda^2 + \lambda - 2 = 0$ and real eigenvalues $\lambda_1 = -2$, $\lambda_2 = 1$ of opposite sign. Hence $(0,n\pi)$ is a saddle point if n is even, as we see in the figure above.

At $(0,n\pi)$, n odd: The Jacobian matrix $\mathbf{J} = \begin{bmatrix} -1 & -1 \\ 2 & 0 \end{bmatrix}$ has characteristic equation

$\lambda^2 + \lambda + 2 = 0$ and complex conjugate eigenvalues $\lambda_1, \lambda_2 = \frac{1}{2}\left(-1 \pm i\sqrt{7}\right)$ with negative

real part. Hence $(0,n\pi)$ is a spiral sink if n is odd, as indicated in the figure.

Problem 5

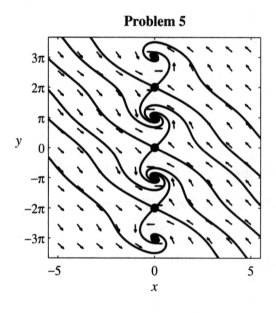

Problem 7

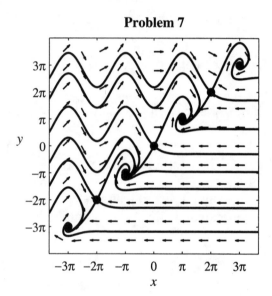

7. The critical points are of the form $(n\pi, n\pi)$, where n is an integer, so we substitute $x = u + n\pi$, $y = v + n\pi$. Then

$$u' = x' = 1 - e^{u-v} = 1 - \left[1 + (u-v) + \frac{1}{2}(u-v)^2 + \cdots\right] \approx -u + v,$$

$$v' = y' = 2\sin(u + n\pi) = 2\sin u \cos n\pi \approx 2(-1)^n u.$$

Hence the linearized system at $(n\pi, n\pi)$ is

$$u' = -u + v, \quad v' = \pm 2u,$$

and has coefficient matrix $\mathbf{A} = \begin{bmatrix} -1 & 1 \\ \pm 2 & 0 \end{bmatrix}$, where we take the plus sign if n is even, the minus sign if n is odd. With n even, the characteristic equation $\lambda^2 + \lambda - 2 = 0$ has real roots

$\lambda_1 = 1$ and $\lambda_2 = -2$ of opposite sign, so $(n\pi, n\pi)$ is an unstable saddle point. With n odd, the characteristic equation $\lambda^2 + \lambda + 2 = 0$ has complex conjugate eigenvalues $\lambda_1, \lambda_2 = \frac{1}{2}\left(-1 \pm i\sqrt{7}\right)$ with negative real part, so $(n\pi, n\pi)$ is a stable spiral point.

Alternatively, we can start by calculating the Jacobian matrix $\mathbf{J}(x, y) = \begin{bmatrix} -e^{x-y} & e^{x-y} \\ 2\cos x & 0 \end{bmatrix}$.

At $(n\pi, n\pi)$, n even: The Jacobian matrix $\mathbf{J} = \begin{bmatrix} -1 & 1 \\ 2 & 0 \end{bmatrix}$ has characteristic equation $\lambda^2 + \lambda - 2 = 0$ and real eigenvalues $\lambda_1 = -2$, $\lambda_2 = 1$ of opposite sign. Hence $(n\pi, n\pi)$ is a saddle point if n is even, as we see in the figure.

At $(n\pi, n\pi)$, n odd: The Jacobian matrix $\mathbf{J} = \begin{bmatrix} -1 & 1 \\ -2 & 0 \end{bmatrix}$ has characteristic equation $\lambda^2 + \lambda + 2 = 0$ and complex conjugate eigenvalues $\lambda_1, \lambda_2 \approx -0.5 \pm 1.3229i$ with negative real part. Hence $(n\pi, n\pi)$ is a spiral sink if n is odd, as we see in the figure.

As preparation for Problems 9–11, we first calculate the Jacobian matrix

$$\mathbf{J}(x, y) = \begin{bmatrix} 0 & 1 \\ -\omega^2 \cos x & -c \end{bmatrix}$$

of the damped pendulum system in (34) in the text. At the critical point $(n\pi, 0)$ we have

$$\mathbf{J}(n\pi, 0) = \begin{bmatrix} 0 & 1 \\ -\omega^2 \cos n\pi & -c \end{bmatrix} = \begin{bmatrix} 0 & 1 \\ \pm\omega^2 & -c \end{bmatrix},$$

where we take the plus sign if n is odd, the minus sign if n is even.

9.   If n is odd then the characteristic equation $\lambda^2 + c\lambda - \omega^2 = 0$ has real roots

$$\lambda_1, \lambda_2 = \frac{-c \pm \sqrt{c^2 + 4\omega^2}}{2}$$

with opposite signs, so $(n\pi, 0)$ is an unstable saddle point.

11.   If n is even and $c^2 < 4\omega^2$, then the two eigenvalues

$$\lambda_1, \lambda_2 = \frac{-c \pm \sqrt{c^2 - 4\omega^2}}{2} = -\frac{c}{2} \pm \frac{i}{2}\sqrt{4\omega^2 - c^2}$$

are complex conjugates with negative real part, so $(n\pi, 0)$ is a stable spiral point.

Problems 12-16 call for us to find and classify the critical points of the first order-system $x' = y$, $y' = -f(x, y)$ that corresponds to the given equation $x'' + f(x, x') = 0$. After finding the critical points $(x, 0)$ where $f(x, 0) = 0$, we first calculate the Jacobian matrix $\mathbf{J}(x, y)$.

13. $\mathbf{J}(x, y) = \begin{bmatrix} 0 & 1 \\ 15x^2 - 20 & -2 \end{bmatrix}$.

At $(0, 0)$: The Jacobian matrix $\mathbf{J} = \begin{bmatrix} 0 & 1 \\ -20 & -2 \end{bmatrix}$ has characteristic equation

$\lambda^2 + 2\lambda + 20 = 0$ and complex conjugate eigenvalues $\lambda_1, \lambda_2 = -1 \pm i\sqrt{19}$ consistent with the spiral node we see at $(0, 0)$ in Fig. 6.4.6 in the textbook.

At $(\pm 2, 0)$: The Jacobian matrix $\mathbf{J} = \begin{bmatrix} 0 & 1 \\ 40 & -2 \end{bmatrix}$ has characteristic equation

$\lambda^2 + 2\lambda - 40 = 0$ and real eigenvalues $\lambda_1, \lambda_2 = -1 \pm \sqrt{41}$ of opposite sign, consistent with the saddle points we see at $(\pm 2, 0)$ in Fig. 6.4.6.

15. $\mathbf{J}(x, y) = \begin{bmatrix} 0 & 1 \\ 2x - 4 & 0 \end{bmatrix}$.

At $(0, 0)$: The Jacobian matrix $\mathbf{J} = \begin{bmatrix} 0 & 1 \\ -4 & 0 \end{bmatrix}$ has characteristic equation $\lambda^2 + 4 = 0$ and pure imaginary eigenvalues $\lambda_1, \lambda_2 = \pm 2i$ consistent with the stable center we see at $(0, 0)$ in Fig. 6.4.13 in the textbook.

At $(4, 0)$: The Jacobian matrix $\mathbf{J} = \begin{bmatrix} 0 & 1 \\ 4 & 0 \end{bmatrix}$ has characteristic equation $\lambda^2 - 4 = 0$ and real eigenvalues $\lambda_1, \lambda_2 = \pm 2$ of opposite sign, consistent with the saddle point we see at $(4, 0)$ in Fig. 6.4.13.

17. $\mathbf{J}(x, y) = \begin{bmatrix} 0 & 1 \\ -5 - \dfrac{15}{4}x^2 & -2 \end{bmatrix}$.

At $(0, 0)$: The Jacobian matrix $\mathbf{J} = \begin{bmatrix} 0 & 1 \\ -5 & -2 \end{bmatrix}$ has characteristic equation

$\lambda^2 + 2\lambda + 5 = 0$ and complex conjugate eigenvalues $\lambda_1, \lambda_2 = -1 \pm 2i$ with negative real part, consistent with the spiral sink we see in the figure.

Problem 17

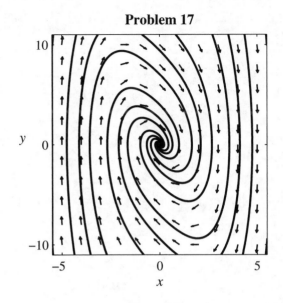

Problem 19

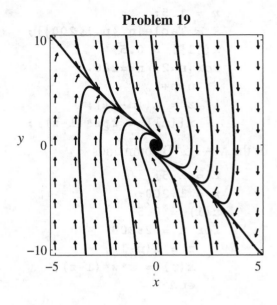

19. $\mathbf{J}(x,y) = \begin{bmatrix} 0 & 1 \\ -5 - \dfrac{15}{4}x^2 & -4|y| \end{bmatrix}.$

At $(0,0)$: The Jacobian matrix $\mathbf{J} = \begin{bmatrix} 0 & 1 \\ -5 & 0 \end{bmatrix}$ has characteristic equation $\lambda^2 + 5 = 0$ and

pure imaginary eigenvalues $\lambda_1, \lambda_2 = \pm i\sqrt{5}$. This corresponds to the indeterminate case of Theorem 2 in Section 6.3, but is not inconsistent with the spiral sink we see in the figure.

The statements of Problems 21–26 in the text include their answers and rather fully outline their solutions, which therefore are omitted here.

SECTION 6.5

CHAOS IN DYNAMICAL SYSTEMS

We list here some programs that may be useful in the projects for this section. Further discussion of these projects can be found in the applications manual that accompanies this text.

As indicated in Fig. 6.5.1 in the text, you can use the Maple commands

```
r := 1.5:
x = array(1..200):
x[1] := 0.5:
for n from 2 to 200 do
    z := x[n-1]:
    x[n] := r*z*(1-z):
    od:
```

the *Mathematica* commands

```
r = 1.5;
x = Table[n,{n,1,200}];
x[[1]] = 0.5;
For[n=2, n<=200,
    n=n+1,
    z = x[[n-1]];
    x[[n]] = r*z*(1-z)];
```

or the MATLAB commands

```
r = 1.5;
x = 1:200;
x(1) = 0.5;
for n = 2:200
    z = x(n-1);
    x(n) = r*z*(1-z);
    end
```

to calculate and assemble a list of the successive iterates given by $x_{n+1} = rx_n(1-x_n)$, as illustrated in Figures 6.5.2 through 6.5.7 in the text. The following BASIC program can be used to investigate periodic cycles for this iteration.

```
100 'Program PERIODS
110 '
120 'The period-doubling iteration
130 '
140 '         x    =    rx(1 - x)
150 '
160 '  r  =  2.75  :  Period  1
170 '  r  =  3.25  :  Period  2
180 '  r  =  3.50  :  Period  4
190 '  r  =  3.55  :  Period  8
200 '  r  =  3.565 :  Period 16
210 '  r  =  3.57  :  CHAOS
220 '  r  =  3.84  :  Period  3
230 '  r  =  3.845 :  Period  6
240 '  r  =  3.848 :  Period 12
250 '
260  DEFDBL R,X
270  INPUT "Value of r"; R
280  INPUT "Print in blocks of k  =  "; K
290  P$  =  "#.####     "
300  X  =  .5                  'Initial seed
310 '
320  FOR I  =  1 TO 500        '500 initial
330     X  =  R*X*(1 - X)      'iterations to
340  NEXT I                    'stabilize.
350 '
360  FOR  I  =  1  TO  K
```

```
370      X  =  R*X*(1 - X)          'Final iterations
380         PRINT USING P$; X;
390   NEXT
400   IF K <> 8 THEN PRINT
410   '
420   'Press any key but Q to continue:
430   A$  =  INKEY$
440   IF A$  =  "" THEN GOTO 430
450   IF A$  =  "q" OR A$  =  "Q" THEN END
460   GOTO 360                     'End of loop
470   '
480   END
```

The next BASIC program below can be used to plot pitchfork diagrams as in Figures 6.5.8 and 6.5.9 in the text. As written, it runs well in Borland TurboBasic (probably now obsolete), but may have to be fine-tuned to run in other dialects of BASIC.

```
100   'Program PICHFORK
110   '
120   'Exhibits the period-doubling toward chaos
130   'generated by the Verhulst iteration
140   '
150   '          x  =  rx(1 - x)
160   '
170   'as the growth parameter  r   is increased
180   'in the range from about  3   to about  4.
190   '
200    DEFDBL H,K,R,X
210    DEFINT I,J,M,N,P,Q
220    INPUT "Rmin,Rmax"; RMIN, RMAX      'Try 2.8 and 4.0
230    INPUT "Xmin,Xmax"; XMIN, XMAX      'Try  0  and  1
240   '
250    KEY OFF   : CLS
260   'SCREEN 1  :  N  =  319          'For med resolution
270    SCREEN 2  :  N  =  639          'For hi  resolution
280    M  =  200                       'Hor rows for either
290    H  =   (RMAX - RMIN)/N
300    K  =   (XMAX - XMIN)/M
310   '
320    LINE (0,0) - (N,0)              'Draws a box
330    LINE - (N,199)                  'around
340    LINE - (0,199)                  'the screen
350    LINE - (  0,0)
360   '
370    FOR P  =   1 TO 9               'Tick marks on
380       Q  =   (P*(N+1)/10) - 1      'top and bottom
390       LINE (Q,0) - (Q,5)           'of box
400       LINE (Q,195) - (Q,199)
```

```
410   NEXT P
420   '
430   FOR J   =   0 TO  N              'Jth vertical column
440       R   =   RMIN + J*H           'of pixels on screen
450       X   =   .5
460       FOR P   =   0 TO 1000        'These iterations
470           X   =   R*X*(1-X)        'to settle down.
480       NEXT P
490       FOR Q   =   0 TO 250         'These iterations
500           X   =   R*X*(1-X)        'are recorded.
510           I   =   INT((X - XMIN)/K)
520           I   =   200 - I
530           IF (0< = I) AND (I<200) THEN PSET (J,I)
540       NEXT Q
550   NEXT J
560   '
570   WHILE INKEY$   =   ""            'Press a key when
580   WEND                            'finished looking.
590   SCREEN 0  :  CLS  :   KEY ON
600   END
```

A more elaborate construction of these pitchfork diagrams is given by the following Mathematica program, a slight elaboration of one found on page 102 of T. Gray and J. Glynn, *Exploring Mathematics with Mathematica*, Addison-Wesley, 1991.

```
g[x_] : =   r x (1 - x);
Clear[r];
a  =  2.8;  b  =  4.0.   (* r-range for Fig 6.5.8 *)
c  =  0;    d  =  1;            (* x-range *)
m  =  250;                      (* no of x-points *)
n  =  500;                      (* no of r-values *)
ListPlot[
   Flatten[Table[
      Transpose[{
         Table[r, {m+1}],
            NestList[g, Nest[q, 0.5, 2m], m] }],
            {r, a, b, (b-a)/n} ],
               1],
   PlotStyle -> PointSize[0.001],
   PlotRange -> {{a,b},{c,d}},
   AspectRatio -> 0.75,
   Frame -> True,
   AxesLabel -> {"r","x"} ]
```

This *Mathematica* program runs slowly, and requires a fast machine with plenty of memory to finish within a reasonable waiting time. The following MATLAB program (which was actually used to construct Figs 6.5.8 and 6.5.9) runs much faster on a comparable computer, and may be easier to understand.

```
% pitchfork diagram script
% for Figures 6.5.8 and 6.5.9

hold off
m  =  400;                      % no of r-subintervals
n  =  400;                      % no of x-subintervals


a  =  2.8;   b  =  4.0;         % r-range for 6.5.9
dr =  (b - a)/m;
R  =  a+dr/2 : dr : b;          % vector of r-values
c  =  0;        d  =  1;        % x-range
dx =  (d - c)/n
X  =  c+dx/2 : dx : d;          % vector of x-values
[rr,xx]  =  meshgrid(R,X);      % matrices of r- and x-coords
                                % of grid points in rx-rect


C = zeros(m,n);
for j  =  1 : m                 % Cycle through r-values
   r  =  a - dr/2 + j*dr;
   x  =  0.5;                   % Initialize x-value
   for k  =  1:1000             % 1000 iterations to stabilize
      x  =  r*x*(1-x);
   end
   for k  =  1:1000             % 1000 more iterations
       x  =  r*x*(1-x);
       i = ceil(x/dx);
       C(i,j) = 1;              % lattice point to plot
       end
   end


C = C + 1;
C = flipud(C);                  % matrix of points for image
image(R,X,C)
colormap([1 1 1; 0 0 0])        % color them black or white
axis square
```

The following MATLAB function defines the forced Duffing equation for Figures 6.5.13 through 6.5.16.

```
function  yp = ypduffing(t,x)
F0 = 0.80;
yp = x;
y = x(2);    x = x(1);
yp(1) = y;
yp(2) = F0*cos(t)-y+x-x.^3;
```

Then the following MATLAB script can be used to construct Fig. 6.5.16.

```
% fig6_5_16.m script
```

```
options = odeset('RelTol',1e-8,'AbsTol',1e-8);
[t,y] = ode45('ypduffing', [0 100], [1;0],options);
n = length(t);
y100 = y(n,:)';
[t,y] = ode45('ypduffing', [100 300], y100,options);
hold off
plot(y(:,1),y(:,2),'b')                    % Fig. 6.5.16(a)
axis([-1.5 1.5 -1.5 1.5])
axis square
hold on
plot([-1.5 1.5],[0 0],'k')
plot([0 0],[-1.5 1.5],'k')
pause
hold off
plot(t,y(:,1),'b')                         % Fig. 6.5.16(b)
axis([100 300 -1.5 1.5])
axis square
hold on
plot([100 300],[0 0],'k')
plot([0 0],[-1.5 1.5],'k')
```

The Lorenz system of Equations (15) in the text is defined by the following MATLAB function.

```
function yp = lorenzeq(t,y)
sigma = 10.;
rho = 28.;
beta = 8./3.;
yp = y;
yp(1) = - sigma*y(1) + sigma *y(2);
yp(2) =      rho*y(1)        - y(2) - y(1)*y(3);
yp(3) =                        y(1)*y(2) - beta*y(3);
```

The Lorenz trajectory in text Figure 6.5.18 was generated by the MATLAB script

```
% fig6_5_18.m
y0 = [-8; 8; 27];   % set initial point
hold off
[t,y] = ode45(@lorenzeq, [0 40], y0);
plot3(y(:,1),y(:,2),y(:,3),'b');
set(gca,'view',[350 -10])
box on;  xlabel('x');  ylabel('y');  zlabel('z')
axis([-20 20 -25 25 5 55])
```

The Rössler system of Equations (16) in the text is defined by the following MATLAB function.

```
function ydot = rosslereq(t,y)
alpha = 0.398;
beta = 2.;
gamma = 4.;
ydot = y;
```

```
ydot(1) =  -y(2) - y(3);
ydot(2) =   y(1) + alpha*y(2);
ydot(3) =   beta + y(3)*(y(1) - gamma);
```

The Rössler band in text Figure 6.5.18 was generated by the MATLAB script

```
% Fig. 6_5_20.m
y0 = [2; 0; 3];        % set initial point
hold off
[t,y] = ode45(@rosslereq, [0 400], y0);
plot3(y(:,1),y(:,2),y(:,3),'b')
set(gca,'view',[-30 15])
box on;  xlabel('x');  ylabel('y');  zlabel('z')
axis([-4 6 -6 4 -2 8])
```

CHAPTER 7

LAPLACE TRANSFORM METHODS

SECTION 7.1

LAPLACE TRANSFORMS AND INVERSE TRANSFORMS

The objectives of this section are especially clear cut. They include familiarity with the definition of the Laplace transform $\mathcal{L}\{f(t)\} = F(s)$ that is given in Equation (1) in the textbook, the direct application of this definition to calculate Laplace transforms of simple functions (as in Examples 1–3), and the use of known transforms (those listed in Figure 7.1.2) to find Laplace transforms and inverse transforms (as in Examples 4-6). Perhaps students need to be told explicitly to memorize the transforms that are listed in the short table that appears in Figure 7.1.2.

1. $\displaystyle \mathcal{L}\{t\} = \int_0^\infty e^{-st} t\, dt \qquad (u = -st, \quad du = -s\, dt)$

$\displaystyle = \int_0^{-\infty} \left[\frac{1}{s^2}\right] u e^u\, du = \frac{1}{s^2}\Big[(u-1)e^u\Big]_0^{-\infty} = \frac{1}{s^2}$

3. $\displaystyle \mathcal{L}\{e^{3t+1}\} = \int_0^\infty e^{-st} e^{3t+1}\, dt = e\int_0^\infty e^{-(s-3)t}\, dt = \frac{e}{s-3}$

5. $\displaystyle \mathcal{L}\{\sinh t\} = \tfrac{1}{2}\mathcal{L}\{e^t - e^{-t}\} = \tfrac{1}{2}\int_0^\infty e^{-st}\left(e^t - e^{-t}\right)dt = \tfrac{1}{2}\int_0^\infty \left(e^{-(s-1)t} - e^{-(s+1)t}\right)dt$

$\displaystyle = \frac{1}{2}\left[\frac{1}{s-1} - \frac{1}{s+1}\right] = \frac{1}{s^2 - 1}$

7. $\displaystyle \mathcal{L}\{f(t)\} = \int_0^1 e^{-st}\, dt = \left[-\frac{1}{s}e^{-st}\right]_0^1 = \frac{1 - e^{-s}}{s}$

9. $\displaystyle \mathcal{L}\{f(t)\} = \int_0^1 e^{-st} t\, dt = \frac{1 - e^{-s} - se^{-s}}{s^2}$

11. $\displaystyle \mathcal{L}\{\sqrt{t} + 3t\} = \frac{\Gamma(3/2)}{s^{3/2}} + 3\cdot\frac{1}{s^2} = \frac{\sqrt{\pi}}{2s^{3/2}} + \frac{3}{s^2}$

13. $\displaystyle \mathcal{L}\{t - 2e^{3t}\} = \frac{1}{s^2} - \frac{2}{s-3}$

15. $\mathcal{L}\{1 + \cosh 5t\} = \dfrac{1}{s} + \dfrac{s}{s^2 - 25}$

17. $\mathcal{L}\{\cos^2 2t\} = \dfrac{1}{2}\mathcal{L}\{1 + \cos 4t\} = \dfrac{1}{2}\left(\dfrac{1}{s} + \dfrac{s}{s^2 + 16}\right)$

19. $\mathcal{L}\left\{(1+t)^3\right\} = \mathcal{L}\left\{1 + 3t + 3t^2 + t^3\right\} = \dfrac{1}{s} + 3 \cdot \dfrac{1!}{s^2} + 3 \cdot \dfrac{2!}{s^3} + \dfrac{3!}{s^4} = \dfrac{1}{s} + \dfrac{3}{s^2} + \dfrac{6}{s^3} + \dfrac{6}{s^4}$

21.   Integration by parts with $u = t$ and $dv = e^{-st}\cos 2t\, dt$ yields

$$\mathcal{L}\{t\cos 2t\} = \int_0^\infty te^{-st}\cos 2t\, dt = -\dfrac{1}{s^2 + 4}\int_0^\infty e^{-st}\left(-s\cos 2t + 2\sin 2t\right)\, dt$$

$$= -\dfrac{1}{s^2 + 4}\left[-s\,\mathcal{L}\{\cos 2t\} + 2\,\mathcal{L}\{\sin 2t\}\right]$$

$$= -\dfrac{1}{s^2 + 4}\left[\dfrac{-s^2}{s^2 + 4} + \dfrac{4}{s^2 + 4}\right] = \dfrac{s^2 - 4}{\left(s^2 + 4\right)^2}.$$

23. $\mathcal{L}^{-1}\left\{\dfrac{3}{s^4}\right\} = \mathcal{L}^{-1}\left\{\dfrac{1}{2} \cdot \dfrac{6}{s^4}\right\} = \dfrac{1}{2}t^3$

25. $\mathcal{L}^{-1}\left\{\dfrac{1}{s} - \dfrac{2}{s^{5/2}}\right\} = \mathcal{L}^{-1}\left\{\dfrac{1}{s} - \dfrac{2}{\Gamma(5/2)} \cdot \dfrac{\Gamma(5/2)}{s^{5/2}}\right\} = 1 - \dfrac{2}{\frac{3}{2} \cdot \frac{1}{2}\sqrt{\pi}} \cdot t^{3/2} = 1 - \dfrac{8t^{3/2}}{3\sqrt{\pi}}$

27. $\mathcal{L}^{-1}\left\{\dfrac{3}{s - 4}\right\} = 3 \cdot \mathcal{L}^{-1}\left\{\dfrac{1}{s - 4}\right\} = 3e^{4t}$

29. $\mathcal{L}^{-1}\left\{\dfrac{5 - 3s}{s^2 + 9}\right\} = \dfrac{5}{3} \cdot \mathcal{L}^{-1}\left\{\dfrac{3}{s^2 + 9}\right\} - 3 \cdot \mathcal{L}^{-1}\left\{\dfrac{s}{s^2 + 9}\right\} = \dfrac{5}{3}\sin 3t - 3\cos 3t$

31. $\mathcal{L}^{-1}\left\{\dfrac{10s - 3}{25 - s^2}\right\} = -10 \cdot \mathcal{L}^{-1}\left\{\dfrac{s}{s^2 - 25}\right\} + \dfrac{3}{5} \cdot \mathcal{L}^{-1}\left\{\dfrac{5}{s^2 - 25}\right\} = -10\cosh 5t + \dfrac{3}{5}\sinh 5t$

33. $\mathcal{L}\{\sin kt\} = \mathcal{L}\left\{\dfrac{e^{ikt} - e^{-ikt}}{2i}\right\} = \dfrac{1}{2i}\left(\dfrac{1}{s - ik} - \dfrac{1}{s + ik}\right)$

$$= \dfrac{1}{2i} \cdot \dfrac{2ik}{(s - ik)(s - ik)} = \dfrac{k}{s^2 + k^2} \qquad \text{(because } i^2 = -1\text{)}$$

35.   Using the given tabulated integral with $a = -s$ and $b = k$, we find that

$$\mathcal{L}\{\cos kt\} = \int_0^\infty e^{-st} \cos kt \, dt = \left[\frac{e^{-st}}{s^2 + k^2} (-s \cos kt + k \sin kt) \right]_{t=0}^\infty$$

$$= \lim_{t \to \infty} \left(\frac{e^{-st}}{s^2 + k^2} (-s \cos kt + k \sin kt) \right) - \frac{e^0}{s^2 + k^2} (-s \cdot 1 + k \cdot 0) = \frac{s}{s^2 + k^2}.$$

37. $f(t) = 1 - u_a(t) = 1 - u(t - a)$ so

$$\mathcal{L}\{f(t)\} = \mathcal{L}\{1\} - \mathcal{L}\{u_a(t)\} = \frac{1}{s} - \frac{e^{-as}}{s} = s^{-1}(1 - e^{-as}).$$

For the graph of f, note that $f(a) = 1 - u(a) = 1 - 1 = 0$.

39.   Use of the geometric series gives

$$\mathcal{L}\{f(t)\} = \sum_{n=0}^\infty \mathcal{L}\{u(t-n)\} = \sum_{n=0}^\infty \frac{e^{-ns}}{s} = \frac{1}{s}\left(1 + e^{-s} + e^{-2s} + e^{-3s} + \cdots\right)$$

$$= \frac{1}{s}\left(1 + (e^{-s}) + (e^{-s})^2 + (e^{-s})^3 + \cdots\right) = \frac{1}{s} \cdot \frac{1}{1 - e^{-s}} = \frac{1}{s(1 - e^{-s})}.$$

41.   By checking values at sample points, you can verify that $g(t) = 2f(t) - 1$ in terms of the square wave function $f(t)$ of Problem 40. Hence

$$\mathcal{L}\{g(t)\} = \mathcal{L}\{2f(t) - 1\} = \frac{2}{s(1 + e^{-s})} - \frac{1}{s} = \frac{1}{s}\left(\frac{2}{1 + e^{-s}} - 1\right) = \frac{1}{s} \cdot \frac{1 - e^{-s}}{1 + e^{-s}}$$

$$= \frac{1}{s} \cdot \frac{1 - e^{-s}}{1 + e^{-s}} \cdot \frac{e^{s/2}}{e^{s/2}} = \frac{1}{s} \cdot \frac{e^{s/2} - e^{-s/2}}{e^{s/2} + e^{-s/2}} = \frac{1}{s} \cdot \frac{\frac{1}{2}\left(e^{s/2} - e^{-s/2}\right)}{\frac{1}{2}\left(e^{s/2} + e^{-s/2}\right)}$$

$$= \frac{1}{s} \cdot \frac{\sinh(s/2)}{\cosh(s/2)} = \frac{1}{s} \tanh\frac{s}{2}.$$

SECTION 7.2

TRANSFORMATION OF INITIAL VALUE PROBLEMS

The focus of this section is on the use of transforms of derivatives (Theorem 1) to solve initial value problems (as in Examples 1 and 2). Transforms of integrals (Theorem 2) appear less frequently in practice, and the extension of Theorem 1 at the end of Section 7.2 may be considered entirely optional (except perhaps for electrical engineering students).

In Problems 1–10 we give first the transformed differential equation, then the transform $X(s)$ of the solution, and finally the inverse transform $x(t)$ of $X(s)$.

1. $[s^2 X(s) - 5s] + 4\{X(s)\} = 0$

$$X(s) = \frac{5s}{s^2 + 4} = 5 \cdot \frac{s}{s^2 + 4}$$

$$x(t) = \mathcal{L}^{-1}\{X(s)\} = 5\cos 2t$$

3. $[s^2 X(s) - 2] - [sX(s)] - 2[X(s)] = 0$

$$X(s) = \frac{2}{s^2 - s - 2} = \frac{2}{(s-2)(s+1)} = \frac{2}{3}\left(\frac{1}{s-2} - \frac{1}{s+1}\right)$$

$$x(t) = (2/3)(e^{2t} - e^{-t})$$

5. $[s^2 X(s)] + [X(s)] = 2/(s^2 + 4)$

$$X(s) = \frac{2}{(s^2 + 1)(s^2 + 4)} = \frac{2}{3} \cdot \frac{1}{s^2 + 1} - \frac{1}{3} \cdot \frac{2}{s^2 + 4}$$

$$x(t) = (2\sin t - \sin 2t)/3$$

7. $[s^2 X(s) - s] + [X(s)] = s/s^2 + 9)$

$$(s^2 + 1)X(s) = s + s/(s^2 + 9) = (s^3 + 10s)/(s^2 + 9)$$

$$X(s) = \frac{s^2 + 10s}{(s^2 + 1)(s^2 + 9)} = \frac{9}{9} \cdot \frac{s}{s^2 + 1} - \frac{1}{8} \cdot \frac{s}{s^2 + 9}$$

$$x(t) = (9\cos t - \cos 3t)/8$$

9. $s^2 X(s) + 4sX(s) + 3X(s) = 1/s$

$$X(s) = \frac{1}{s(s^2 + 4s + 3)} = \frac{1}{s(s+1)(s+3)} = \frac{1}{3} \cdot \frac{1}{s} - \frac{1}{2} \cdot \frac{1}{s+1} + \frac{1}{6} \cdot \frac{1}{s+3}$$

$$x(t) = (2 - 3e^{-t} + e^{-3t})/6$$

11. The transformed equations are

$$sX(s) - 1 = 2X(s) + Y(s)$$

$$sY(s) + 2 = 6X(s) + 3Y(s).$$

We solve for the Laplace transforms

$$X(s) = \frac{s-5}{s(s-5)} = \frac{1}{s}$$

$$Y(s) = X(s) = \frac{-2s+10}{s(s-5)} = -\frac{2}{s}.$$

Hence the solution is given by

$$x(t) = 1, \qquad\qquad y(t) = -2.$$

13. The transformed equations are

$$sX(s) + 2[sY(s) - 1] + X(s) = 0$$
$$sX(s) - [sY(s) - 1] + Y(s) = 0,$$

which we solve for the transforms

$$X(s) = -\frac{2}{3s^2-1} = -\frac{2}{3}\cdot\frac{1}{s^2-1/3} = -\frac{2}{\sqrt{3}}\cdot\frac{1/\sqrt{3}}{s^2-\left(1/\sqrt{3}\right)^2}$$

$$Y(s) = \frac{3s+1}{3s^2-1} = \frac{s+1/3}{s^2-1/3} = \frac{s}{s^2-\left(1/\sqrt{3}\right)^2} + \frac{1}{\sqrt{3}}\cdot\frac{1/\sqrt{3}}{s^2-\left(1/\sqrt{3}\right)^2}.$$

Hence the solution is

$$x(t) = -\left(2/\sqrt{3}\right)\sinh\left(t/\sqrt{3}\right)$$
$$y(t) = \cosh\left(t/\sqrt{3}\right) + \left(1/\sqrt{3}\right)\sinh\left(t/\sqrt{3}\right).$$

15. The transformed equations are

$$[s^2X - s] + [sX - 1] + [sY - 1] + 2X - Y = 0$$
$$[s^2Y - s] + [sX - 1] + [sY - 1] + 4X - 2Y = 0,$$

which we solve for

$$X(s) = \frac{s^2+3s+2}{s^3+3s^2+3s} = \frac{1}{3}\left(\frac{2}{s} + \frac{s+3}{s^2+3s+3}\right) = \frac{1}{3}\left(\frac{2}{s} + \frac{s+3}{(s+3/2)^2 + (3/4)}\right)$$

$$= \frac{1}{3}\left(\frac{2}{s} + \frac{s+3/2}{(s+3/2)^2 + (\sqrt{3}/2)^2} + \sqrt{3}\cdot\frac{\sqrt{3}/2}{(s+3/2)^2 + (\sqrt{3}/2)^2}\right)$$

$$Y(s) = \frac{-s^3 - 2s^2 + 2s + 4}{s^3 + 3s^2 + 3s} = \frac{1}{21}\left(\frac{28}{s} - \frac{9}{s-1} + \frac{2s+15}{s^2 + 3s + 3}\right)$$

$$= \frac{1}{21}\left(\frac{28}{s} - \frac{9}{s-1} + \frac{2s+15}{(s+3/2)^2 + 3/4}\right)$$

$$= \frac{1}{21}\left(\frac{28}{s} - \frac{9}{s-1} + 2 \cdot \frac{s+3/2}{(s+3/2)^2 + (\sqrt{3}/2)^2} + 8\sqrt{3} \cdot \frac{\sqrt{3}/2}{(s+3/2)^2 + (\sqrt{3}/2)^2}\right).$$

Here we've used some fairly heavy-duty partial fractions (Section 7.3). The transforms

$$\mathcal{L}\{e^{at}\cos kt\} = \frac{s-a}{(s-a)^2 + k^2}, \quad \mathcal{L}\{e^{at}\sin kt\} = \frac{k}{(s-a)^2 + k^2}$$

from the inside-front-cover table (with $a = -3/2$, $k = \sqrt{3}/2$) finally yield

$$x(t) = \frac{1}{3}\left\{2 + e^{-3t/2}\left[\cos\left(\sqrt{3}t/2\right) + \sqrt{3}\sin\left(\sqrt{3}t/2\right)\right]\right\}$$

$$y(t) = \frac{1}{21}\left\{28 - 9e^t + e^{-3t/2}\left[2\cos\left(\sqrt{3}t/2\right) + 8\sqrt{3}\sin\left(\sqrt{3}t/2\right)\right]\right\}.$$

17.　$f(t) = \displaystyle\int_0^t e^{3\tau}\,d\tau = \left[\frac{1}{3}e^{3\tau}\right]_{\tau=0}^t = \frac{1}{3}\left(e^{3t} - 1\right)$

19.　$f(t) = \displaystyle\int_0^t \frac{1}{2}\sin 2\tau\,d\tau = \left[-\frac{1}{4}\cos 2\tau\right]_{\tau=0}^t = \frac{1}{4}\left(1 - \cos 2t\right)$

21.　$f(t) = \displaystyle\int_0^t\left[\int_0^\tau \sin t\,dt\right]d\tau = \int_0^t (1-\cos\tau)\,d\tau = \left[\tau - \sin\tau\right]_{\tau=0}^t = t - \sin t$

23.　$f(t) = \displaystyle\int_0^t\left[\int_0^\tau \sinh t\,dt\right]d\tau = \int_0^t (\cosh\tau - 1)\,d\tau = \left[\sinh\tau - \tau\right]_{\tau=0}^t = \sinh t - t$

25.　With $f(t) = \cos kt$ and $F(s) = s/(s^2 + k^2)$, Theorem 1 in this section yields

$$\mathcal{L}\{-k\sin kt\} = \mathcal{L}\{f'(t)\} = sF(s) - 1 = s \cdot \frac{s}{s^2 + k^2} - 1 = -\frac{k^2}{s^2 + k^2},$$

so division by $-k$ yields $\mathcal{L}\{\sin kt\} = k/(s^2 + k^2)$.

27.　**(a)**　With $f(t) = t^n e^{at}$ and $f'(t) = nt^{n-1}e^{at} + at^n e^{at}$, Theorem 1 yields

$$\mathcal{L}\{nt^{n-1}e^{at} + at^n e^{at}\} = s\,\mathcal{L}\{t^n e^{at}\}$$

so

$$n\,\mathcal{L}\{t^{n-1}e^{at}\} = (s-a)\mathcal{L}\{t^n e^{at}\}$$

and hence

$$\mathcal{L}\{t^n e^{at}\} = \frac{n}{s-a}\mathcal{L}\{t^{n-1}e^{at}\}.$$

(b) $n=1$: $\mathcal{L}\{t e^{at}\} = \dfrac{1}{s-a}\mathcal{L}\{e^{at}\} = \dfrac{1}{s-a}\cdot\dfrac{1}{s-a} = \dfrac{1}{(s-a)^2}$

       $n=2$: $\mathcal{L}\{t^2 e^{at}\} = \dfrac{2}{s-a}\mathcal{L}\{t e^{at}\} = \dfrac{2}{s-a}\cdot\dfrac{1}{(s-a)^2} = \dfrac{2!}{(s-a)^3}$

       $n=3$: $\mathcal{L}\{t^3 e^{at}\} = \dfrac{3}{s-a}\mathcal{L}\{t^2 e^{at}\} = \dfrac{3}{s-a}\cdot\dfrac{2!}{(s-a)^3} = \dfrac{3!}{(s-a)^4}$

And so forth.

29.   Let $f(t) = t \sinh kt$, so $f(0) = 0$. Then

$$f'(t) = \sinh kt + kt \cosh kt$$

$$f''(t) = 2k \cosh kt + k^2 t \sinh kt,$$

and thus $f'(0) = 0$, so Formula (5) in this section yields

$$\mathcal{L}\{2k \cosh kt + k^2 t \sinh kt\} = s^2 \mathcal{L}\{t \sinh kt\},$$

$$2k\cdot\frac{s}{s^2-k^2} + k^2 F(s) = s^2 F(s).$$

We readily solve this last equation for

$$\mathcal{L}\{t \sinh kt\} = F(s) = \frac{2ks}{\left(s^2-k^2\right)^2}.$$

31.   Using the known transform of $\sin kt$ and the Problem 28 transform of $t \cos kt$, we obtain

$$\mathcal{L}\left\{\frac{1}{2k^3}(\sin kt - kt\cos kt)\right\} = \frac{1}{2k^3}\cdot\frac{k}{s^2+k^2} - \frac{k}{2k^3}\cdot\frac{s^2-k^2}{\left(s^2+k^2\right)^2}$$

$$= \frac{1}{2k^2}\left(\frac{1}{s^2+k^2} - \frac{s^2-k^2}{\left(s^2+k^2\right)^2}\right) = \frac{1}{2k^2}\cdot\frac{2k^2}{\left(s^2+k^2\right)^2} = \frac{1}{\left(s^2+k^2\right)^2}$$

33. $f(t) = u_a(t) - u_b(t) = u(t-a) - u(t-b)$, so the result of Problem 32 gives

$$\mathcal{L}\{f(t)\} = \mathcal{L}\{u(t-a)\} - \mathcal{L}\{u(t-b)\} = \frac{e^{-as}}{s} - \frac{e^{-bs}}{s} = \frac{e^{-as} - e^{-bs}}{s}.$$

35.   Let's write $g(t)$ for the on-off function of this problem to distinguish it from the square wave function of Problem 34. Then comparison of Figures 7.2.9 and 7.2.10 makes it clear that $g(t) = \frac{1}{2}(1 + f(t))$, so (using the result of Problem 34) we obtain

$$G(s) = \frac{1}{2s} + \frac{1}{2}F(s) = \frac{1}{2s} + \frac{1}{2s}\tanh\frac{s}{2} = \frac{1}{2s}\left(1 + \frac{e^{s/2} - e^{-s/2}}{e^{s/2} + e^{-s/2}} \cdot \frac{e^{-s/2}}{e^{-s/2}}\right)$$

$$= \frac{1}{2s}\left(1 + \frac{1 - e^{-s}}{1 + e^{-s}}\right) = \frac{1}{2s} \cdot \frac{2}{1 + e^{-s}} = \frac{1}{s(1 + e^{-s})}.$$

37.   We observe that $f(0) = 0$ and that the sawtooth function has jump -1 at each of the points $t_n = n = 1, 2, 3, \cdots$. Also, $f'(t) \equiv 1$ wherever the derivative is defined. Hence Eq. (21) in this section gives

$$\frac{1}{s} = sF(s) + \sum_{n=1}^{\infty} e^{-ns} = sF(s) - 1 + \sum_{n=0}^{\infty} e^{-ns} = sF(s) - 1 + \frac{1}{1 - e^{-ns}},$$

using the geometric series $\sum_{n=0}^{\infty} x^n = 1/(1-x)$ with $x = e^{-s}$. Solution for $F(s)$ gives

$$F(s) = \frac{1}{s^2} + \frac{1}{s} - \frac{1}{s(1 - e^{-s})} = \frac{1}{s^2} - \frac{e^{-s}}{s(1 - e^{-s})}.$$

SECTION 7.3

TRANSLATION AND PARTIAL FRACTIONS

This section is devoted to the computational nuts and bolts of the staple technique for the inversion of Laplace transforms — partial fraction decompositions. If time does not permit going further in this chapter, Sections 7.1–7.3 provide a self-contained introduction to Laplace transforms that suffices for the most common elementary applications.

1. $\mathcal{L}\{t^4\} = \dfrac{24}{s^5}$,   so $\mathcal{L}\{t^4 e^{\pi t}\} = \dfrac{24}{(s - \pi)^5}$

3. $\mathcal{L}\{\sin 3\pi t\} = \dfrac{3\pi}{s^2 + 9\pi^2}$,   so $\mathcal{L}\{e^{-2t}\sin 3\pi t\} = \dfrac{3\pi}{(s + 2)^2 + 9\pi^2}.$

5. $F(s) = \dfrac{3}{2s - 4} = \dfrac{3}{2} \cdot \dfrac{1}{s - 2}$,   so $f(t) = \dfrac{3}{2}e^{2t}$

7. $F(s) = \dfrac{1}{(s+2)^2}$, so $f(t) = t\,e^{-2t}$

9. $F(s) = 3 \cdot \dfrac{s-3}{(s-3)^2+16} + \dfrac{7}{2} \cdot \dfrac{4}{(s-3)^2+16}$, so $f(t) = e^{3t}[3\cos 4t + (7/2)\sin 4t]$

11. $F(s) = \dfrac{1}{4} \cdot \dfrac{1}{s-2} - \dfrac{1}{4} \cdot \dfrac{1}{s+2}$, so $f(t) = \dfrac{1}{4}\left(e^{2t} - e^{-2t}\right) = \dfrac{1}{2}\sinh 2t$

13. $F(s) = 3 \cdot \dfrac{1}{s+2} - 5 \cdot \dfrac{1}{s+5}$, so $f(t) = 3e^{-2t} - 5e^{-5t}$

15. $F(s) = \dfrac{1}{25}\left(-1 \cdot \dfrac{1}{s} - 5 \cdot \dfrac{1}{s^2} + \dfrac{1}{s-5}\right)$, so $f(t) = \dfrac{1}{25}\left(-1 - 5t + e^{5t}\right)$

17. $F(s) = \dfrac{1}{8}\left(\dfrac{1}{s^2-4} - \dfrac{1}{s^2+4}\right) = \dfrac{1}{16}\left(\dfrac{2}{s^2-4} - \dfrac{2}{s^2+4}\right)$

$f(t) = \dfrac{1}{16}\left(\sinh 2t - \sin 2t\right)$

19. $F(s) = \dfrac{s^2-2s}{(s^2+1)(s^2+4)} = \dfrac{1}{3}\left(\dfrac{-2s-1}{s^2+1} + \dfrac{2s+4}{s^2+4}\right)$

$f(t) = \dfrac{1}{3}\left(-2\cos t - \sin t + 2\cos 2t + 2\sin 2t\right)$

21. First we need to find A, B, C, D so that

$$\frac{s^2+3}{\left(s^2+2s+2\right)^2} = \frac{As+B}{s^2+2s+2} + \frac{Cs+D}{\left(s^2+2s+2\right)^2}.$$

When we multiply both sides by the quadratic factor s^2+2s+2 and collect coefficients, we get the linear equations

$$-2B - D + 3 = 0$$
$$-2A - 2B - C = 0$$
$$-2A - B + 1 = 0$$
$$-A = 0$$

which we solve for $A = 0$, $B = 1$, $C = -2$, $D = 1$. Thus

$$F(s) = \frac{1}{(s+1)^2+1} + \frac{-2s+1}{\left[(s+1)^2+1\right]^2} = \frac{1}{(s+1)^2+1} - 2\cdot\frac{s+1}{\left[(s+1)^2+1\right]^2} + 3\cdot\frac{1}{\left[(s+1)^2+1\right]^2}.$$

We now use the inverse Laplace transforms given in Eq. (16) and (17) of Section 7.3 — supplying the factor e^{-t} corresponding to the translation $s \to s+1$ — and get

$$f(t) = e^{-t}\left[\sin t - 2\cdot\frac{1}{2}t\sin t + 3\cdot\frac{1}{2}(\sin t - t\cos t)\right] = \frac{1}{2}e^{-t}(5\sin t - 2t\sin t - 3t\cos t).$$

23. $\dfrac{s^3}{s^4+4a^4} = \dfrac{1}{2}\left(\dfrac{s-a}{s^2-2as+2a^2} + \dfrac{s+a}{s^2+2as+2a^2}\right),$

and $s^2 \pm 2as + 2a^2 = (s\pm a)^2 + a^2$, so it follows that

$$\mathcal{L}^{-1}\left\{\frac{s^3}{s^4+4a^4}\right\} = \frac{1}{2}\left(e^{at} + e^{-at}\right)\cos at = \cosh at \cos at.$$

25. $\dfrac{s}{s^4+4a^4} = \dfrac{1}{4a}\left(\dfrac{s}{s^2-2as+2a^2} - \dfrac{s}{s^2+2as+2a^2}\right)$

$$= \frac{1}{4a}\left(\frac{s-a}{s^2-2as+2a^2} + \frac{a}{s^2-2as+2a^2} - \frac{s+a}{s^2+2as+2a^2} + \frac{a}{s^2+2as+2a^2}\right),$$

and $s^2 \pm 2as + 2a^2 = (s\pm a)^2 + a^2$, so it follows that

$$\mathcal{L}^{-1}\left\{\frac{s}{s^4+4a^4}\right\} = \frac{1}{4a}\left[e^{at}(\cos at + \sin at) - e^{-at}(\cos at - \sin at)\right]$$

$$= \frac{1}{2a}\left[\frac{1}{2}\left(e^{at}+e^{-at}\right)\sin at + \frac{1}{2}\left(e^{at}-e^{-at}\right)\cos at\right]$$

$$= \frac{1}{2a}\left(\cosh at \sin at + \sinh at \cos at\right).$$

In Problems 27–40 we give first the transformed equation, then the Laplace transform $X(s)$ of the solution, and finally the desired solution $x(t)$.

27. $[s^2X(s) - 2s - 3] + 6[sX(s) - 2] + 25X(s) = 0$

$$X(s) = \frac{2s+15}{s^2+6s+25} = 2\cdot\frac{s+3}{(s+3)^2+16} + \frac{9}{4}\cdot\frac{4}{(s+3)^2+16}$$

$x(t) = e^{-3t}[2\cos 4t + (9/4)\sin 4t]$

29. $s^2 X(s) - 4X(s) = \dfrac{3}{s^2}$

$$X(s) = \frac{3}{s^2(s^2-4)} = \frac{3}{4}\left(\frac{1}{s^2-4} - \frac{1}{s^2}\right)$$

$$x(t) = \frac{3}{8}\sinh 2t - \frac{3}{4}t = \frac{3}{8}\left(\sinh 2t - 2t\right)$$

31. $[s^3 X(s) - s - 1] + [s^2 X(s) - 1] - 6[sX(s)] = 0$

$$X(s) = \frac{s+2}{s^3+s^2-6s} = \frac{1}{15}\left(-\frac{5}{s} - \frac{1}{s+3} + \frac{6}{s-2}\right)$$

$$x(t) = \frac{1}{15}\left(-5 - e^{-3t} + 6e^{2t}\right)$$

33. $[s^4 X(s) - 1] + X(s) = 0$

$$X(s) = \frac{1}{s^4+1}$$

It therefore follows from Problem 26 with $a = \sqrt[4]{1/4} = 1/\sqrt{2}$ that

$$x(t) = \frac{1}{\sqrt{2}}\left(\cosh\frac{t}{\sqrt{2}}\sin\frac{t}{\sqrt{2}} - \sinh\frac{t}{\sqrt{2}}\cos\frac{t}{\sqrt{2}}\right).$$

35. $\left[s^4 X(s) - 1\right] + 8s^2 X(s) + 16 X(s) = 0$

$$X(s) = \frac{1}{s^4+8s^2+16} = \frac{1}{\left(s^2+4\right)^2}$$

$$x(t) = \frac{1}{16}\left(\sin 2t - 2t\cos 2t\right) \qquad \text{(by Eq. (17) in Section 7.3)}$$

37. $\left[s^2 X(s) - 2\right] + 4sX(s) + 13X(s) = \dfrac{1}{(s+1)^2}$

$$X(s) = \frac{2 + 1/(s+1)^2}{s^2+4s+13} = \frac{2s^2+4s+13}{(s+1)^2\left(s^2+4s+3\right)}$$

$$= \frac{1}{50}\left[-\frac{1}{s+1} + \frac{5}{(s+1)^2} + \frac{s+98}{(s+2)^2+9} \right]$$

$$= \frac{1}{50}\left[-\frac{1}{s+1} + \frac{5}{(s+1)^2} + \frac{s+2}{(s+2)^2+9} + 32\cdot\frac{3}{(s+2)^2+9} \right]$$

$$x(t) = \frac{1}{50}\left[(-1+5t)e^{-t} + e^{-2t}(\cos 3t + 32\sin 3t) \right]$$

39. $x'' + 9x = 6\cos 3t, \quad x(0) = x'(0) = 0$

$$s^2 X(s) + 9X(s) = \frac{6s}{s^2+9}$$

$$X(s) = \frac{6s}{\left(s^2+9\right)^2}$$

$$x(t) = 6\cdot\frac{1}{2\cdot 3}t\sin 3t = t\sin 3t \qquad \text{(by Eq. (16) in Section 7.3)}$$

The graph of this resonance is shown in the figure below.

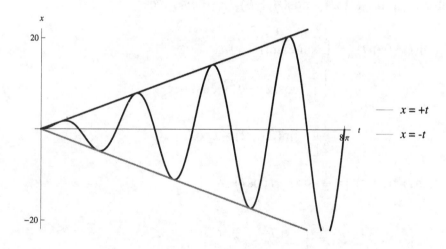

SECTION 7.4

DERIVATIVES, INTEGRALS, AND PRODUCTS OF TRANSFORMS

This section completes the presentation of the standard "operational properties" of Laplace transforms, the most important one here being the convolution property $\mathcal{L}\{f*g\} = \mathcal{L}\{f\}\cdot\mathcal{L}\{g\}$, where the **convolution** $f*g$ is defined by

$$f*g(t) = \int_0^t f(x)g(t-x)\,dx.$$

Here we use x rather than τ as the variable of integration; compare with Eq. (3) in Section 7.4 of the textbook.

1.   With $f(t)=t$ and $g(t)=1$ we calculate

$$t*1 = \int_0^t x\cdot 1\,dx = \left[\frac{1}{2}x^2\right]_{x=0}^{x=t} = \frac{1}{2}t^2.$$

3.   To compute $(\sin t)*(\sin t) = \int_0^t \sin x \sin(t-x)\,dx,$ we first apply the identity $\sin A \sin B = [\cos(A-B) - \cos(A+B)]/2.$ This gives

$$\begin{aligned}
(\sin t)*(\sin t) &= \int_0^t \sin x \sin(t-x)\,dx \\
&= \frac{1}{2}\int_0^t [\cos(2x-t) - \cos t]\,dx \\
&= \frac{1}{2}\left[\frac{1}{2}\sin(2x-t) - x\cos t\right]_{x=0}^{x=t} \\
(\sin t)*(\sin t) &= \frac{1}{2}(\sin t - t\cos t).
\end{aligned}$$

5. $e^{at}*e^{at} = \int_0^t e^{ax}e^{a(t-x)}\,dx = \int_0^t e^{at}\,dx = e^{at}\left[x\right]_{x=0}^{x=t} = t\,e^{at}$

7. $f(t) = 1*e^{3t} = e^{3t}*1 = \int_0^t e^{3x}\cdot 1\,dx = \frac{1}{3}\left(e^{3t} - 1\right)$

9. $f(t) = \frac{1}{9}\sin 3t * \sin 3t = \frac{1}{9}\int_0^t \sin 3x \sin 3(t-x)\,dx$

$$= \frac{1}{9}\int_0^t \sin 3x \left[\sin 3t \cos 3x - \cos 3t \sin 3x\right] dx$$

$$= \frac{1}{9}\sin 3t \int_0^t \sin 3x \cos 3x \, dx - \frac{1}{9}\cos 3t \int_0^t \sin^2 3x \, dx$$

$$= \frac{1}{9}\sin 3t \left[\frac{1}{6}\sin^2 3x\right]_{x=0}^{x=t} - \frac{1}{9}\cos 3t \left[\frac{1}{2}\left(x - \frac{1}{6}\sin 6x\right)\right]_{x=0}^{x=t}$$

$$f(t) = \frac{1}{54}\left(\sin 3t - 3t \cos 3t\right)$$

11. $$f(t) = \cos 2t * \cos 2t = \int_0^t \cos 2x \cos 2(t-x) \, dx$$

$$= \int_0^t \cos 2x \left(\cos 2t \cos 2x + \sin 2t \sin 2x\right) dx$$

$$= (\cos 2t)\int_0^t \cos^2 2x \, dx + (\sin 2t)\int_0^t \cos 2x \sin 2x \, dx$$

$$= (\cos 2t)\left[\frac{1}{2}\left(x + \frac{1}{4}\sin 4x\right)\right]_{x=0}^{x=t} + (\sin 2t)\left[\frac{1}{4}\sin^2 2x\right]_{x=0}^{x=t}$$

$$f(t) = \frac{1}{4}\left(\sin 2t + 2t \cos 2t\right)$$

13. $$f(t) = e^{3t} * \cos t = \int_0^t (\cos x) e^{3(t-x)} dx$$

$$= e^{3t}\int_0^t e^{-3x} \cos x \, dx$$

$$= e^{3t}\left[\frac{e^{-3x}}{10}(-3\cos x + \sin x)\right]_{x=0}^{x=t} \qquad \text{(by integral formula \#50)}$$

$$f(t) = \frac{1}{10}\left(3e^{3t} - 3\cos t + \sin t\right)$$

15. $$\mathcal{L}\{t \sin 3t\} = -\frac{d}{ds}\left(\mathcal{L}\{\sin 3t\}\right) = -\frac{d}{ds}\left(\frac{3}{s^2+9}\right) = \frac{6s}{\left(s^2+9\right)^2}$$

17. $$\mathcal{L}\{e^{2t}\cos 3t\} = (s-2)/(s^2 - 4s + 13)$$

$$\mathcal{L}\{te^{2t}\cos 3t\} = -(d/ds)[(s-2)/(s^2 - 4s + 13)] = (s^2 - 4s - 5)/(s^2 - 4s + 13)^2$$

19. $$\mathcal{L}\left\{\frac{\sin t}{t}\right\} = \int_s^\infty \frac{ds}{s^2+1} = \left[\tan^{-1} s\right]_s^\infty = \frac{\pi}{2} - \tan^{-1} s = \tan^{-1}\left(\frac{1}{s}\right)$$

21. $\mathcal{L}\left\{e^{3t}-1\right\} = \dfrac{1}{s-3} - \dfrac{1}{s}, \text{ so}$

$$\mathcal{L}\left\{\dfrac{e^{3t}-1}{t}\right\} = \int_{s}^{\infty}\left(\dfrac{1}{s-3}-\dfrac{1}{s}\right)ds = \left[\ln\left(\dfrac{s-3}{s}\right)\right]_{s}^{\infty} = \ln\left(\dfrac{s}{s-3}\right)$$

23. $f(t) = -\dfrac{1}{t}\mathcal{L}^{-1}\{F'(s)\} = -\dfrac{1}{t}\mathcal{L}^{-1}\left\{\dfrac{1}{s-2}-\dfrac{1}{s+2}\right\} = -\dfrac{1}{t}\left(e^{2t}-e^{-2t}\right) = -\dfrac{2\sinh 2t}{t}$

25. $f(t) = -\dfrac{1}{t}\mathcal{L}^{-1}\{F'(s)\} = -\dfrac{1}{t}\mathcal{L}^{-1}\left\{\dfrac{2s}{s^2+1}-\dfrac{1}{s+2}-\dfrac{1}{s-3}\right\} = \dfrac{1}{t}\left(e^{-2t}+e^{3t}-2\cos t\right)$

27. $f(t) = -\dfrac{1}{t}\mathcal{L}^{-1}\{F'(s)\} = -\dfrac{1}{t}\mathcal{L}^{-1}\left\{\dfrac{-2/s^3}{1+1/s^2}\right\}$

$$= \dfrac{2}{t}\mathcal{L}^{-1}\left\{\dfrac{1}{s^3+s}\right\} = \dfrac{2}{t}\mathcal{L}^{-1}\left\{\dfrac{1}{s}-\dfrac{s}{s^2+1}\right\} = \dfrac{2}{t}(1-\cos t)$$

29. $-[s^2 X(s) - x'(0)]' - [s\,X(s)]' - 2[s\,X(s)] + X(s) = 0$

$s(s+1)X'(s) + 4s\,X(s) = 0$ (separable)

$X(s) = \dfrac{A}{(s+1)^4}$ with $A \neq 0$

$x(t) = Ct^3 e^{-t}$ with $C \neq 0$

31. $-[s^2 X(s) - x'(0)]' + 4[s\,X(s)]' - [s\,X(s)] - 4[X(s)]' + 2X(s) = 0$

$(s^2 - 4s + 4)X'(s) + (3s - 6)X(s) = 0$ (separable)

$(s - 2)X'(s) + 3X(s) = 0$

$X(s) = \dfrac{A}{(s-2)^3}$ with $A \neq 0$

$x(t) = Ct^2 e^{2t}$ with $C \neq 0$

33. $-[s^2 X(s) - x(0)]' - 2[s\,X(s)] - [X(s)]' = 0$

$(s^2 + 1)X'(s) + 4s\,X(s) = 0$ (separable)

$X(s) = \dfrac{A}{(s^2+1)^2}$ with $A \neq 0$

$x(t) = C(\sin t - t \cos t)$ with $C \neq 0$

35. $\mathcal{L}^{-1}\left\{\dfrac{1}{(s-1)\sqrt{s}}\right\} = e^t * \dfrac{1}{\sqrt{\pi t}} = \displaystyle\int_0^t \dfrac{1}{\sqrt{\pi x}} \cdot e^{t-x}\, dx$

$$= \dfrac{e^t}{\sqrt{\pi}}\int_0^{\sqrt{t}} \dfrac{1}{u}\cdot e^{-u^2}\cdot 2u\, du = \dfrac{2e^t}{\sqrt{\pi}}\int_0^{\sqrt{t}} e^{-u^2}\, du = e^t \mathrm{erf}\left(\sqrt{t}\right)$$

37. $s^2 X(s) + 2sX(s) + X(s) = F(s)$

$X(s) = F(s)\cdot \dfrac{1}{(s+1)^2}$

$x(t) = te^{-t} * f(t) = \displaystyle\int_0^t \tau e^{-\tau} f(t-\tau)\, d\tau$

SECTION 7.5

PERIODIC AND PIECEWISE CONTINUOUS INPUT FUNCTIONS

In Problems 1 through 10, we first derive the inverse Laplace transform $f(t)$ of $F(s)$ and then show the graph of $f(t)$.

1. $F(s) = e^{-3s}\mathcal{L}\{t\}$ so Eq. (3b) in Theorem 1 gives

$$f(t) = u(t-3)\cdot(t-3) = \begin{cases} 0 & \text{if } t < 3, \\ t-3 & \text{if } t \geq 3. \end{cases}$$

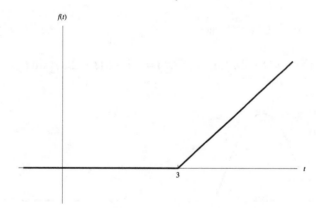

3. $F(s) = e^{-s}\mathcal{L}\{e^{-2t}\}$ so $f(t) = u(t-1)\cdot e^{-2(t-1)} = \begin{cases} 0 & \text{if } t < 1, \\ e^{-2(t-1)} & \text{if } t \geq 1. \end{cases}$

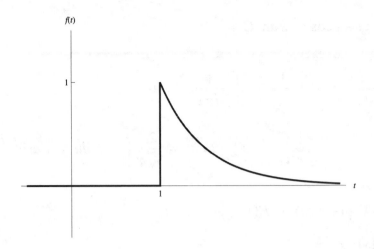

5. $F(s) = e^{-\pi s}\mathcal{L}\{\sin t\}$ so

$$f(t) = u(t-\pi)\cdot\sin(t-\pi) = -u(t-\pi)\sin t = \begin{cases} 0 & \text{if } t < \pi, \\ -\sin t & \text{if } t \geq \pi. \end{cases}$$

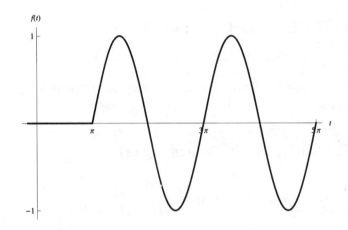

7. $F(s) = \mathcal{L}\{\sin t\} - e^{-2\pi s}\mathcal{L}\{\sin t\}$ so

$$f(t) = \sin t - u(t-2\pi)\sin(t-2\pi) = \left[1-u(t-2\pi)\right]\sin t = \begin{cases} \sin t & \text{if } t < 2\pi, \\ 0 & \text{if } t \geq 2\pi. \end{cases}$$

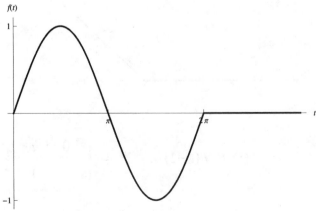

9. $F(s) = \mathcal{L}\{\cos\pi t\} + e^{-3s}\mathcal{L}\{\cos\pi t\}$ so

$$f(t) = \cos\pi t + u(t-3)\cos\pi(t-3) = [1-u(t-3)]\cos\pi t = \begin{cases} \cos\pi t & \text{if } t < 3, \\ 0 & \text{if } t \geq 3. \end{cases}$$

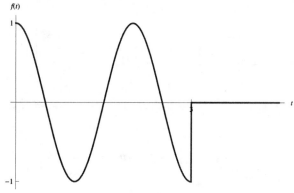

11. $f(t) = 2 - u(t-3)\cdot 2$ so $F(s) = \dfrac{2}{s} - e^{-3s}\dfrac{2}{s} = \dfrac{2}{s}\left(1 - e^{-3s}\right).$

13. $f(t) = [1-u(t-2\pi)]\sin t = \sin t - u(t-2\pi)\sin(t-2\pi)$ so

$$F(s) = \frac{1}{s^2+1} - e^{-2\pi s}\cdot\frac{1}{s^2+1} = \frac{1-e^{-2\pi s}}{s^2+1}.$$

15. $f(t) = [1-u(t-3\pi)]\sin t = \sin t + u(t-3\pi)\sin(t-3\pi)$ so

$$F(s) = \frac{1}{s^2+1} + \frac{e^{-3\pi s}}{s^2+1} = \frac{1+e^{-3\pi s}}{s^2+1}.$$

17. $f(t) = [u(t-2) - u(t-3)]\sin\pi t = u(t-2)\sin\pi(t-2) + u(t-3)\sin\pi(t-3)$ so

$$F(s) = \left(e^{-2s} + e^{-3s}\right)\cdot\frac{\pi}{s^2+\pi^2} = \frac{\pi\left(e^{-2s} + e^{-3s}\right)}{s^2+\pi^2}.$$

19. If $g(t) = t+1$ then $f(t) = u(t-1)\cdot t = u(t-1)\cdot g(t-1)$ so

$$F(s) = e^{-s}G(s) = e^{-s}L\{t+1\} = e^{-s}\cdot\left(\frac{1}{s^2} + \frac{1}{s}\right) = \frac{e^{-s}(s+1)}{s^2}.$$

21. If $g(t) = t+1$ and $h(t) = t+2$ then

$$\begin{aligned} f(t) &= t[1-u(t-1)] + (2-t)[u(t-1)-u(t-2)] \\ &= t - 2t\,u(t-1) + 2u(t-1) - 2u(t-2) + t\,u(t-2) \\ &= t - 2u(t-1)g(t-1) + 2u(t-1) - 2u(t-2) + u(t-2)h(t-2) \end{aligned}$$

so

$$F(s) = \frac{1}{s^2} - 2e^{-s}\left(\frac{1}{s^2} + \frac{1}{s}\right) + \frac{2e^{-s}}{s} - \frac{2e^{-2s}}{s} + e^{-2s}\left(\frac{1}{s^2} + \frac{2}{s}\right) = \frac{\left(1 - e^{-s}\right)^2}{s^2}.$$

23. With $f(t) = 1$ and $p = 1$, Formula (12) in the text gives

$$\mathcal{L}\{1\} = \frac{1}{1 - e^{-s}} \int_0^1 e^{-st} \cdot 1\, dt = \frac{1}{1 - e^{-s}}\left[-\frac{e^{-st}}{s}\right]_{t=0}^{t=1} = \frac{1}{s}.$$

25. With $p = 2a$ and $f(t) = 1$ if $0 \le t \le a$, $f(t) = 0$ if $a < t \le 2a$, Formula (12) gives

$$\mathcal{L}\{f(t)\} = \frac{1}{1 - e^{-2as}} \int_0^a e^{-st} \cdot 1\, dt = \frac{1}{1 - e^{-2as}}\left[-\frac{e^{-st}}{s}\right]_{t=0}^{t=a}$$

$$= \frac{1 - e^{-as}}{s\left(1 - e^{-as}\right)\left(1 + e^{-as}\right)} = \frac{1}{s\left(1 + e^{-as}\right)}.$$

27. $G(s) = \mathcal{L}\{t/a - f(t)\} = (1/as^2) - F(s)$. Now substitution of the result of Problem 26 in place of $F(s)$ immediately gives the desired transform.

29. With $p = 2\pi/k$ and $f(t) = \sin kt$ for $0 \le t \le \pi/k$ while $f(t) = 0$ for $\pi/k \le t \le 2\pi/k$, Formula (12) and the integral formula

$$\int e^{at} \sin bt\, dt = e^{at}\left[\frac{a \sin bt - b \cos bt}{a^2 + b^2}\right] + C$$

give

$$\mathcal{L}\{f(t)\} = \frac{1}{1 - e^{-2\pi s/k}} \int_0^{\pi/k} e^{-st} \cdot \sin kt\, dt$$

$$= \frac{1}{1 - e^{-2\pi s/k}}\left[e^{-st}\left(\frac{-s \sin kt - k \cos kt}{s^2 + k^2}\right)\right]_{t=0}^{t=\pi/k}$$

$$= \frac{1}{1 - e^{-2\pi s/k}}\left[\frac{e^{-\pi s/k}\left(k\right) - \left(-k\right)}{s^2 + k^2}\right]$$

$$= \frac{k\left(1 + e^{-\pi s/k}\right)}{\left(1 - e^{-\pi s/k}\right)\left(1 + e^{-\pi s/k}\right)\left(s^2 + k^2\right)} = \frac{k}{\left(s^2 + k^2\right)\left(1 - e^{-\pi s/k}\right)}.$$

In Problems 31-42, we first write and transform the appropriate differential equation. Then we solve for the transform of the solution, and finally inverse transform to find the desired solution.

31. $x'' + 4x = 1 - u(t - \pi)$

$$s^2 X(s) + 4X(s) = \frac{1 - e^{-\pi s}}{s}$$

$$X(s) = \frac{1 - e^{-\pi s}}{s(s^2 + 4)} = \frac{1}{4}(1 - e^{-\pi s})\left(\frac{1}{s} - \frac{s}{s^2 + 4}\right)$$

$$x(t) = (1/4)[1 - u(t - \pi)][1 - \cos 2(t - \pi)] = (1/2)[1 - u(t - \pi)]\sin^2 t$$

The graph of the position function $x(t)$ is shown at the top of the next page.

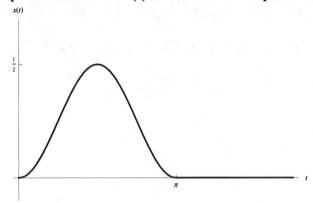

33. $x'' + 9x = [1 - u(t - 2\pi)]\sin t$

$$X(s) = \frac{1 - e^{-2\pi s}}{(s^2 + 1)(s^2 + 9)} = \frac{1}{8}(1 - e^{-2\pi s})\left(\frac{1}{s^2 + 1} - \frac{1}{s^2 + 9}\right)$$

$$x(t) = \frac{1}{8}[1 - u(t - 2\pi)]\left(\sin t - \frac{1}{3}\sin 3t\right)$$

The figure below show the graph of this position function.

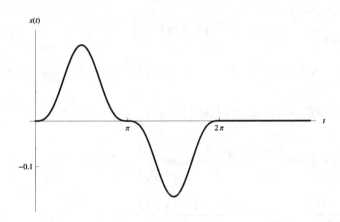

35. $x'' + 4x' + 4x = [1 - u(t - 2)]t = t - u(t - 2)g(t - 2)$ where $g(t) = t + 2$

$$(s+2)^2 X(s) = \frac{1}{s^2} - e^{-2s}\left(\frac{2}{s} + \frac{1}{s^2}\right)$$

$$X(s) = \frac{1}{s^2(s+2)^2} - e^{-2s}\frac{2s+1}{s^2(s+2)^2}$$

$$= \frac{1}{4}\left(-\frac{1}{s} + \frac{1}{s^2} + \frac{1}{s+2} + \frac{1}{(s+2)^2}\right) - \frac{1}{4}e^{-2s}\left(\frac{1}{s} + \frac{1}{s^2} - \frac{1}{s+2} - \frac{3}{(s+2)^2}\right)$$

$$x(t) = (1/4)\{-1 + t + (1+t)e^{-2t} + u(t-2)[1 - t + (3t-5)e^{-2(t-2)}]\}$$

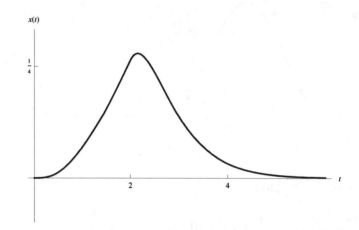

37. $i'(t) + 10^4 \int i(t)\, dt = 100[1 - u(t - 2\pi)]$

$$sI(s) + 10^4 \frac{I(s)}{s} = 100\frac{1 - e^{-2\pi s}}{s}$$

$$I(s) = \frac{100\left(1 - e^{-2\pi s}\right)}{s^2 + 10^4} = \left(1 - e^{-2\pi s}\right)\mathcal{L}\{\sin 100t\}$$

$$i(t) = \sin 100t - u(t - 2\pi)\sin 100(t - 2\pi) = [1 - u(t - 2\pi)]\sin 100t$$

39. $i'(t) + 150\, i(t) + 5000 \int i(t)\, dt = 100t[1 - u(t - 1)]$

$$sI(s) + 150I(s) + 5000\frac{I(s)}{s} = \frac{100}{s^2} - 100e^{-s}\left(\frac{1}{s} + \frac{1}{s^2}\right)$$

$$I(s) = \frac{100}{s(s+50)(s+100)} - e^{-s}\cdot\frac{100(s+1)}{s(s+50)(s+100)}$$

$$= \frac{1}{50}\left(\frac{1}{s} - \frac{2}{s+50} + \frac{1}{s+100}\right) - \frac{1}{50}e^{-s}\left(\frac{1}{s} + \frac{98}{s+50} - \frac{99}{s+100}\right)$$

$$i(t) = (1/50)[1 - 2e^{-50t} + e^{-100t}] - (1/50)u(t-1)[1 + 98e^{-50(t-1)} - 99e^{-100(t-1)}]$$

41. $x'' + 4x = f(t), \qquad x(0) = x'(0) = 0$

$$\left(s^2 + 4\right)X(s) = \frac{4\left(1 - e^{-\pi s}\right)}{s\left(1 + e^{-\pi s}\right)} \qquad \text{(by Example 6 of Section 7.5)}$$

$$\left(s^2 + 4\right)X(s) = \frac{4}{s} + \frac{8}{s}\sum_{n=1}^{\infty}(-1)^n e^{-n\pi s} \qquad \text{(as in Eq. (16) of Section 7.5)}$$

Now let

$$g(t) = \mathcal{L}^{-1}\left\{\frac{4}{s(s^2 + 4)}\right\} = 1 - \cos 2t = 2\sin^2 t.$$

Then it follows that

$$x(t) = g(t) + 2\sum_{n=1}^{\infty}(-1)^n u_{n\pi}(t)g(t - n\pi) = 2\sin^2 t + 4\sum_{n=1}^{\infty}(-1)^n u_{n\pi}(t)\sin^2 t.$$

Hence

$$x(t) = \begin{cases} 2\sin^2 t & \text{if } 2n\pi \le t < (2n+1)\pi, \\ -2\sin^2 t & \text{if } (2n-1)\pi \le t < 2n\pi. \end{cases}$$

Consequently the complete solution

$$x(t) = 2|\sin t|\sin t$$

is periodic, so the transient solution is zero. The graph of $x(t)$:

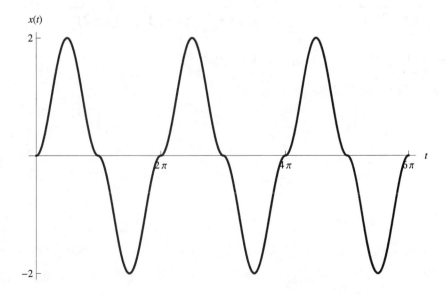

SECTION 7.6

IMPULSES AND DELTA FUNCTIONS

Among the several ways of introducing delta functions, we consider the physical approach of the first two pages of this section to be the most tangible one for elementary students. Whatever the approach, however, the practical consequences are the same — as described in the discussion associated with equations (11)–(19) in the text. That is, in order to solve a differential equation of the form

$$a\,x''(t) + b\,x'(t) + c\,x(t) \; = \; f(t)$$

where $f(t)$ involves delta functions, we transform the equation using the operational principle $\mathcal{L}\{\delta_a(t)\} \; = \; e^{-as}$, then solve for $X(s)$, and finally invert as usual to find the formal solution $x(t)$. Then we show the graph of $x(t)$.

1. $\quad s^2 X(s) + 4X(s) \; = \; 1$

$$X(s) \; = \; \frac{1}{s^2 + 4}$$

$$x(t) \; = \; \frac{1}{2}\sin 2t$$

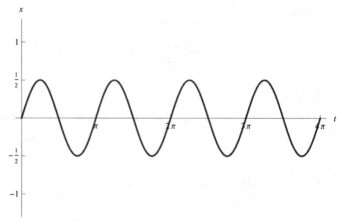

3. $\quad s^2 X(s) + 4sX(s) + 4X(s) \; = \; \dfrac{1}{s} + e^{-2s}$

$$X(s) \; = \; \frac{1}{s(s+2)^2} + \frac{e^{-2s}}{(s+2)^2} = \frac{1}{4}\left(\frac{1}{s} - \frac{1}{s+2} - \frac{2}{(s+2)^2}\right) + \frac{e^{-2s}}{(s+2)^2}$$

$$x(t) \; = \; \frac{1}{4}\left[1 - e^{-2t} - 2t\,e^{-2t}\right] + u(t-2)(t-2)e^{-2(t-2)}$$

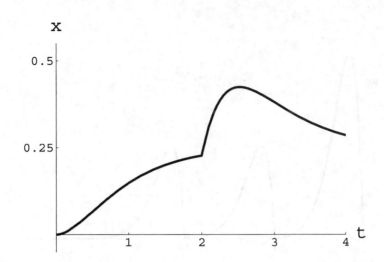

5. $(s^2 + 2s + 2)X(s) = 2e^{-\pi s}$

$$X(s) = \frac{2e^{-\pi s}}{(s+1)^2 + 1}$$

$$x(t) = 2u(t - \pi)e^{-(t-\pi)}\sin(t - \pi) = \begin{cases} 0 & \text{if} \quad 0 \le t \le \pi, \\ -2e^{-(t-\pi)}\sin t & \text{if} \quad t \ge \pi. \end{cases}$$

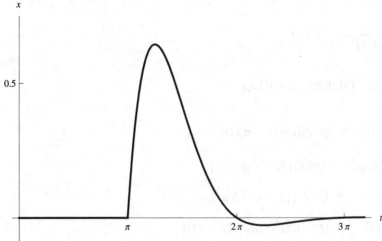

7. $[s^2 X(s) - 2] + 4sX(s) + 5X(s) = e^{-\pi s} + e^{-2\pi s}$

$$X(s) = \frac{2 + e^{-\pi s} + e^{-2\pi s}}{(s+2)^2 + 1}$$

$x(t) = 2e^{-2t}\sin t + u_\pi(t)e^{-2(t-\pi)}\sin(t - \pi) + u_{2\pi}(t)e^{-2(t-2\pi)}\sin(t - 2\pi)$

$\quad\quad = [2 - e^{2\pi}u(t - \pi) + e^{4\pi}u(t - 2\pi)]\,e^{-2t}\sin t$

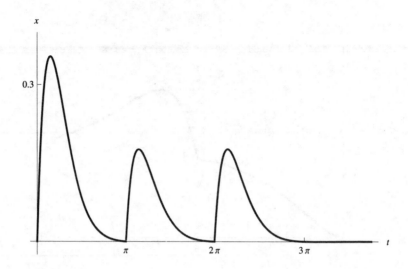

9. $s^2 X(s) + 4X(s) = F(s)$

$$X(s) = \frac{1}{s^2 + 4} \cdot F(s)$$

$$x(t) = \tfrac{1}{2} \int_0^t (\sin 2u)\, f(t - u)\, du$$

11. $(s^2 + 6s + 8)X(s) = F(s)$

$$X(s) = \frac{1}{(s+3)^2 - 1} \cdot F(s)$$

$$x(t) = \int_0^t e^{-3u} (\sinh u)\, f(t - u)\, du$$

13. **(a)** $mx_\varepsilon''(t) = (p/\varepsilon)[u_0(t) - u_\varepsilon(t)]$

$ms^2 X_\varepsilon(s) = (p/\varepsilon)[1/s - e^{-\varepsilon s}/s]$

$mX_\varepsilon(s) = (p/\varepsilon)[(1 - e^{-\varepsilon s})/s^3]$

$mx_\varepsilon(t) = (p/2\varepsilon)[t^2 - u_\varepsilon(t)(t - \varepsilon)^2]$

(b) If $t \geq \varepsilon$ then

$$mx_\varepsilon(t) = (p/2\varepsilon)[t^2 - (t^2 - 2\varepsilon t + \varepsilon^2)] = (p/2\varepsilon)(2\varepsilon t - \varepsilon^2).$$

Hence $mx_\varepsilon(t) \to pt$ as $\varepsilon \to 0$.

(c) $mv = (mx)' = (pt)' = p.$

15. Each of the two given initial value problems transforms to

$$(ms^2 + k)X(s) = mv_0 = p_0.$$

17.   **(b)** $i' + 100i = \delta_1(t) - \delta_2(t), \quad i(0) = 0$

$$I(s) = \frac{e^{-s} - e^{-2s}}{s + 100} I(s)$$

$$i(t) = u_1(t)e^{-100(t-1)} - u_2(t)e^{-100(t-2)}$$

19. $\left(s^2 + 100\right)I(s) = 10\sum_{n=0}^{\infty}(-1)^n e^{-n\pi s/10}$

$$I(s) = \frac{10\sum_{n=0}^{\infty}(-1)^n e^{-n\pi s/10}}{s^2 + 100} = \sum_{n=0}^{\infty}\left((-1)^n e^{-n\pi s/10} \cdot \frac{10}{s^2 + 100}\right)$$

$$i(t) = \sum_{n=0}^{\infty}(-1)^n u_{n\pi/10}(t)\sin 10(t - n\pi/10) = \sum_{n=0}^{\infty}u(t - n\pi/10)\sin 10t$$

because $\sin(10t - n\pi) = (-1)^n \sin 10t$. Hence

$$i(t) = (n+1)\sin 10t$$

if $n\pi/10 < t < (n+1)\pi/10$.

21. $\left(s^2 + 60s + 1000\right)I(s) = 10\sum_{n=0}^{\infty}(-1)^n e^{-n\pi s/10}$

$$I(s) = \frac{10\sum_{n=0}^{\infty}(-1)^n e^{-n\pi s/10}}{s^2 + 60s + 1000} = \sum_{n=0}^{\infty}\left((-1)^n e^{-n\pi s/10} \cdot \frac{10}{(s+30)^2 + 100}\right)$$

$$i(t) = \Sigma\,(-1)^n u_{n\pi/10}(t)\,g(t - n\pi/10)$$

where $g(t) = e^{-30t}\sin 10t$, and so

$$g(t - n\pi/10) = \exp[-30(t - n\pi/10)]\sin 10(t - n\pi/10)$$

$$= e^{3n\pi}e^{-30t}\cdot(-1)^n\sin 10t$$

Therefore

$$i(t) = \sum_{n=0}^{\infty}u\left(t - \frac{n\pi}{10}\right)e^{3n\pi}e^{-30t}\sin 10t\,.$$

If $n\pi/10 < t < (n+1)\pi/10$ then it follows that

$$i(t) = (1 + e^{3\pi} + \cdots + e^{3n\pi})e^{-30t}\sin 10t = \frac{e^{3(n+1)\pi} - 1}{e^{3\pi} - 1}e^{-30t}\sin 10t.$$

The graph of $i(t)$ is shown.

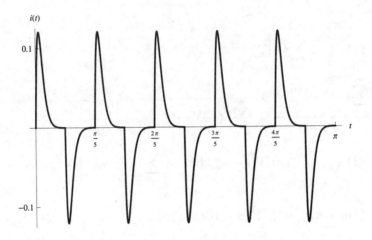

CHAPTER 8

POWER SERIES METHODS

SECTION 8.1

INTRODUCTION AND REVIEW OF POWER SERIES

The power series method consists of substituting a series $y = \Sigma c_n x^n$ into a given differential equation in order to determine what the coefficients $\{c_n\}$ must be in order that the power series will satisfy the equation. It might be pointed out that, if we find a recurrence relation in the form $c_{n+1} = \phi(n)c_n$, then we can determine the radius of convergence ρ of the series solution directly from the recurrence relation

$$\rho = \lim_{n \to \infty} \left| \frac{c_n}{c_{n+1}} \right| = \lim_{n \to \infty} \left| \frac{1}{\phi(n)} \right|.$$

In Problems 1–10 we give first a recurrence relation that can be used to find the radius of convergence and to calculate the succeeding coefficients $c_1, c_2, c_3, \cdots$ in terms of the arbitrary constant c_0. Then we give the series itself.

1. $\quad c_{n+1} = \dfrac{c_n}{n+1}$; it follows that $c_n = \dfrac{c_0}{n!}$ and $\rho = \lim_{n \to \infty}(n+1) = \infty$.

$$y(x) = c_0\left(1 + x + \frac{x^2}{2} + \frac{x^3}{6} + \frac{x^4}{24} + \cdots\right) = c_0\left(1 + \frac{x}{1!} + \frac{x^2}{2!} + \frac{x^3}{3!} + \frac{x^4}{4!} + \cdots\right) = c_0 e^x$$

3. $\quad c_{n+1} = -\dfrac{3c_n}{2(n+1)}$; it follows that $c_n = \dfrac{(-1)^n 3^n c_0}{2^n n!}$ and $\rho = \lim_{n \to \infty} \dfrac{2(n+1)}{3} = \infty$.

$$y(x) = c_0\left(1 - \frac{3x}{2} + \frac{9x^2}{8} - \frac{9x^3}{16} + \frac{27x^4}{128} - \cdots\right)$$

$$= c_0\left(1 - \frac{3x}{1!2} + \frac{3^2 x^2}{2!2^2} - \frac{3^3 x^3}{3!2^3} + \frac{3^4 x^4}{4!2^4} - \cdots\right) = c_0 e^{-3x/2}$$

5. When we substitute $y = \Sigma c_n x^n$ into the equation $y' = x^2 y$, we find that

$$c_1 + 2c_2 x + \sum_{n=0}^{\infty}\left[(n+3)c_{n+3} - c_n\right]x^{n+2} = 0.$$

Hence $c_1 = c_2 = 0$ — which we see by equating constant terms and x-terms on the two sides of this equation — and $c_{n+3} = \dfrac{c_n}{n+3}$. It follows that

$$c_{3k+1} = c_{3k+2} = 0 \quad \text{and} \quad c_{3k} = \frac{c_0}{3 \cdot 6 \cdots (3k)} = \frac{c_0}{k! \, 3^k}.$$

Hence

$$y(x) = c_0\left(1 + \frac{x^3}{3} + \frac{x^6}{18} + \frac{x^9}{162} + \cdots\right) = c_0\left(1 + \frac{x^3}{1!3} + \frac{x^6}{2!3^2} + \frac{x^9}{3!3^3} + \cdots\right) = c_0 e^{(x^3/3)}$$

and $\rho = \infty$.

7. $c_{n+1} = 2c_n;$ it follows that $c_n = 2^n c_0$ and $\rho = \lim\limits_{n\to\infty} \dfrac{1}{2} = \dfrac{1}{2}.$

$$y(x) = c_0\left(1 + 2x + 4x^2 + 8x^3 + 16x^4 + \cdots\right)$$
$$= c_0\left[1 + (2x) + (2x)^2 + (2x)^3 + (2x)^4 + \cdots\right] = \frac{c_0}{1-2x}$$

9. $c_{n+1} = \dfrac{(n+2)c_n}{n+1};$ it follows that $c_n = (n+1)c_0$ and $\rho = \lim\limits_{n\to\infty} \dfrac{n+1}{n+2} = 1.$

$$y(x) = c_0\left(1 + 2x + 3x^2 + 4x^3 + 5x^4 + \cdots\right)$$

Separation of variables gives $y(x) = \dfrac{c_0}{(1-x)^2}.$

In Problems 11–14 the differential equations are second-order, and we find that the two initial coefficients c_0 and c_1 are both arbitrary. In each case we find the even-degree coefficients in terms of c_0 and the odd-degree coefficients in terms of c_1. The solution series in these problems are all recognizable power series that have infinite radii of convergence.

11. $c_{n+1} = \dfrac{c_n}{(n+1)(n+2)};$ it follows that $c_{2k} = \dfrac{c_0}{(2k)!}$ and $c_{2k+1} = \dfrac{c_1}{(2k+1)!}.$

$$y(x) = c_0\left(1 + \frac{x^2}{2!} + \frac{x^4}{4!} + \frac{x^6}{6!} + \cdots\right) + c_1\left(x + \frac{x^3}{3!} + \frac{x^5}{5!} + \frac{x^7}{7!} + \cdots\right) = c_0 \cosh x + c_1 \sinh x$$

13. $c_{n+2} = -\dfrac{9c_n}{(n+1)(n+2)};$ it follows that $c_{2k} = \dfrac{(-1)^k 3^{2k} c_0}{(2k)!}$ and $c_{2k+1} = \dfrac{(-1)^k 3^{2k} c_1}{(2k+1)!}.$

$$y(x) = c_0\left(1 - \frac{9x^2}{2} + \frac{27x^4}{8} - \frac{81x^6}{80} + \cdots\right) + c_1\left(x - \frac{3x^3}{2} + \frac{27x^5}{40} - \frac{81x^7}{560} + \cdots\right)$$

$$= c_0\left(1 - \frac{(3x)^2}{2!} + \frac{(3x)^4}{4!} - \frac{(3x)^6}{6!} + \cdots\right) + \frac{c_1}{3}\left((3x) - \frac{(3x)^3}{3!} + \frac{(3x)^5}{5!} - \frac{(3x)^7}{7!} + \cdots\right)$$

$$= c_0 \cos 3x + \frac{c_1}{3}\sin 3x$$

15. Assuming a power series solution of the form $y = \Sigma c_n x^n$, we substitute it into the differential equation $xy' + y = 0$ and find that $(n+1)c_n = 0$ for all $n \geq 0$. This implies that $c_n = 0$ for all $n \geq 0$, which means that the only power series solution of our differential equation is the trivial solution $y(x) \equiv 0$. Therefore the equation has no *non-trivial* power series solution.

17. Assuming a power series solution of the form $y = \Sigma c_n x^n$, we substitute it into the differential equation $x^2 y' + y = 0$. We find that $c_0 = c_1 = 0$ and that $c_{n+1} = -nc_n$ for $n \geq 1$, so it follows that $c_n = 0$ for all $n \geq 0$. Just as in Problems 15 and 16, this means that the equation has no *non-trivial* power series solution.

In Problems 19–22 we first give the recurrence relation that results upon substitution of an assumed power series solution $y = \Sigma c_n x^n$ into the given second-order differential equation. Then we give the resulting general solution, and finally apply the initial conditions $y(0) = c_0$ and $y'(0) = c_1$ to determine the desired particular solution.

19. $c_{n+2} = -\dfrac{2^2 c_n}{(n+1)(n+2)}$ for $n \geq 0$, so $c_{2k} = \dfrac{(-1)^k 2^{2k} c_0}{(2k)!}$ and $c_{2k+1} = \dfrac{(-1)^k 2^{2k} c_1}{(2k+1)!}$.

$$y(x) = c_0\left(1 - \frac{2^2 x^2}{2!} + \frac{2^4 x^4}{4!} - \frac{2^6 x^6}{6!} + \cdots\right) + c_1\left(x - \frac{2^2 x^3}{3!} + \frac{2^4 x^5}{5!} - \frac{2^6 x^7}{7!} + \cdots\right)$$

$c_0 = y(0) = 0$ and $c_1 = y'(0) = 3$, so

$$y(x) = 3\left(x - \frac{2^2 x^3}{3!} + \frac{2^4 x^5}{5!} - \frac{2^6 x^7}{7!} + \cdots\right)$$

$$= \frac{3}{2}\left[(2x) - \frac{(2x)^3}{3!} + \frac{(2x)^5}{5!} - \frac{(2x)^7}{7!} + \cdots\right] = \frac{3}{2}\sin 2x.$$

21. $c_{n+1} = \dfrac{2nc_n - c_{n-1}}{n(n+1)}$ for $n \geq 1$; with $c_0 = y(0) = 0$ and $c_1 = y'(0) = 1$, we obtain

$c_2 = 1$, $c_3 = \dfrac{1}{2}$, $c_4 = \dfrac{1}{6} = \dfrac{1}{3!}$, $c_5 = \dfrac{1}{24} = \dfrac{1}{4!}$, $c_6 = \dfrac{1}{120} = \dfrac{1}{5!}$. Evidently $c_n = \dfrac{1}{(n-1)!}$, so

$$y(x) = x + x^2 + \frac{x^3}{2!} + \frac{x^4}{3!} + \frac{x^5}{4!} + \cdots = x\left(1 + x + \frac{x^2}{2!} + \frac{x^3}{3!} + \frac{x^4}{4!} + \cdots\right) = xe^x.$$

23. $c_0 = c_1 = 0$ and the recursion relation

$$(n^2 - n + 1)c_n + (n - 1)c_{n-1} = 0$$

for $n \geq 2$ imply that $c_n = 0$ for $n \geq 0$. Thus any assumed power series solution $y = \Sigma c_n x^n$ must reduce to the trivial solution $y(x) \equiv 0$.

25. Substitution of $\sum_{n=0}^{\infty} c_n x^n$ into the differential equation $y'' = y' + y$ leads routinely — via shifts of summation to exhibit x^n-terms throughout — to the recurrence formula

$$(n+2)(n+1)c_{n+2} = (n+1)c_{n+1} + c_n,$$

and the given initial conditions yield $c_0 = 0 = F_0$ and $c_1 = 1 = F_1$. But instead of proceeding immediately to calculate explicit values of further coefficients, let us first multiply the recurrence relation by $n!$. This trick provides the relation

$$(n+2)!c_{n+2} = (n+1)!c_{n+1} + n!c_n,$$

that is, the Fibonacci-defining relation $F_{n+2} = F_{n+1} + F_n$ where $F_n = n!c_n$, so we see that $c_n = F_n / n!$ as desired.

27. **(b)** The roots of the characteristic equation $r^3 = 1$ are $r_1 = 1$, $r_2 = \alpha = (-1 + i\sqrt{3})/2$, and $r_3 = \beta = (-1 - i\sqrt{3})/2$. Then the general solution is

$$y(x) = Ae^x + Be^{\alpha x} + Ce^{\beta x}. \tag{*}$$

Imposing the initial conditions, we get the equations

$$A + B + C = 1$$
$$A + \alpha B + \beta C = 1$$
$$A + \alpha^2 B + \beta^2 C = -1.$$

The solution of this system is $A = 1/3$, $B = (1 - i\sqrt{3})/3$, $C = (1 + i\sqrt{3})/3$. Substitution of these coefficients in (*) and use of Euler's relation $e^{i\theta} = \cos\theta + i\sin\theta$ finally yields the desired result.

SECTION 8.2

SERIES SOLUTIONS NEAR ORDINARY POINTS

Instead of deriving in detail the recurrence relations and solution series for Problems 1 through 15, we indicate where some of these problems and answers originally came from. Each of the differential equations in Problems 1–10 is of the form

$$(Ax^2 + B)y'' + Cxy' + Dy = 0$$

with selected values of the constants A, B, C, D. When we substitute $y = \Sigma c_n x^n$, shift indices where appropriate, and collect coefficients, we get

$$\sum_{n=0}^{\infty} \left[An(n-1)c_n + B(n+1)(n+2)c_{n+2} + Cnc_n + Dc_n \right] x^n = 0.$$

Thus the recurrence relation is

$$c_{n+2} = -\frac{An^2 + (C-A)n + D}{B(n+1)(n+2)} c_n \quad \text{for } n \ge 0.$$

It yields a solution of the form

$$y = c_0\, y_{\text{even}} + c_1\, y_{\text{odd}}$$

where y_{even} and y_{odd} denote series with terms of even and odd degrees, respectively. The even-degree series $c_0 + c_2 x^2 + c_4 x^4 + \cdots$ converges (by the ratio test) provided that

$$\lim_{n \to \infty} \left| \frac{c_{n+2} x^{n+2}}{c_n x^n} \right| = \left| \frac{Ax^2}{B} \right| < 1.$$

Hence its radius of convergence is at least $\rho = \sqrt{|B/A|}$, as is that of the odd-degree series $c_1 x + c_3 x^3 + c_5 x^4 + \cdots$. (See Problem 6 for an example in which the radius of convergence is, surprisingly, greater than $\sqrt{|B/A|}$.)

In Problems 1–15 we give first the recurrence relation and the radius of convergence, then the resulting power series solution.

1. $c_{n+2} = c_n;$ $\rho = 1;$ $c_0 = c_2 = c_4 = \cdots;$ $c_1 = c_3 = c_5 = \cdots$

$$y(x) = c_0 \sum_{n=0}^{\infty} x^{2n} + c_1 \sum_{n=0}^{\infty} x^{2n+1} = \frac{c_0 + c_1 x}{1 - x^2}$$

3. $\qquad c_{n+2} = -\dfrac{c_n}{(n+2)}; \qquad \rho = \infty;$

$$c_{2n} = \frac{(-1)^n c_0}{(2n)(2n-2)\cdots 4\cdot 2} = \frac{(-1)^n c_0}{n!\,2^n};$$

$$c_{2n+1} = \frac{(-1)^n c_1}{(2n+1)(2n-1)\cdots 5\cdot 3} = \frac{(-1)^n c_1}{(2n+1)!!}$$

$$y(x) = c_0 \sum_{n=0}^{\infty} (-1)^n \frac{x^{2n}}{n!\,2^n} + c_1 \sum_{n=0}^{\infty} (-1)^n \frac{x^{2n+1}}{(2n+1)!!}$$

5. $\qquad c_{n+2} = \dfrac{n c_n}{3(n+2)}; \qquad \rho = \sqrt{3}; \qquad c_2 = c_4 = c_6 = \cdots = 0$

$$c_{2n+1} = \frac{2n-1}{3(2n+1)} \cdot \frac{2n-3}{3(2n-1)} \cdots \frac{3}{3(5)} \cdot \frac{1}{3(3)} c_1 = \frac{c_1}{(2n+1)3^n}$$

$$y(x) = c_0 + c_1 \sum_{n=0}^{\infty} \frac{x^{2n+1}}{(2n+1)3^n}$$

7. $\qquad c_{n+2} = -\dfrac{(n-4)^2}{3(n+1)(n+2)} c_n; \qquad \rho \geq \sqrt{3}$

The factor $(n-4)$ yields $c_6 = c_8 = c_{10} = \cdots = 0$, so y_{even} is a 4th-degree polynomial.

We find first that $c_3 = -c_1/2$ and $c_5 = c_1/120$, and then for $n \geq 3$ that

$$c_{2n+1} = \left(-\frac{(2n-5)^2}{3(2n)(2n+1)} \right)\left(-\frac{(2n-7)^2}{3(2n-2)(2n-1)} \right) \cdots \left(-\frac{1^2}{3(6)(7)} \right) c_5 =$$

$$= (-1)^{n-2} \frac{[(2n-5)!!]^2}{3^{n-2}(2n+1)(2n-1)\cdots 7\cdot 6} \cdot \frac{c_1}{120} = 9\cdot(-1)^n \frac{[(2n-5)!!]^2}{3^n(2n+1)!} c_1$$

$$y(x) = c_0\left(1 - \frac{8}{3}x^2 + \frac{8}{27}x^4 \right) + c_1\left[x - \frac{1}{2}x^3 + \frac{1}{120}x^5 + 9\sum_{n=3}^{\infty} \frac{[(2n-5)!!]^2(-1)^n}{(2n+1)!\,3^n} x^{2n+1} \right]$$

9. $\qquad c_{n+2} = \dfrac{(n+3)(n+4)}{(n+1)(n+2)} c_n; \qquad \rho = 1$

$$c_{2n} = \frac{(2n+1)(2n+2)}{(2n-1)(2n)} \cdot \frac{(2n-1)(2n)}{(2n-3)(2n-2)} \cdots \frac{3\cdot 4}{1\cdot 2} c_0 = (n+1)(2n+1)c_0$$

$$c_{2n+1} = \frac{(2n+2)(2n+3)}{(2n)(2n+1)} \cdot \frac{(2n)(2n+1)}{(2n-2)(2n-1)} \cdots \frac{4\cdot 5}{2\cdot 3} c_1 = \frac{1}{3}(n+1)(2n+3)c_1$$

$$y(x) = c_0 \sum_{n=0}^{\infty} (n+1)(2n+1)x^{2n} + \frac{1}{3}c_1 \sum_{n=0}^{\infty} (n+1)(2n+3)x^{2n+1}$$

11. $c_{n+2} = \dfrac{2(n-5)}{5(n+1)(n+2)}c_n; \qquad \rho = \infty$

The factor $(n-5)$ yields $c_7 = c_9 = c_{11} = \cdots = 0$, so y_{odd} is a 5th-degree polynomial. We find first that $c_2 = -c_1$, $c_4 = c_0/10$ and $c_6 = c_0/750$, and then for $n \geq 4$ that

$$c_{2n} = \frac{2(2n-7)}{5(2n)(2n-1)} \cdot \frac{2(2n-5)}{5(2n-2)(2n-3)} \cdots \frac{2(1)}{5(8)(7)}c_6$$

$$= \frac{2^{n-3}(2n-7)!!}{5^{n-3}(2n)(2n-1)\cdots(8)(7)} \cdot \frac{c_0}{750} =$$

$$= \frac{5^3 \cdot 6!}{2^3 \cdot 750} \cdot \frac{2^n(2n-7)!!}{5^n(2n)(2n)\cdots(8)(7)\cdot 6!} \cdot c_1 = 15 \cdot \frac{2^n(2n-7)!!}{5^n(2n)!}c_0$$

$$y(x) = c_1\left(x - \frac{4x^3}{15} + \frac{4x^5}{375}\right) + c_0\left[1 - x^2 + \frac{x^4}{10} + \frac{x^6}{750} + 15\sum_{n=4}^{\infty} \frac{(2n-7)!! \, 2^n}{(2n)! \, 5^n}x^{2n}\right]$$

13. $c_{n+3} = -\dfrac{c_n}{n+3}; \qquad \rho = \infty$

When we substitute $y = \Sigma c_n x^n$ into the given differential equation, we find first that $c_2 = 0$, so the recurrence relation yields $c_5 = c_8 = c_{11} = \cdots = 0$ also.

$$y(x) = c_0\sum_{n=0}^{\infty} \frac{(-1)^n x^{3n}}{n! \, 3^n} + c_1\sum_{n=0}^{\infty} \frac{(-1)^n x^{3n+1}}{1 \cdot 4 \cdots (3n+1)}$$

15. $c_{n+4} = -\dfrac{c_n}{(n+3)(n+4)}; \qquad \rho = \infty$

When we substitute $y = \Sigma c_n x^n$ into the given differential equation, we find first that $c_2 = c_3 = 0$, so the recurrence relation yields $c_6 = c_{10} = \cdots = 0$ and $c_7 = c_{11} = \cdots = 0$ also. Then

$$c_{4n} = \frac{-1}{(4n)(4n-1)} \cdot \frac{-1}{(4n-4)(4n-5)} \cdots \frac{-1}{4 \cdot 3}c_0 = \frac{(-1)^n c_0}{4^n n! \cdot (4n-1)(4n-5)\cdots 5 \cdot 3},$$

$$c_{3n+1} = \frac{-1}{(4n+1)(4n)} \cdot \frac{-1}{(4n-3)(4n-4)} \cdots \frac{-1}{5 \cdot 4}c_1 = \frac{(-1)^n c_1}{4^n n! \cdot (4n+1)(4n-3)\cdots 9 \cdot 5}.$$

$$y(x) = c_0\left[1 + \sum_{n=1}^{\infty} \frac{(-1)^n x^{4n}}{4^n n! \cdot 3 \cdot 7 \cdots (4n-1)}\right] + c_1\left[x + \sum_{n=1}^{\infty} \frac{(-1)^n x^{4n+1}}{4^n n! \cdot 5 \cdot 9 \cdots (4n+1)}\right]$$

17. The recurrence relation

$$c_{n+2} = -\frac{(n-2)c_n}{(n+1)(n+2)}$$

yields $c_2 = c_0 = y(0) = 1$ and $c_4 = c_6 = \cdots = 0$. Because $c_1 = y'(0) = 0$, it follows also that $c_1 = c_3 = c_5 = \cdots = 0$. Thus the desired particular solution is $y(x) = 1 + x^2$.

19.   The substitution $t = x - 1$ yields $(1 - t^2)y'' - 6ty' - 4y = 0$, where primes now denote differentiation with respect to t. When we substitute $y = \Sigma c_n t^n$ we get the recurrence relation

$$c_{n+2} = \frac{n+4}{n+2}c_n.$$

for $n \geq 0$, so the solution series has radius of convergence $\rho = 1$, and therefore converges if $-1 < t < 1$. The initial conditions give $c_0 = 0$ and $c_1 = 1$, so $c_{\text{even}} = 0$ and

$$c_{2n+1} = \frac{2n+3}{2n+1} \cdot \frac{2n+1}{2n-1} \cdots \frac{7}{5} \cdot \frac{5}{3} c_1 = \frac{2n+3}{3}.$$

Thus

$$y = \frac{1}{3}\sum_{n=0}^{\infty}(2n+3)t^{2n+1} = \frac{1}{3}\sum_{n=0}^{\infty}(2n+3)(x-1)^{2n+1},$$

and the x-series converges if $0 < x < 2$.

21.   The substitution $t = x + 2$ yields $(4t^2 + 1)y'' = 8y$, where primes now denote differentiation with respect to t. When we substitute $y = \Sigma c_n t^n$ we get the recurrence relation

$$c_{n+2} = -\frac{4(n-2)}{(n+2)}c_n$$

for $n \geq 0$. The initial conditions give $c_0 = 1$ and $c_1 = 0$. It follows that $c_{\text{odd}} = 0$, $c_2 = 4$ and $c_4 = c_6 = \cdots = 0$, so the solution reduces to

$$y = 2 + 4t^2 = 1 + 4(x+2)^2.$$

In Problems 23–26 we first derive the recurrence relation, and then calculate the solution series $y_1(x)$ with $c_0 = 1$ and $c_1 = 0$ as well as the solution series $y_2(x)$ with $c_0 = 0$ and $c_1 = 1$.

23.   Substitution of $y = \Sigma c_n x^n$ yields

$$c_0 + 2c_2 + \sum_{n=1}^{\infty}\left[c_{n-1} + c_n + (n+1)(n+2)c_{n+2}\right]x^n = 0,$$

so

$$c_2 = -\frac{1}{2}c_0, \qquad c_{n+2} = -\frac{c_{n-1} + c_n}{(n+1)(n+2)} \quad \text{for} \quad n \geq 1.$$

$$y_1(x) = 1 - \frac{x^2}{2} - \frac{x^3}{6} + \frac{x^4}{24} + \cdots; \qquad y_2(x) = x - \frac{x^3}{6} - \frac{x^4}{12} + \frac{x^5}{120} + \cdots$$

25.   Substitution of $y = \Sigma c_n x^n$ yields

$$2c_2 + 6c_3 x + \sum_{n=2}^{\infty} \left[c_{n-2} + (n-1)c_{n-1} + (n+1)(n+2)c_{n+2} \right] x^n = 0,$$

so

$$c_2 = c_3 = 0, \qquad c_{n+2} = -\frac{c_{n-2} + (n-1)c_{n-1}}{(n+1)(n+2)} \quad \text{for} \quad n \ge 2.$$

$$y_1(x) = 1 - \frac{x^4}{12} + \frac{x^7}{126} + \frac{x^8}{672} + \cdots; \qquad y_2(x) = x - \frac{x^4}{12} - \frac{x^5}{20} + \frac{x^7}{126} + \cdots$$

27.   Substitution of $y = \Sigma c_n x^n$ yields

$$c_0 + 2c_2 + (2c_1 + 6c_3)x + \sum_{n=2}^{\infty} \left[2c_{n-2} + (n+1)c_n + (n+1)(n+2)c_{n+2} \right] x^n = 0,$$

so

$$c_2 = -\frac{c_0}{2}, \qquad c_3 = -\frac{c_1}{3}, \qquad c_{n+2} = -\frac{2c_{n-2} + (n+1)c_n}{(n+1)(n+2)} \quad \text{for} \quad n \ge 2.$$

With $c_0 = y(0) = 1$ and $c_1 = y'(0) = -1$, we obtain

$$y(x) = 1 - x - \frac{x^2}{2} + \frac{x^3}{3} - \frac{x^4}{24} + \frac{x^5}{30} + \frac{29x^6}{720} - \frac{13x^7}{630} - \frac{143x^8}{40320} + \frac{31x^9}{22680} + \cdots.$$

Finally, $x = 0.5$ gives

$$y(0.5) = 1 - 0.5 - 0.125 + 0.041667 - 0.002604 + 0.001042$$
$$+ 0.000629 - 0.000161 - 0.000014 + 0.000003 + \cdots$$
$$y(0.5) \approx 0.415562 \approx 0.4156.$$

29.   When we substitute $y = \Sigma c_n x^n$ and $\cos x = \sum (-1)^n x^{2n} /(2n)!$ and then collect coefficients of the terms involving $1, x, x^2, \cdots, x^6$, we obtain the equations

$$c_0 + 2c_2 = 0, \quad c_1 + 6c_3 = 0, \quad 12c_4 = 0, \quad -2c_3 + 20c_5 = 0,$$

$$\frac{1}{12}c_2 - 5c_4 + 30c_6 = 0, \quad \frac{1}{4}c_3 - 9c_5 + 42c_6 = 0,$$

$$-\frac{1}{360}c_2 + \frac{1}{2}c_4 - 14c_6 + 56c_8 = 0.$$

Given c_0 and c_1, we can solve easily for $c_2, c_3, \cdots, c_8$ in turn. With the choices $c_0 = 1$, $c_1 = 0$ and $c_0 = 0$, $c_1 = 1$ we obtain the two series solutions

$$y_1(x) = 1 - \frac{x^2}{2} + \frac{x^6}{720} + \frac{13x^8}{40320} + \cdots \quad \text{and} \quad y_2(x) = x - \frac{x^3}{6} - \frac{x^5}{60} - \frac{13x^7}{5040} + \cdots.$$

33. Substitution of $y = \Sigma c_n x^n$ in Hermite's equation leads in the usual way to the recurrence formula

$$c_{n+2} = -\frac{2(\alpha - n)c_n}{(n+1)(n+2)}.$$

Starting with $c_0 = 1$, this formula yields

$$c_2 = -\frac{2\alpha}{2!}, \quad c_4 = +\frac{2^2 \alpha(\alpha - 2)}{4!}, \quad c_6 = -\frac{2^3 \alpha(\alpha - 2)(\alpha - 4)}{6!}, \quad \ldots.$$

Starting with $c_1 = 1$, it yields

$$c_3 = -\frac{2(\alpha - 1)}{3!}, \quad c_5 = +\frac{2^2(\alpha - 1)(\alpha - 3)}{5!}, \quad c_7 = -\frac{2^3(\alpha - 1)(\alpha - 3)(\alpha - 5)}{7!}, \quad \ldots.$$

This gives the desired even-term and odd-term series y_1 and y_2. If α is an integer, then obviously one series or the other has only finitely many non-zero terms. For instance, with $\alpha = 4$ we get

$$y_1(x) = 1 - \frac{2 \cdot 4}{2}x^2 + \frac{2^2 \cdot 4 \cdot 2}{24}x^4 = 1 - 4x^2 + \frac{4}{3}x^4 = \frac{1}{12}\left(16x^4 - 48x^2 + 12\right),$$

and with $\alpha = 5$ we get

$$y_2(x) = x - \frac{2 \cdot 4}{6}x^3 + \frac{2^2 \cdot 4 \cdot 2}{120}x^5 = x - \frac{4}{3}x^3 + \frac{4}{15}x^5 = \frac{1}{120}\left(32x^5 - 160x^3 + 120\right).$$

The figure below shows the interlaced zeros of the 4th and 5th Hermite polynomials.

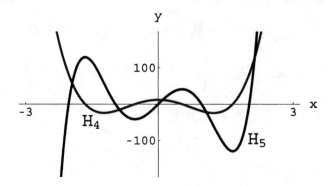

35.   **(a)** If

$$y_0 = 1 + \sum_{n=1}^{\infty} \frac{(2n-1)!!}{2^{3n} n!} x^{2n} = 1 + \sum_{n=1}^{\infty} a_n z^n$$

where $a_n = \dfrac{(2n-1)!!}{2^{3n} n!}$, then the radius of convergence of the series in $z = x^2$ is

$$\rho = \lim_{n \to \infty} \left| \frac{a_n}{a_{n+1}} \right| = \lim_{n \to \infty} \frac{(2n-1)!!/2^{3n} n!}{(2n+1)!!/2^{3n+3}(n+1)!} = \lim_{n \to \infty} \frac{2^3(n+1)}{2n+1} = 4.$$

Thus the series in z converges if $-4 < z = x^2 < 4$, so the series $y_0(x)$ converges if $-2 < x < 2$, and thus has radius of convergence equal to 2.

(b) If

$$y_1 = x \left(1 + \sum_{n=1}^{\infty} \frac{n!}{2^n (2n+1)!!} x^{2n} \right) = x \left(1 + \sum_{n=1}^{\infty} b_n z^n \right)$$

where $b_n = \dfrac{n!}{2^n (2n+1)!!}$, then the radius of convergence of the series in z is

$$\rho = \lim_{n \to \infty} \left| \frac{b_n}{b_{n+1}} \right| = \lim_{n \to \infty} \frac{n!/2^n (2n+1)!!}{(n+1)!/2^{n+1}(2n+3)!!} = \lim_{n \to \infty} \frac{2(2n+3)}{n+1} = 4.$$

Hence it follows as in part (a) that the series $y_1(x)$ has radius of convergence equal to 2.

SECTION 8.3

REGULAR SINGULAR POINTS

1.   Upon division of the given differential equation by x we see that $P(x) = 1 - x^2$ and $Q(x) = (\sin x)/x$. Because both are analytic at $x = 0$ — in particular, $(\sin x)/x \to 1$ as $x \to 0$ because

$$\frac{\sin x}{x} = \frac{1}{x} \sum_{n=0}^{\infty} \frac{(-1)^n x^{2n+1}}{(2n+1)!} = \sum_{n=1}^{\infty} \frac{(-1)^n x^{2n}}{(2n+1)!} = 1 - \frac{x^2}{3!} + \frac{x^4}{5!} - \frac{x^6}{7!} + \cdots$$

— it follows that $x = 0$ is an ordinary point.

3.   When we rewrite the given equation in the standard form of Equation (3) in this section, we see that $p(x) = (\cos x)/x$ and $q(x) = x$. Because $(\cos x)/x \to \infty$ as $x \to 0$ it follows that $p(x)$ is not analytic at $x = 0$, so $x = 0$ is an irregular singular point.

5.   In the standard form of Equation (3) we have $p(x) = 2/(1+x)$ and $q(x) = 3x^2/(1+x)$. Both are analytic $x = 0$, so $x = 0$ is a regular singular point. The indicial equation is

$$r(r-1) + 2r = r^2 + r = r(r+1) = 0,$$

so the exponents are $r_1 = 0$ and $r_2 = -1$.

7.   In the standard form of Equation (3) we have $p(x) = (6 \sin x)/x$ and $q(x) = 6$, so $x = 0$ is a regular singular point with $p_0 = q_0 = 6$. The indicial equation is $r^2 + 5r + 6 = 0$, so the exponents are $r_1 = -2$ and $r_2 = -3$.

9.   The only singular point of the differential equation $y'' + \dfrac{x}{1-x} y' + \dfrac{x^2}{1-x} y = 0$ is $x = 1$.

Upon substituting $t = x - 1$, $x = t + 1$ we get the transformed equation

$y'' - \dfrac{t+1}{t} y' - \dfrac{(t+1)^2}{t} y = 0$, where primes now denote differentiation with respect to t.

In the standard form of Equation (3) we have $p(t) = -(1+t)$ and $q(t) = -t(1+t)^2$. Both these functions are analytic, so it follows that $x = 1$ is a regular singular point of the original equation.

11.   The only singular points of the differential equation $y'' - \dfrac{2x}{1-x^2} y' + \dfrac{12}{1-x^2} y = 0$ are $x = +1$ and $x = -1$.

$x = +1$: Upon substituting $t = x - 1$, $x = t + 1$ we get the transformed equation

$y'' + \dfrac{2(t+1)}{t(t+2)} y' - \dfrac{12}{t(t+2)} y = 0$, where primes now denote differentiation with respect to

t. In the standard form of Equation (3) we have $p(t) = \dfrac{2(t+1)}{t+2}$ and $q(t) = -\dfrac{12t}{t+2}$.

Both these functions are analytic at $t = 0$, so it follows that $x = +1$ is a regular singular point of the original equation.

$x = -1$: Upon substituting $t = x + 1$, $x = t - 1$ we get the transformed equation

$y'' + \dfrac{2(t-1)}{t(t-2)} y' - \dfrac{12}{t(t-2)} y = 0$, where primes now denote differentiation with respect to

t. In the standard form of Equation (3) we have $p(t) = \dfrac{2(t-1)}{t-2}$ and $q(t) = -\dfrac{12t}{t-2}$.

Both these functions are analytic at $t = 0$, so it follows that $x = -1$ is a regular singular point of the original equation.

13.   The only singular points of the differential equation $y'' + \dfrac{1}{x+2}y' + \dfrac{1}{x-2}y = 0$ are $x = +2$ and $x = -2$.

$x = +2$: Upon substituting $t = x - 2$, $x = t + 2$ we get the transformed equation $y'' + \dfrac{1}{t+4}y' + \dfrac{1}{t}y = 0$, where primes now denote differentiation with respect to t. In the standard form of Equation (3) we have $p(t) = \dfrac{t}{t+4}$ and $q(t) = t$. Both these functions are analytic at $t = 0$, so it follows that $x = +2$ is a regular singular point of the original equation.

$x = -2$: Upon substituting $t = x + 2$, $x = t - 2$ we get the transformed equation $y'' + \dfrac{1}{t}y' + \dfrac{1}{t-4}y = 0$, where primes now denote differentiation with respect to t. In the standard form of Equation (3) we have $p(t) \equiv 1$ and $q(t) = \dfrac{t^2}{t-4}$. Both these functions are analytic at $t = 0$, so it follows that $x = -2$ is a regular singular point of the original equation.

15.   The only singular point of the differential equation $y'' - \dfrac{x^2 - 4}{(x-2)^2}y' + \dfrac{x+2}{(x-2)^2}y = 0$ is $x = 2$. Upon substituting $t = x - 2$, $x = t + 2$ we get the transformed equation $y'' - \dfrac{t+4}{t}y' + \dfrac{t+4}{t^2}y = 0$, where primes now denote differentiation with respect to t. In the standard form of Equation (3) we have $p(t) = -(t+4)$ and $q(t) = t+4$. Both these functions are analytic, so it follows that $x = 2$ is a regular singular point of the original equation.

Each of the differential equations in Problems 17–20 is of the form

$$Axy'' + By' + Cy = 0$$

with indicial equation $Ar^2 + (B - A)r = 0$. Substitution of $y = \Sigma c_n x^{n+r}$ into the differential equation yields the recurrence relation

$$c_n = -\frac{Cc_{n-1}}{A(n+r)^2 + (B-A)(n+r)}$$

for $n \geq 1$. In these problems the exponents $r_1 = 0$ and $r_2 = (A - B)/A$ do *not* differ by an integer, so this recurrence relation yields two linearly independent Frobenius series solutions when we apply it separately with $r = r_1$ and with $r = r_2$.

17. With exponent $r_1 = 0$: $c_n = -\dfrac{c_{n-1}}{4n^2 - 2n}$

$$y_1(x) = x^0\left(1 - \frac{x}{2} + \frac{x^2}{24} - \frac{x^3}{720} + \cdots\right) = \sum_{n=0}^{\infty} \frac{(-1)^n \left(\sqrt{x}\right)^{2n}}{(2n)!} = \cos\sqrt{x}$$

With exponent $r_2 = \dfrac{1}{2}$: $c_n = -\dfrac{c_{n-1}}{4n^2 + 2n}$

$$y_2(x) = x^{1/2}\left(1 - \frac{x}{6} + \frac{x^2}{120} - \frac{x^3}{5040} + \cdots\right) = \sum_{n=0}^{\infty} \frac{(-1)^n \left(\sqrt{x}\right)^{2n+1}}{(2n+1)!} = \sin\sqrt{x}$$

19. With exponent $r_1 = 0$: $c_n = \dfrac{c_{n-1}}{2n^2 - 3n}$

$$y_1(x) = x^0\left(1 - x - \frac{x^2}{2} - \frac{x^3}{18} - \frac{x^4}{360} - \cdots\right) = 1 - x - \sum_{n=2}^{\infty} \frac{x^n}{n!(2n-3)!!}$$

With exponent $r_2 = \dfrac{3}{2}$: $c_n = \dfrac{c_{n-1}}{2n^2 + 3n}$

$$y_2(x) = x^{3/2}\left(1 + \frac{x}{5} + \frac{x^2}{70} + \frac{x^3}{1890} + \frac{x^4}{83160} + \cdots\right) = x^{3/2}\left[1 + 3\sum_{n=1}^{\infty} \frac{x^n}{n!(2n+3)!!}\right]$$

The differential equations in Problems 21–24 are all of the form

$$Ax^2y'' + Bxy' + (C + Dx^2)y = 0 \tag{1}$$

with indical equation

$$\phi(r) = Ar^2 + (B - A)r + C = 0. \tag{2}$$

Substitution of $y = \Sigma c_n x^{n+r}$ into the differential equation yields

$$\phi(r)c_0 x^r + \phi(r+1)c_1 x^{r+1} + \sum_{n=2}^{\infty}\left[\phi(r+n)c_n + Dc_{n-2}\right]x^{n+r} = 0. \tag{3}$$

In each of Problems 21–24 the exponents r_1 and r_2 do *not* differ by an integer. Hence when we substitute either $r = r_1$ or $r = r_2$ into Equation (*) above, we find that c_0 is arbitrary because $\phi(r)$ is then zero, that $c_1 = 0$ — because its coefficient $\phi(r+1)$ is then nonzero — and that

$$c_n = -\frac{Dc_{n-2}}{\phi(r+n)} = -\frac{Dc_{n-2}}{A(n+r)^2 + (B-A)(n+r) + C} \tag{4}$$

for $n \geq 2$. Thus this recurrence formula yields two linearly independent Frobenius series solutions when we apply it separately with $r = r_1$ and with $r = r_2$.

21. With exponent $r_1 = 1$: $c_1 = 0$, $c_n = \dfrac{2c_{n-2}}{n(2n+3)}$

$$y_1(x) = x^1\left(1 + \frac{x^2}{7} + \frac{x^4}{154} + \frac{x^6}{6930} + \cdots\right) = x\left[1 + \sum_{n=1}^{\infty} \frac{x^{2n}}{n!\cdot 7\cdot 11\cdots\cdots(4n+3)}\right]$$

With exponent $r_2 = -\dfrac{1}{2}$: $c_1 = 0$, $c_n = \dfrac{2c_{n-2}}{n(2n-3)}$

$$y_2(x) = x^{-1/2}\left(1 + x^2 + \frac{x^4}{10} + \frac{x^6}{270} + \cdots\right) = \frac{1}{\sqrt{x}}\left[1 + \sum_{n=1}^{\infty} \frac{x^{2n}}{n!\cdot 1\cdot 5\cdots\cdots(4n-3)}\right]$$

23. With exponent $r_1 = \dfrac{1}{2}$: $c_1 = 0$, $c_n = \dfrac{c_{n-2}}{n(6n+7)}$

$$y_1(x) = x^{1/2}\left(1 + \frac{x^2}{38} + \frac{x^4}{4712} + \frac{x^6}{1215696} + \cdots\right) = \sqrt{x}\left[1 + \sum_{n=1}^{\infty} \frac{x^{2n}}{2^n n!\cdot 19\cdot 31\cdots\cdots(12n+7)}\right]$$

With exponent $r_2 = -\dfrac{2}{3}$: $c_1 = 0$, $c_n = \dfrac{c_{n-2}}{n(6n-7)}$

$$y_2(x) = x^{-2/3}\left(1 + \frac{x^2}{10} + \frac{x^4}{680} + \frac{x^6}{118320} + \cdots\right) = x^{-2/3}\left[1 + \sum_{n=1}^{\infty} \frac{x^{2n}}{2^n n!\cdot 5\cdot 17\cdots\cdots(12n-7)}\right]$$

25. With exponent $r_1 = \dfrac{1}{2}$: $c_n = -\dfrac{c_{n-1}}{2n}$

$$y_1(x) = x^{1/2}\left(1 - \frac{x}{2} + \frac{x^2}{8} - \frac{x^3}{48} + \frac{x^4}{384} - \cdots\right) = \sqrt{x}\sum_{n=0}^{\infty} \frac{(-1)^n x^n}{n!\, 2^n} = \sqrt{x}\, e^{-x/2}$$

With exponent $r_2 = 0$: $c_n = -\dfrac{c_{n-1}}{2n-1}$

$$y_2(x) = x^0\left(1 - x + \frac{x^2}{3} - \frac{x^3}{15} + \frac{x^4}{105} - \cdots\right) = 1 + \sum_{n=1}^{\infty} \frac{(-1)^n x^n}{(2n-1)!!}$$

The differential equations in Problems 27–29 (after multiplication by x) and the one in Problem 31 are of the same form (1) above as those in Problems 21–24. However, now the exponents r_1 and $r_2 = r_1 - 1$ *do* differ by an integer. Hence when we substitute the smaller exponent $r = r_2$ into Equation (3), we find that c_0 and c_1 are *both* arbitrary, and that c_n is given (for $n \geq 2$) by the recurrence relation in (4). Thus the *smaller* exponent r_2 yields the general solution $y(x) = c_0 y_1(x) + c_1 y_2(x)$ in terms of the two linearly independent Frobenius series solutions $y_1(x)$ and $y_2(x)$.

27. Exponents $r_1 = 0$ and $r_2 = -1$; with $r = -1$: $c_n = -\dfrac{9c_{n-2}}{n(n-1)}$

$$y(x) = \frac{c_0}{x}\left(1 - \frac{9x^2}{2} + \frac{27x^4}{8} - \frac{81x^6}{80} + \cdots\right) + \frac{c_1}{x}\left(x - \frac{3x^3}{2} + \frac{27x^5}{40} - \frac{81x^7}{560} + \cdots\right)$$

$$= \frac{c_0}{x}\left(1 - \frac{9x^2}{2} + \frac{81x^4}{24} - \frac{729x^6}{720} + \cdots\right) + \frac{c_1}{3x}\left(3x - \frac{27x^3}{6} + \frac{243x^5}{120} - \frac{2187x^7}{5040} + \cdots\right)$$

$$y(x) = c_0\frac{\cos 3x}{x} + \frac{1}{3}c_1\frac{\sin 3x}{x}$$

The figure below shows the graphs of the independent solutions $y_1(x) = \dfrac{\cos 3x}{x}$ and

$y_2(x) = \dfrac{\sin 3x}{x}.$

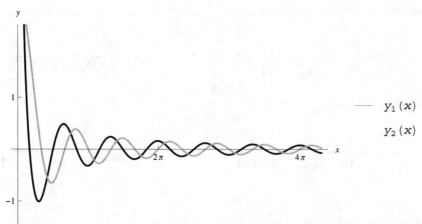

29.   Exponents $r_1 = 0$ and $r_2 = -1$; with $r = -1$: $\quad c_n = -\dfrac{c_{n-2}}{4n(n-1)}$

$$y(x) = \frac{c_0}{x}\left(1 - \frac{x^2}{8} + \frac{x^4}{384} - \frac{x^6}{46080} + \cdots\right) + \frac{c_1}{x}\left(x - \frac{x^3}{24} + \frac{x^5}{1920} - \frac{x^7}{322560} + \cdots\right)$$

$$= \frac{c_0}{x}\left(1 - \frac{x^2}{2^2 \cdot 2} + \frac{x^4}{2^4 \cdot 24} - \frac{x^6}{2^6 \cdot 720} + \cdots\right) + \frac{2c_1}{x}\left(\frac{x}{2} - \frac{x^3}{2^3 \cdot 6} + \frac{x^5}{2^5 \cdot 120} - \frac{x^7}{2^7 \cdot 5040} + \cdots\right)$$

$$y(x) = \frac{c_0}{x}\cos\frac{x}{2} + \frac{2c_1}{x}\sin\frac{x}{2}$$

The figure below shows the graphs of the independent solutions

$y_1(x) = \dfrac{\cos x/2}{x}$ and $y_2(x) = \dfrac{\sin x/2}{x}.$

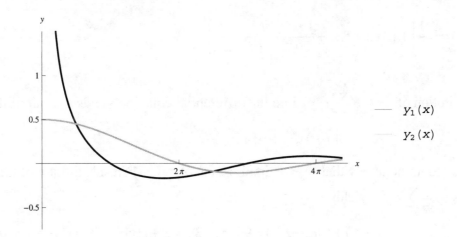

31.   The given differential equation $4x^2y'' - 4xy' + (3 - 4x^2)y = 0$ has indicial equation $4r^2 - 8r + 3 = (2r - 3)(2r - 1) = 0$, so its exponents are $r_1 = 3/2$ and $r_2 = 1/2$. With $r = 3/2$, the recurrence relation $c_n = c_{n-2}/n(n-1)$ yields the general solution

$$y(x) = c_0 x^{1/2}\left(1 + \frac{x^2}{2} + \frac{x^4}{24} + \frac{x^6}{720} + \cdots\right) + c_1 x^{1/2}\left(x + \frac{x^3}{6} + \frac{x^5}{120} + \frac{x^7}{5040} + \cdots\right)$$

$$y(x) = c_0\sqrt{x}\cosh x + c_1\sqrt{x}\sinh x.$$

The figure below shows the graphs of the independent solutions $y_1(x) = \sqrt{x}\cosh x$ and $y_2(x) = \sqrt{x}\sinh x$.

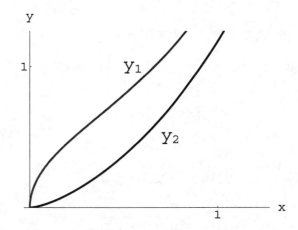

33.   Exponents $r_1 = 1/2$ and $r_2 = -1$. With each exponent we find that c_0 is arbitrary and we can solve recursively for c_n in terms of c_{n-1}.

$$y_1(x) = \sqrt{x}\left(1 + \frac{11x}{20} - \frac{11x^2}{224} + \frac{671x^3}{24192} - \frac{9577x^4}{387072} + \cdots\right)$$

$$y_2(x) = \frac{1}{x}\left(1 + 10x + 5x^2 + \frac{10x^3}{9} - \frac{7x^4}{18} + \cdots\right)$$

35. Substitution of $y = x^r \sum c_n x^n$ into the differential equation yields a result of the form

$$-rc_0 x^{r-1} + (\cdots)x^r + (\cdots)x^{r+1} + \cdots = 0,$$

so we see immediately that $c_0 \neq 0$ implies that $r = 0$. Then substitution of the power series $y = \sum c_n x^n$ yields

$$(c_0 - c_1) + (4c_1 - 2c_2)x + (9c_2 - 3c_3)x^2 + (16c_3 - 4c_4)x^4 + \cdots = 0$$

Evidently $c_n = nc_{n-1}$, so if $c_0 = 1$ it follows that $c_n = n!$ for $n \geq 1$. But the series $\sum n! x^n$ has zero radius of convergence, and hence converges only if $x = 0$. We therefore conclude that the given differential equation has *no* nontrivial Frobenius series solution.

37. Substitution of $y = x^r \sum c_n x^n$ into the differential equation $x^3 y'' - xy' + y = 0$ yields a result of the form

$$(r-1)^2 c_0 x^r + (\cdots)x^{r+1} + (\cdots)x^{r+2} + \cdots = 0,$$

so it follows that $r = 1$. But then substitution of $y = x \sum c_n x^n$ into the differential equation yields

$$c_1 x^2 + 4c_2 x^3 + 9c_3 x^4 + 16c_4 x^5 + 25c_5 x^6 + \cdots = 0,$$

so it follows that $c_1 = c_2 = c_3 = c_4 = \cdots = 0$. Hence $y(x) = c_0 x$.

39. Exponents $r_1 = 1$ and $r_2 = -1$; with $r = +1$: $c_1 = 0$, $c_n = -\dfrac{c_{n-2}}{n(n+2)}$

$$y(x) = c_0 x \left(1 - \frac{x^2}{8} + \frac{x^4}{192} - \frac{x^6}{9216} + \frac{x^8}{737280} - \cdots\right)$$

$$= c_0 x \left(1 - \frac{x^2}{2^2 1! 2!} + \frac{x^4}{2^4 2! 3!} - \frac{x^6}{2^6 3! 4!} + \frac{x^8}{2^8 4! 5!} - \cdots\right)$$

If $c_0 = 1/2$, then

$$y(x) = J_1(x) = \frac{x}{2} \sum_{n=0}^{\infty} \frac{(-1)^n}{n!(n+1)} \left(\frac{x}{2}\right)^{2n}.$$

Now, consider the smaller exponent $r_2 = -1$. A Frobenius series with $r = -1$ is of the form $y = x^{-1} \sum_{n=0}^{\infty} c_n x^n$ with $c_0 \neq 0$. However, substitution of this series into Bessel's equation of order 1 gives

$$-c_1 + c_0 x + (c_1 + 3c_3)x^2 + (c_2 + 8c_4)x^3 + (c_3 + 15c_5)x^5 + \cdots = 0,$$

so it follows that $c_0 = 0$, after all. Thus Bessel's equation of order 1 does not have a Frobenius series solution with leading term $c_0 x^{-1}$. However, there is a little more here that meets the eye. We see further that c_2 is arbitrary and that $c_1 = 0$ and $c_n = c_{n-2}/n(n-2)$ for $n > 2$. It follows that our assumed Frobenius series

$$y = x^{-1} \sum_{n=0}^{\infty} c_n x^n \quad \text{actually reduces to}$$

$$y(x) = c_2 x \left(1 - \frac{x^2}{8} + \frac{x^4}{192} - \frac{x^6}{9216} + \frac{x^8}{737280} - \cdots \right).$$

But this is the same as our series solution obtained above using the larger exponent $r = +1$ (calling the arbitrary constant c_2 rather than c_0).

SECTION 8.4

METHOD OF FROBENIUS: THE EXCEPTIONAL CASES

Each of the differential equations in Problems 1–6 is of (or can be written in) the form

$$xy'' + (A + Bx)y' + Cy = 0.$$

The origin is a regular singular point with exponents $r = 0$ and $r = 1 - A$, so if A is an integer then we have an exceptional case of the method of Frobenius. When we substitute $y = \Sigma c_n x^{n+r}$ in the differential equation we find that the coefficient of x^{n+r} is

$$[(n + r)^2 + (A - 1)(n + r)]c_n + [B(n + r) + C - B]c_{n-1} = 0. \qquad (*)$$

Case 1:   In each of Problems 1–4 we have $A \geq 2$ and $B = C$, so the larger exponent $r_1 = 0$ and the smaller exponent $r_2 = 1 - A = -N$ differ by a positive integer. When we substitute the smaller exponent $r = -N$ in Equation (*) above, it simplifies to

$$n(n - N)c_n + B(n - N)c_{n-1} = 0. \qquad (1)$$

This equation determines $c_1, c_2, \cdots, c_{N-1}$ in terms of c_0, thereby yielding the solution

$$y_1(x) = x^{-N}(c_0 + c_1 x + \cdots + c_{N-1} x^{n-1}), \tag{2}$$

provided it is possible to choose $c_N = 0$. But when $n = N$, Equation (1) reduces to

$$0 \cdot c_N + 0 \cdot c_{N-1} = 0,$$

so c_N may be chosen arbitrarily. With $C_N = 0$ we get the terminating Frobenius series solution in (2). For $n > N$, Equation (1) yields the recurrence formula $c_n = -B c_{n-1}/n$, which if $C_N \neq 0$ gives a second (non-terminating) Frobenius series solution of the form

$$y_2(x) = c_N + c_{N+1} x + c_{N+2} x^2 + \cdots. \tag{3}$$

Case 2: If $A \leq 0$ then the larger exponent $r_1 = 1 - A = N$ and the smaller exponent $r_2 = 0$ again differ by a positive integer. In Problems 5 and 6 we have this case with $B = -1$. When we substitute the smaller exponent $r = 0$ in Equation (*), it simplifies to

$$n(n - N)c_n - (n - C - 1)c_{n-1} = 0. \tag{4}$$

This equation determines $c_1, c_2, \cdots, c_{N-1}$ in terms of c_0. When $n = N$ it reduces to

$$0 \cdot c_N - (N - C - 1)c_{N-1} = 0. \tag{5}$$

If either $N - C - 1 = 0$ or $c_{N-1} = 0$ (the latter happens in Problem 5) then c_N can be chosen arbitrarily, and finally $c_{N+1}, c_{N+2}, \cdots$ are determined in terms of c_N. Thus we get *two* Frobenius series solutions

$$\begin{aligned} y_1 &= c_0 + c_1 x + \cdots + c_{N-1} x^{N-1}, &&\text{(terminating)} \\ y_2 &= c_N x^N + c_{N+1} x^{N+1} + \cdots. &&\text{(not terminating)} \end{aligned}$$

On the other hand, if (as in Problem 6) neither $N - C - 1 = 0$ nor $c_{N-1} = 0$, then c_N cannot be chosen so as to satisfy Equation (5), and hence there is no Frobenius series solution corresponding to the smaller exponent $r_2 = 0$. We therefore find the *single* Frobenius series solution by substituting the larger exponent $r_1 = N$ in Equation (*) and using the resulting recurrence relation to determine $c_1, c_2, c_3, \cdots$ in terms of c_0.

Problems 1–4 correspond to case 1 above. We give first the indicial roots and the critical index N, then the recurrence relation that defines c_n in terms of c_{n-1}, for both the N-term solution $y_1(x)$ in (2) and the non-terminating series solution $y_2(x)$ in (3).

1.　　$r_1 = 0, \ r_2 = -2, \ N = 2, \ c_n = \dfrac{c_{n-1}}{n}; \qquad y_1(x) = x^{-2}\left(1 + x\right);$

$$y_2(x) = 1 + \frac{x}{3} + \frac{x^2}{3 \cdot 4} + \frac{x^3}{3 \cdot 4 \cdot 5} + \cdots = 1 + 2 \sum_{n=1}^{\infty} \frac{x^n}{(n+2)!}$$

3. $r_1 = 0,\ r_2 = -4,\ N = 4,\ c_n = -\dfrac{3c_{n-1}}{n};$ $y_1(x) = x^{-4}\left(1 - 3x + \dfrac{9}{2}x^2 - \dfrac{9}{2}x^3\right)$

$$y_2(x) = 1 - \frac{3x}{5} + \frac{3^2 x^2}{5 \cdot 6} - \frac{3^3 x^3}{5 \cdot 6 \cdot 7} + \cdots = 1 + 24\sum_{n=1}^{\infty} \frac{(-1)^n 3^n x^n}{(n+4)!}$$

Problems 5 and 6 correspond to case 2 described above.

5. $r_1 = 5,\ r_2 = 0,\ N = 5,\ c_n = \dfrac{(n-4)c_{n-1}}{n(n-5)}$ for $n \neq 5$

$$y_1(x) = 1 + \frac{3}{4}x + \frac{1}{4}x^2 + \frac{1}{24}x^3$$

With $n = 5$ the recurrence relation is $0 \cdot c_5 - c_4 = 0$. Because $c_4 = 0$ we can choose $c_5 = 1$ arbitrarily and proceed:

$$y_2(x) = x^5 + \frac{2x^6}{6} + \frac{3x^7}{6 \cdot 7} + \frac{4x^8}{6 \cdot 7 \cdot 8} + \cdots = x^5\left[1 + 120\sum_{n=1}^{\infty} \frac{(n+1)x^n}{(n+5)!}\right]$$

7.   The indicial exponents are $r = -2,\ 1$. Substitution of $y = x^{-2}\sum_{n=0}^{\infty} c_n x^n$ in the differential equation leads to the recurrence relation

$$n(n-3)c_n + 3(n-3)c_{n-1} = 0$$

that reduces to $0 \cdot c_3 + 0 \cdot c_2 = 0$ when $n = 3$ so — having found c_1 and c_2 — c_3 can be chosen arbitrarily. With $c_0 = 2$ and $c_3 = 0$ we get the terminating Frobenius series

$$y_1(x) = x^{-2}(2 - 6x + 9x^2).$$

Starting afresh with $c_3 = 3/3! = 1/2$, the recurrence relation $c_n = -3c_{n-1}/n$ for $n > 3$ yields the second Frobenius series solution

$$y_2(x) = x^{-2}\left(\frac{3x^3}{3!} - \frac{3^2 x^4}{4!} + \frac{3^3 x^5}{5!} - \cdots\right) = \sum_{n=1}^{\infty} \frac{(-1)^{n-1} 3^n x^n}{(n+2)!}.$$

In Problems 11–15, we give first the Frobenius series solution $y_1(x)$ corresponding to the larger indicial exponent r_1 of the given differential equation. Then, writing the equation in the form $y'' + P(x)y' + Q(x)y = 0$, we apply the reduction of order formula

$$y_2(x) = \int \frac{\exp\left(-\int P(x)\,dx\right)}{y_1(x)^2}\,dx$$

to derive a second independent solution $y_2(x)$.

9. $r_1 = r_2 = 0$

$$y_1 = 1 + \frac{x^2}{4} + \frac{x^4}{64} + \frac{x^6}{2304} + \frac{x^8}{147456} + \cdots$$

$$P(x) = 1/x; \qquad \exp\left(-\int P(x)\,dx\right) = 1/x$$

$$y_2 = y_1 \int x^{-1} \cdot \left(1 + \frac{x^2}{4} + \frac{x^4}{64} + \frac{x^6}{2304} + \frac{x^8}{147456} + \cdots\right)^{-2} dx$$

$$= y_1 \int x^{-1}\left(1 + \frac{x^2}{2} + \frac{3x^4}{32} + \frac{5x^6}{576} + \frac{35x^8}{73728} + \cdots\right)^{-1} dx$$

$$= y_1 \int x^{-1}\left(1 - \frac{x^2}{2} + \frac{5x^4}{32} - \frac{23x^6}{576} + \frac{677x^8}{73728} + \cdots\right) dx$$

$$y_2 = y_1\left(\ln x - \frac{x^2}{4} + \frac{5x^4}{128} - \frac{23x^6}{3456} + \frac{677x^6}{589824} - \cdots\right)$$

11. $r_1 = r_2 = 2$

$$y_1 = x^2\left(1 - 2x + \frac{3x^2}{2} - \frac{2x^3}{3} + \frac{5x^4}{24} - \cdots\right)$$

$$P(x) = 1 - 3/x; \qquad \exp\left(-\int P(x)\,dx\right) = x^3 e^{-x}$$

$$y_2 = y_1 \int x^3 e^{-x} \cdot x^{-4}\left(1 - 2x + \frac{3x^2}{2} - \frac{2x^3}{3} + \frac{5x^4}{24} - \cdots\right)^{-2} dx$$

$$= y_1 \int x^{-1} e^{-x}\left(1 - 4x + 7x^2 - \frac{22x^3}{3} + \frac{16x^4}{3} - \cdots\right)^{-1} dx$$

$$= y_1 \int x^{-1}\left(1 - x + \frac{x^2}{2} - \frac{x^3}{6} + \frac{x^4}{24} - \cdots\right)\left(1 + 4x + 9x^2 + \frac{46x^3}{3} + \frac{67x^4}{3} + \cdots\right) dx$$

$$= y_1 \int x^{-1}\left(1 + 3x + \frac{11x^2}{2} + \frac{49x^3}{6} + \frac{87x^4}{8} + \cdots\right) dx$$

$$y_2 = y_1\left(\ln x + 3x + \frac{11x^2}{4} + \frac{49x^3}{18} + \frac{87x^4}{32} + \cdots\right)$$

13. $r_1 = 3, \quad r_2 = 1$

$$y_1 = x^3 \left(1 - 2x + 2x^2 - \frac{4x^3}{3} + \frac{2x^4}{3} - \cdots \right)$$

$$P(x) = 2 - 3/x; \qquad \exp\left(-\int P(x)\,dx \right) = x^3 e^{-2x}$$

$$y_2 = y_1 \int x^3 e^{-2x} \cdot x^{-6} \left(1 - 2x + 2x^2 - \frac{4x^3}{3} + \frac{2x^4}{3} - \cdots \right)^{-2} dx$$

$$= y_1 \int x^{-3} e^{-2x} \left(1 - 4x + 8x^2 - \frac{32x^3}{3} + \frac{32x^4}{3} - \cdots \right)^{-1} dx$$

$$= y_1 \int x^{-3} \left(1 - 2x + 2x^2 - \frac{4x^3}{3} + \frac{2x^4}{3} - \cdots \right) \left(1 + 4x + 8x^2 + \frac{32x^3}{3} + \frac{32x^4}{3} + \cdots \right) dx$$

$$= y_1 \int x^{-3} \left(1 + 2x + 2x^2 + \frac{4x^3}{3} + \frac{2x^4}{3} + \cdots \right) dx$$

$$y_2 = y_1 \left(2\ln x - \frac{1}{2x^2} - \frac{2}{x} + \frac{4x}{3} + \frac{x^2}{3} + \cdots \right)$$

15. $r_1 = r_2 = 0$

$$J_0(x) = 1 - \frac{x^2}{4} + \frac{x^4}{64} - \frac{x^6}{2304} + \frac{x^8}{147456} - \cdots$$

$$P(x) = 1/x; \qquad \exp\left(-\int P(x)\,dx \right) = 1/x$$

$$y_2(x) = J_0(x) \int x^{-1} \cdot \left(1 - \frac{x^2}{4} + \frac{x^4}{64} - \frac{x^6}{2304} + \frac{x^8}{147456} - \cdots \right)^{-2} dx$$

$$= J_0(x) \int x^{-1} \left(1 - \frac{x^2}{2} + \frac{3x^4}{32} - \frac{5x^6}{576} + \frac{35x^8}{73728} - \cdots \right)^{-1} dx$$

$$= J_0(x) \int x^{-1} \left(1 + \frac{x^2}{2} + \frac{5x^4}{32} + \frac{23x^6}{576} + \frac{677x^8}{73728} + \cdots \right) dx$$

$$= J_0(x) \left(\ln x + \frac{x^2}{4} + \frac{5x^4}{128} + \frac{23x^6}{3456} + \frac{677x^6}{589824} - \cdots \right)$$

$$= J_0(x)\ln x +$$

$$\left(1 - \frac{x^2}{4} + \frac{x^4}{64} - \frac{x^6}{2304} + \frac{x^8}{147456} - \cdots \right) \left(\frac{x^2}{4} + \frac{5x^4}{128} + \frac{23x^6}{3456} + \frac{677x^6}{589824} - \cdots \right)$$

$$y_2(x) = J_0(x)\ln x + \frac{x^2}{4} - \frac{3x^4}{128} + \frac{11x^6}{13284} - \cdots$$

17. The given first solution

$$y_1(x) = xe^x = x\left(1 + x + \frac{x^2}{2} + \frac{x^3}{6} + \frac{x^4}{24} + \frac{x^5}{120} + \cdots\right)$$

can be derived by starting with the single exponent $r = 1$, substituting $y = x\sum_{n=0}^{\infty} c_n x^n$ into the differential equation, and calculating successive coefficients recursively as usual. We can verify the alleged second solution by applying the method of reduction of order as in Problems 9–14:

$$P(x) = -1 - \frac{1}{x}; \qquad \exp\left(-\int P(x)\,dx\right) = xe^x$$

$$y_2 = y_1 \int xe^x \cdot \left(xe^x\right)^{-2} dx = y_1 \int x^{-1} e^{-x}\, dx$$

$$= y_1 \int x^{-1}\left(1 - x + \frac{x^2}{2} - \frac{x^3}{6} + \frac{x^4}{24} - \frac{x^5}{120} + \cdots\right) dx$$

$$= y_1\left(\ln x - x + \frac{x^2}{4} - \frac{x^3}{18} + \frac{x^4}{96} - \frac{x^5}{600} + \cdots\right)$$

$$= y_1 \ln x + \left(1 - x + \frac{x^2}{2} - \frac{x^3}{6} + \frac{x^4}{24} - \frac{x^5}{120} + \cdots\right)\left(-x + \frac{x^2}{4} - \frac{x^3}{18} + \frac{x^4}{96} - \frac{x^5}{600} + \cdots\right)$$

$$= y_1 \ln x - \left(x^2 + \frac{3x^3}{4} + \frac{11x^4}{36} + \frac{25x^5}{288} + \frac{137x^6}{7200} + \cdots\right)$$

$$y_2(x) = xe^x \ln x - \sum_{n=1}^{\infty} \frac{H_n x^{n+1}}{n!}$$

SECTION 8.5

BESSEL'S EQUATION

Of course Bessel's equation is the most important special ordinary differential equation in mathematics, and every student should be exposed at least to Bessel functions of the first kind. Though Bessel functions of integral order can be treated without the gamma function, the subsection on the gamma function is also needed for Chapter 7 on Laplace transforms. The final subsections on Bessel function identities and the parametric Bessel equation will not be needed until Section 10.4, and therefore may be considered optional at this point in the course.

1. $$J_0'(x) = D_x\left(1 + \sum_{m=1}^{\infty} \frac{(-1)^m x^{2m}}{2^{2m}(m!)^2}\right) = \sum_{m=1}^{\infty} \frac{(-1)^m 2m\, x^{2m-1}}{2^{2m}(m!)^2}$$

$$= \sum_{m=1}^{\infty} \frac{(-1)^m x^{2m-1}}{2^{2m-1}(m-1)!(m!)} = \sum_{m=0}^{\infty} \frac{(-1)^{m+1} x^{2m+1}}{2^{2m+1}(m)!(m+1)!}$$

$$= -\sum_{m=0}^{\infty} \frac{(-1)^m x^{2m+1}}{2^{2m+1}(m)!(m+1)!} = -J_1(x)$$

3.   **(a)** $\Gamma\left(m + \frac{2}{3}\right) = \Gamma\left(\frac{3m+2}{3}\right) = \frac{3m-1}{3} \cdot \frac{3m-4}{3} \cdot \Gamma\left(\frac{3m-4}{3}\right)$

$$= \frac{3m-1}{3} \cdot \frac{3m-4}{3} \cdots \cdot \frac{5}{3} \cdot \frac{2}{3} \cdot \Gamma\left(\frac{2}{3}\right) = \frac{2 \cdot 5 \cdot 8 \cdots (3m-1)}{3^m} \Gamma\left(\frac{2}{3}\right)$$

(b) $J_{-1/3}(x) = \sum_{m=0}^{\infty} \frac{(-1)^m}{m! \Gamma(m+2/3)} \left(\frac{x}{2}\right)^{2m-1/3} = \frac{(x/2)^{-1/3}}{\Gamma(2/3)} \sum_{m=0}^{\infty} \frac{(-1)^m 3^m x^{2m}}{m! \cdot 2 \cdot 3 \cdot 8 \cdots (3m-1)}$

5.   Starting with $p = 3$ in Equation (26) we get

$$J_4(x) = \frac{6}{x} J_3(x) - J_2(x) = \frac{6}{x} \left[\frac{4}{x} J_2(x) - J_1(x)\right] - J_2(x)$$

$$= \left(\frac{24}{x^2} - 1\right)\left[\frac{2}{x} J_1(x) - J_0(x)\right] - \frac{6}{x} J_1(x)$$

$$= \frac{x^2 - 24}{x^2} J_0(x) + \frac{8(6 - x^2)}{x^3} J_1(x)$$

11. $\Gamma(p + m + 1) = (p + m)(p + m - 1) \cdots (p + 2)(p + 1)\Gamma(p + 1)$, so

$$J_p(x) = \sum_{m=0}^{\infty} \frac{(-1)^m}{m! \Gamma(p + m + 1)} \left(\frac{x}{2}\right)^{2m+p}$$

$$= \frac{(x/2)^p}{\Gamma(p + 1)} \sum_{m=0}^{\infty} \frac{(-1)^m}{m!(p+1)(p+2) \cdots (p+m)} \left(\frac{x}{2}\right)^{2m}.$$

In Problems 13–21 we use a conspicuous dot • to indicate our choice of u and dv in the integration by parts formula $\int u \cdot dv = uv - \int v \, du$. We use repeatedly the facts (from Example 1) that $\int x J_0(x) \, dx = x J_1(x) + C$ and $\int J_1(x) \, dx = -J_0(x) + C$.

13. $\int x^2 J_0(x) \, dx = \int x \cdot x J_0(x) \, dx$

$$= x^2 J_1(x) - \int x \cdot J_1(x) \, dx$$

$$= x^2 J_1(x) - \left(-x J_0(x) + \int J_0(x) \, dx\right)$$

$$= x^2 J_1(x) + x J_0(x) - \int J_0(x) \, dx + C$$

15. $\displaystyle\int x^4 J_0(x)\,dx = \int x^3 \cdot x J_0(x)\,dx$

$\qquad = x^4 J_1(x) - 3\int x^3 \cdot J_1(x)\,dx$

$\qquad = x^4 J_1(x) - 3\left(-x^3 J_0(x) + 3\int x \cdot x J_0(x)\,dx\right)$

$\qquad = x^4 J_1(x) + 3x^3 J_0(x) - 9\left(x^2 J_1(x) - \int x \cdot J_1(x)\,dx\right)$

$\qquad = x^4 J_1(x) + 3x^3 J_0(x) - 9x^2 J_1(x) + 9\left(-x J_0(x) + \int J_0(x)\,dx\right)$

$\qquad = (x^4 - 9x^2)J_1(x) + (3x^3 - 9x)J_0(x) + 9\int J_0(x)\,dx + C$

17. $\displaystyle\int x^2 J_1(x)\,dx = \int x^2 \cdot J_1(x)\,dx$

$\qquad = -x^2 J_0(x) + 2\int x J_0(x)\,dx = -x^2 J_0(x) + 2x J_1(x) + C$

19. $\displaystyle\int x^4 J_1(x)\,dx = \int x^4 \cdot J_1(x)\,dx$

$\qquad = -x^4 J_0(x) + 4\int x^2 \cdot x J_0(x)\,dx$

$\qquad = -x^4 J_0(x) + 4\left(x^3 J_1(x) - 2\int x^2 \cdot J_1(x)\,dx\right)$

$\qquad = -x^4 J_0(x) + 4x^3 J_1(x) - 8\left(-x^2 J_0(x) + 2\int x J_0(x)\,dx\right)$

$\qquad = (-x^4 + 8x^2)J_0(x) + (4x^3 - 16x)J_1(x) + C$

21.   With $p = 2$, Eq. (23) in the text gives $\displaystyle\int x^{-2} J_3(x)\,dx = -x^{-2} J_2(x) + C$.   Hence

$\displaystyle\int J_3(x)\,dx = \int x^2 \cdot x^{-2} J_3(x)\,dx$

$\qquad = x^2\left(-x^{-2} J_2(x)\right) + 2\int x^{-1} J_2(x)\,dx$

$\qquad = -J_2(x) - \dfrac{2}{x} J_1(x) + C$ $\qquad$ (by Example 3)

$\qquad = -\left(\dfrac{2}{x} J_1(x) - J_0(x)\right) - \dfrac{2}{x} J_1(x) + C$ $\qquad$ (By Eq. (26) with $p = 1$)

$\qquad = J_0(x) - \dfrac{4}{x} J_1(x) + C.$

23.   This is a special case of the discussion below in Problem 24.

24.   Given an integer $n \geq 1$, let us define

$$g_n(x) = \int_0^\pi \cos(n\theta - x\sin\theta)\,d\theta.$$

Differentiation yields

$$g_n'(x) = \int_0^\pi \sin(n\theta - x\sin\theta)\sin\theta\, d\theta.$$

Integration by parts with $u = \sin(n\theta - x\sin\theta)$ and $dv = \sin\theta\, d\theta$ yields

$$g_n'(x) = n\int_0^\pi \cos\theta\,\cos(n\theta - x\sin\theta)\, d\theta - x\int_0^\pi \cos^2\theta\,\cos(n\theta - x\sin\theta)\, d\theta.$$

But differentiation of the first equation for $g_n'(x)$ yields

$$g_n''(x) = -\int_0^\pi \sin^2\theta\,\cos(n\theta - x\sin\theta)\, d\theta.$$

It follows that

$$g_n''(x) + \frac{1}{x}g_n'(x) = -g_n(x) + \frac{n}{x}\int_0^\pi \cos\theta\,\cos(n\theta - x\sin\theta)\, d\theta$$

$$= -g_n(x) - \frac{n}{x^2}\int_0^\pi \left[(n - x\cos\theta) - n\right]\cos(n\theta - x\sin\theta)\, d\theta$$

$$= -g_n(x) - \frac{n}{x^2}\left[\sin(n\theta - x\sin\theta)\right]_0^\pi + \frac{n^2}{x^2}g_n(x) = -\left(1 - \frac{n^2}{x^2}\right)g_n(x).$$

Upon equating the first and last members of this continued inequality and multiplying by x^2, we see that $y = g_n(x)$ satisfies Bessel's equation of order $n \geq 1$. The initial values of $g_n(x)$ are

$$g_n(0) = \int_0^\pi \cos(n\theta)\, d\theta = 0 \quad \text{and} \quad g_n'(0) = \int_0^\pi \sin(\theta)\sin(n\theta)\, d\theta = 0.$$

If $n = 1$ then $g_1'(0) = \pi/2$, whereas $g_n'(0) = 0$ if $n \geq 1$. In either case the values of $g_n(0)$ and $g_n'(0)$ are π times those of $J_n(0)$ and $J_n'(0)$, respectively. Now we know from the general solution of Bessel's equation that $g_n(x) = c\,J_n(x)$ for some constant c. If $n = 1$ then the fact that

$$\pi/2 = g_n'(0) = c\,J_1'(0) = c/2$$

implies that $c = \pi$, as desired. But if $n > 1$ the fact that

$$0 = g_n'(0) = c\,J_n'(0) = c\cdot 0$$

does not suffice to determine c.

SECTION 8.6

APPLICATIONS OF BESSEL FUNCTIONS

Problems 1–12 are routine applications of the theorem in this section. In each case it is necessary only to identify the coefficients A, B, C and the exponent q in the differential equation

$$x^2 y'' + Axy' + (B + Cx^q)y = 0. \tag{1}$$

Then we can calculate the values

$$\alpha = \frac{1-A}{2}, \quad \beta = \frac{q}{2}, \quad k = \frac{2\sqrt{C}}{q}, \quad p = \frac{\sqrt{(1-A)^2 - 4B}}{q} \tag{2}$$

and finally write the general solution

$$y(x) = x^\alpha \left[c_1 J_p(kx^\beta) + c_2 J_{-p}(kx^\beta) \right] \tag{3}$$

specified in Theorem 1 of this section. This is a "template procedure" that we illustrate only in a couple of problems.

1. We have $A = -1, B = 1, C = 1, q = 2$ so

$$\alpha = \frac{1-(-1)}{2} = 1, \quad \beta = \frac{2}{2} = 1, \quad k = \frac{2\sqrt{1}}{2} = 1, \quad p = \frac{\sqrt{(1-(-1))^2 - 4(1)}}{2} = 0,$$

so our general solution is $y(x) = x[c_1 J_0(x) + c_2 Y_0(x)]$, using $Y_0(x)$ because $p = 0$ is an integer.

3. $y(x) = x[c_1 J_{1/2}(3x^2) + c_2 J_{-1/2}(3x^2)]$

5. To match the given equation with Eq. (1) above, we first divide through by the leading coefficient 16 to obtain the equation

$$x^2 y'' + \frac{5}{3}xy' + \left(-\frac{5}{36} + \frac{1}{4}x^3\right)y = 0$$

with $A = 5/3, B = -5/36, C = 1/4,$ and $q = 3$. Then

$$\alpha = \frac{1-5/3}{3} = -\frac{1}{3}, \quad \beta = \frac{3}{2}, \quad k = \frac{2\sqrt{1/4}}{3} = \frac{1}{3}, \quad p = \frac{\sqrt{(1-5/3)^2 - 4(-5/36)}}{3} = \frac{1}{3},$$

so our general solution is $y(x) = x^{-1/3}[c_1 J_{1/3}(x^{3/2}/3) + c_2 J_{-1/3}(x^{3/2}/3)]$.

7. $y(x) = x^{-1}[c_1 J_0(x) + c_2 Y_0(x)]$

9. $y(x) = x^{1/2}[c_1 J_{1/2}(2x^{3/2}) + c_2 J_{-1/2}(2x^{3/2})]$

11. $y(x) = x^{1/2}[c_1 J_{1/6}(x^3/3) + c_2 J_{-1/6}(x^3/3)]$

13. We want to solve the equation $xy'' + 2y' + xy = 0$. If we rewrite it as

$$x^2 y'' + 2xy' + x^2 y = 0$$

then we have the form in Equation (1) with $A = 2$, $B = 0$, $C = 1$, and $q = 2$. Then Equation (2) gives $\alpha = -1/2$, $\beta = 1$, $k = 1$, and $p = 1/2$, so by Equation (3) the general solution is

$$y(x) = x^{-1/2}\left[c_1 J_{1/2}(x) + c_1 J_{-1/2}(x)\right]$$

$$= x^{-1/2}\left[c_1 \sqrt{\frac{2}{\pi x}}\cos x + c_2 \sqrt{\frac{2}{\pi x}}\sin x\right] = \frac{1}{x}\left(a_1 \cos x + a_2 \sin x\right),$$

(with $a_i = c_i \sqrt{2/\pi}$) using Equations (19) in Section 3.5.

15.　The substitution

$$y = -\frac{u'}{u}, \quad y' = \frac{(u')^2}{u^2} - \frac{u''}{u}$$

immediately transforms $y' = x^2 + y^2$ to $u'' + x^2 u = 0$. The equivalent equation

$$x^2 u'' + x^4 u = 0$$

is of the form in (1) with $A = B = 0$, $C = 1$, and $q = 4$. Equations (2) give $\alpha = 1/2$, $\beta = 2$, $k = 1/2$, and $p = 1/4$, so the general solution is

$$u(x) = x^{1/2}\left[c_1 J_{1/4}(x^2/2) + c_2 J_{-1/4}(x^2/2)\right].$$

To compute $u'(x)$, let $z = x^2/2$ so $x = 2^{1/2} z^{1/2}$. Then Equation (22) in Section 8.5 with $p = 1/4$ yields

$$\frac{d}{dx}\left(x^{1/2} J_{1/4}(x^2/2)\right) = \frac{d}{dz}\left(2^{1/4} z^{1/4} J_{1/4}(z)\right) \cdot \frac{dz}{dx}$$

$$= 2^{1/4} z^{1/4} J_{-3/4}(z) \cdot \frac{dz}{dx}$$

$$= 2^{1/4} \cdot \frac{x^{1/2}}{2^{1/4}} J_{-3/4}(x^2/2) \cdot x = x^{3/2} J_{-3/4}(x^2/2).$$

Similarly, Equation (23) in Section 8.5 with $p = -1/4$ yields

$$\frac{d}{dx}\left(x^{1/2} J_{-1/4}(x^2/2)\right) = \frac{d}{dz}\left(2^{1/4} z^{1/4} J_{-1/4}(z)\right) \cdot \frac{dz}{dx} = -x^{3/2} J_{3/4}(x^2/2).$$

Therefore

$$u'(x) = x^{3/2}\left[c_1 J_{-3/4}(x^2/2) - c_2 J_{3/4}(x^2/2)\right].$$

It follows finally that the general solution of the Riccati equation $y' = x^2 + y^2$ is

$$y(x) = -\frac{u'}{u} = x \cdot \frac{J_{3/4}(\frac{1}{2}x^2) - cJ_{-3/4}(\frac{1}{2}x^2)}{cJ_{1/4}(\frac{1}{2}x^2) + J_{-1/4}(\frac{1}{2}x^2)}$$

where the arbitrary constant is $c = c_1/c_2$.

17. If we write the equation $x^4 y'' + \gamma^2 y = 0$ in the form

$$x^2 y'' + \gamma^2 x^{-2} y = 0,$$

then we see that it is of the form in Equation (3) of this section with $A = B = 0$, $C = \gamma^2$, and $q = -2$. Then Equations (5) give $\alpha = 1/2$, $\beta = -1$, $k = \gamma$, and $p = -1/2$, so the theorem yields the general solution

$$y(x) = x^{1/2}[c_1 J_{1/2}(\gamma/x) + c_2 J_{-1/2}(\gamma/x)] = x[A\cos(\gamma/x) + B\sin(\gamma/x)],$$

using Equations (19) in Section 8.5 for $J_{1/2}(x)$ and $J_{-1/2}(x)$. With a and b both nonzero, the initial conditions $y(a) = y(b) = 0$ yield the equations

$$A\cos(\gamma/a) + B\sin(\gamma/a) = 0$$

$$A\cos(\gamma/b) + B\sin(\gamma/b) = 0.$$

These equations have a nontrivial solution for A and B only if the coefficient determinant

$$\Delta = \sin(\gamma/b)\cos(\gamma/a) - \sin(\gamma/a)\cos(\gamma/b)$$

$$= \sin(\gamma/b - \gamma/a) = \sin(\gamma L/ab)$$

is nonzero. Hence $\gamma L/ab$ must be an integral multiple $n\pi$ of π, and then the nth buckling force is

$$P_n = \frac{EI_0 \gamma_n^2}{b^4} = \frac{EI_0}{b^4}\left(\frac{n\pi ab}{L}\right)^2 = EI_0\left(\frac{n\pi}{L}\right)^2\left(\frac{a}{b}\right)^2.$$

CHAPTER 9

FOURIER SERIES METHODS
AND PARTIAL DIFFERENTIAL EQUATIONS

SECTION 9.1

PERIODIC FUNCTIONS AND TRIGONOMETRIC SERIES

The basic trigonometric functions $\cos(t)$ and $\sin(t)$ have period $P = 2\pi$, so the sine or cosine of ωt (as in Problems 1–4) completes its first period when $\omega t = 2\pi$; hence $P = 2\pi / \omega$.

1. Smallest period $P = 2\pi/3$ (left-hand figure below)

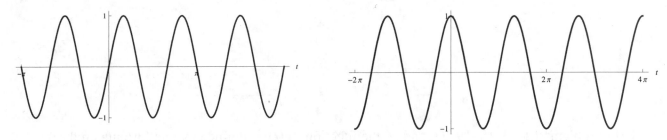

3. Smallest period $P = 4\pi/3$ (right-hand figure above)

However, the basic tangent and cotangent functions have period π (instead of 2π), so $P = \pi / \omega$ in Problems 5 and 6.

5. Smallest period $P = \pi$; see the left-hand figure below.

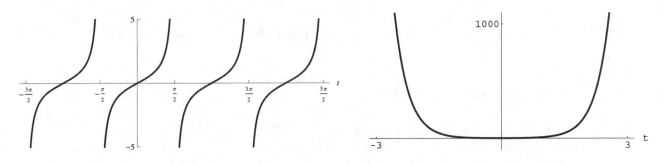

The hyperbolic sine and cosine functions of Problems 7 and 8 are steadily increasing (for $t > 0$), and hence are not periodic.

7. Not periodic (right-hand figure above)

9. Smallest period $P = \pi$.

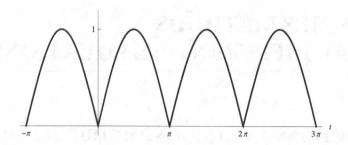

11. With $f(t) = 1$ the integral formulas of Eqs. (16) and (17) in the text give $a_0 = 2$ and $a_n = b_n = 0$ for $n \geq 0$. Thus the Fourier series of f is the single term series $f(t) = 1$:

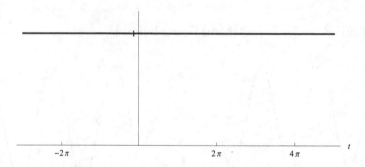

In Problems 12–13, 18–19, 23, and 26 the function $f(t)$ is defined by one formula on the interval $(-\pi, 0)$ and by another formula on $(0, \pi)$. The coefficient integrals must therefore be split accordingly, and the appropriate formula substituted in each integral:

$$a_0 = \frac{1}{\pi}\int_{-\pi}^{0} f(t)\, dt + \frac{1}{\pi}\int_{0}^{\pi} f(t)\, dt,$$

$$a_n = \frac{1}{\pi}\int_{-\pi}^{0} f(t)\cos nt\, dt + \frac{1}{\pi}\int_{0}^{\pi} f(t)\cos nt\, dt, \qquad (n > 0)$$

$$b_n = \frac{1}{\pi}\int_{-\pi}^{0} f(t)\sin nt\, dt + \frac{1}{\pi}\int_{0}^{\pi} f(t)\sin nt\, dt.$$

13.

$$a_0 = \frac{1}{\pi}\int_{-\pi}^{0} (0)\, dt + \frac{1}{\pi}\int_{0}^{\pi} (1)\, dt = 1$$

$$a_n = \frac{1}{\pi}\int_{-\pi}^{0} (0)\cos nt\, dt + \frac{1}{\pi}\int_{0}^{\pi} (1)\cos nt\, dt = \frac{\sin n\pi}{n\pi} = 0$$

$$b_n = \frac{1}{\pi}\int_{-\pi}^{0} (0)\sin nt\, dt + \frac{1}{\pi}\int_{0}^{\pi} (1)\sin nt\, dt =$$

$$= \frac{1 - \cos n\pi}{n\pi} = \begin{cases} 0 & \text{for } n \text{ even} \\ 2/n\pi & \text{for } n \text{ odd} \end{cases}$$

$$f(t) \sim \frac{1}{2} + \frac{2}{\pi}\left[\frac{\sin t}{1} + \frac{\sin 3t}{3} + \frac{\sin 5t}{5} + \frac{\sin 7t}{7} + \cdots\right]$$ (figure below)

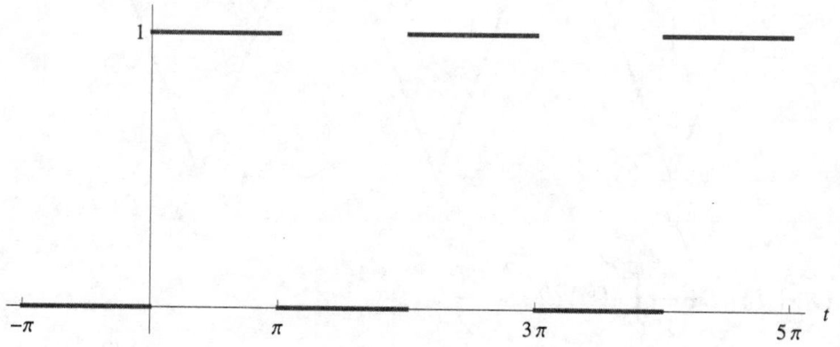

15. $$a_0 = \frac{1}{\pi}\int_{-\pi}^{\pi} t\, dt = 0, \qquad a_n = \frac{1}{\pi}\int_{-\pi}^{\pi} t\cos nt\, dt = 0$$

$$b_n = \frac{1}{\pi}\int_{-\pi}^{\pi} t\sin nt\, dt = \frac{2\sin n\pi - 2n\pi\cos n\pi}{n^2\pi} = \begin{cases} -2/n & \text{for } n \text{ even} \\ +2/n & \text{for } n \text{ odd} \end{cases}$$

$$f(t) \sim 2\left[\frac{\sin t}{1} - \frac{\sin 2t}{2} + \frac{\sin 3t}{3} - \frac{\sin 4t}{4} + \cdots\right]$$ (figure at top of next page)

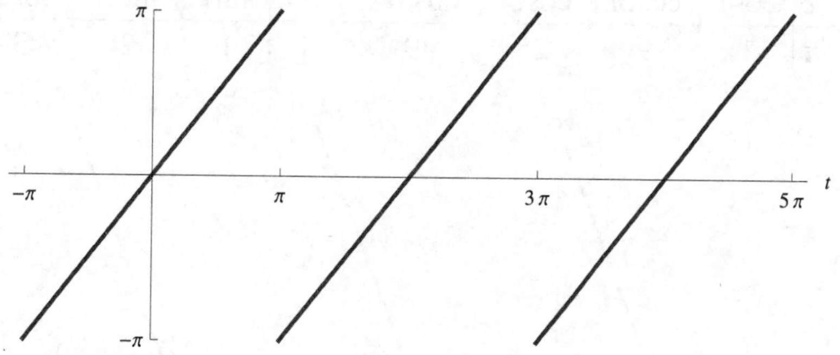

17. $$a_0 = \frac{1}{\pi}\int_{-\pi}^{0} (-t)\, dt + \frac{1}{\pi}\int_{0}^{\pi} (t)\, dt = \pi$$

$$a_n = \frac{1}{\pi}\int_{-\pi}^{0} (-t)\cos nt\, dt + \frac{1}{\pi}\int_{0}^{\pi} (t)\cos nt\, dt$$

$$= \frac{2(\cos n\pi + n\pi\sin n\pi - 1)}{n^2\pi} = \begin{cases} 0 & \text{for } n \text{ even} \\ -4/n^2\pi & \text{for } n \text{ odd} \end{cases}$$

$$b_n = \frac{1}{\pi}\int_{-\pi}^{0} (-t)\sin nt\, dt + \frac{1}{\pi}\int_{0}^{\pi} (t)\sin nt\, dt = 0$$

$$f(t) \sim \frac{\pi}{2} - \frac{4}{\pi}\left[\frac{\cos t}{1} + \frac{\cos 3t}{9} + \frac{\cos 5t}{25} + \frac{\cos 7t}{49} + \cdots\right]$$

See the figure at the top of the next page.

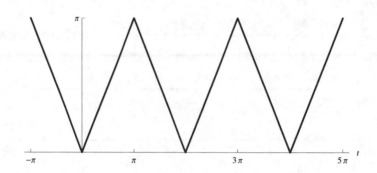

19.

$$a_0 = \frac{1}{\pi}\int_{-\pi}^{0}(\pi+t)\,dt + \frac{1}{\pi}\int_{0}^{\pi}(0)\,dt = \frac{\pi}{2}$$

$$a_n = \frac{1}{\pi}\int_{-\pi}^{0}(\pi+t)\cos nt\,dt + \frac{1}{\pi}\int_{0}^{\pi}(0)\cos nt\,dt$$

$$= \frac{1-\cos n\pi}{n^2\pi} = \begin{cases} 0 & \text{for } n \text{ even} \\ 2/n^2\pi & \text{for } n \text{ odd} \end{cases}$$

$$b_n = \frac{1}{\pi}\int_{-\pi}^{0}(\pi+t)\sin nt\,dt + \frac{1}{\pi}\int_{0}^{\pi}(0)\sin nt\,dt = \frac{\sin n\pi - n\pi}{n^2\pi} = -\frac{1}{n}$$

$$f(t) \sim \frac{\pi}{4} + \frac{2}{\pi}\left[\frac{\cos t}{1} + \frac{\cos 3t}{9} + \frac{\cos 5t}{25} + \frac{\cos 7t}{49} + \cdots\right] - \left[\frac{\sin t}{1} + \frac{\sin 2t}{2} + \frac{\sin 3t}{3} + \frac{\sin 4t}{4} + \cdots\right]$$

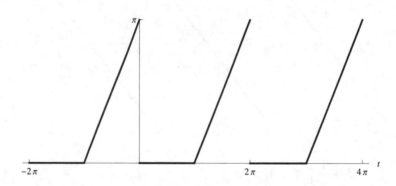

21.

$$a_0 = \frac{1}{\pi}\int_{-\pi}^{\pi}t^2\,dt = \frac{2\pi^2}{3}$$

$$a_n = \frac{1}{\pi}\int_{-\pi}^{\pi}t^2\cos nt\,dt = \frac{4n\pi\cos n\pi - 2(n^2\pi^2-2)\sin n\pi}{n^3\pi} = \begin{cases} +4/n^2 & \text{for } n \text{ even} \\ -4/n^2 & \text{for } n \text{ even} \end{cases}$$

$$b_n = \frac{1}{\pi}\int_{-\pi}^{\pi}t^2\sin nt\,dt = 0$$

$$f(t) \sim \frac{\pi^2}{3} - 4\left[\frac{\cos t}{1} - \frac{\cos 2t}{4} + \frac{\cos 3t}{9} - \frac{\cos 4t}{16} + \cdots\right] \qquad \text{(figure below)}$$

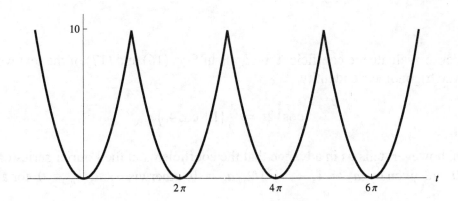

23. $a_0 = \dfrac{1}{\pi}\displaystyle\int_0^{\pi} t^2\,dt = \dfrac{\pi^2}{3}$

$a_n = \dfrac{1}{\pi}\displaystyle\int_0^{\pi} t^2\cos nt\,dt = \dfrac{2n\pi\cos n\pi + (n^2\pi^2 - 2)\sin n\pi}{n^3\pi} = \dfrac{2(-1)^n}{n^2}$

$b_n = \dfrac{1}{\pi}\displaystyle\int_0^{2\pi} t^2\sin nt\,dt$

$\quad = \dfrac{(2 - n^2\pi^2)\cos n\pi + 2n\pi\sin n\pi - 2}{n^3\pi} = \begin{cases} -\pi/n & \text{for } n \text{ even} \\ (n^2\pi^2 - 4)/\pi n^3 & \text{for } n \text{ odd} \end{cases}$

$f(t) \sim \dfrac{\pi^2}{6} - 2\left[\dfrac{\cos t}{1} - \dfrac{\cos 2t}{4} + \dfrac{\cos 3t}{9} - \dfrac{\cos 4t}{16} + \cdots\right]$ (figure below)

$\quad + \pi\left[\dfrac{\sin t}{1} - \dfrac{\sin 2t}{2} + \dfrac{\sin 3t}{3} - \dfrac{\sin 4t}{4} + \cdots\right] - \dfrac{4}{\pi}\left[\dfrac{\sin t}{1} + \dfrac{\sin 3t}{27} + \dfrac{\sin 5t}{125} + \dfrac{\sin 7t}{343} + \cdots\right]$

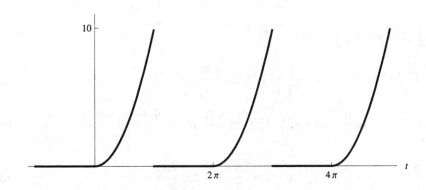

The trigonometric identities

$$2\cos A\cos B = \cos(A+B) + \cos(A-B)$$
$$2\sin A\cos B = \sin(A+B) + \sin(A-B)$$
$$2\sin A\sin B = \cos(A-B) - \cos(A+B)$$

are needed to evaluate the integrals that appear in Problems 24–26.

25. In order to evaluate the coefficient integrals in Eqs. (16) and (17) of the text we would need the trigonometric identity

$$\cos^2 2t = \frac{1}{2}(1 + \cos 4t)$$

which, however, tells us in advance that the coefficients in the Fourier series of $f(t) = \cos^2 2t$ are given by $a_0 = 1$, $a_4 = 1/2$, $a_n = 0$ otherwise, and $b_n = 0$ for all $n \geq 1$.

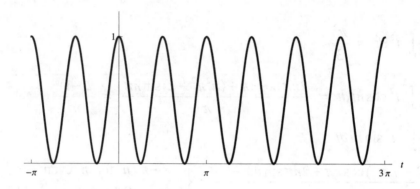

SECTION 9.2

GENERAL FOURIER SERIES AND CONVERGENCE

1.

$$a_0 = \frac{1}{3}\int_{-3}^{0}(-2)\,dt + \frac{1}{3}\int_{0}^{3}(2)\,dt = 0$$

$$a_n = \frac{1}{3}\int_{-3}^{0}(-2)\cos\frac{n\pi t}{3}\,dt + \frac{1}{3}\int_{0}^{3}(2)\cos\frac{n\pi t}{3}\,dt = 0$$

$$b_n = \frac{1}{3}\int_{-3}^{0}(-2)\sin\frac{n\pi t}{3}\,dt + \frac{1}{3}\int_{0}^{3}(2)\sin\frac{n\pi t}{3}\,dt = \frac{4(1-\cos n\pi)}{n\pi} = \frac{4}{n\pi}\Big[1-(-1)^n\Big]$$

$$f(t) = \frac{8}{\pi}\left[\sin\frac{\pi t}{3} + \frac{1}{3}\sin\frac{3\pi t}{3} + \frac{1}{5}\sin\frac{5\pi t}{3} + \frac{1}{7}\sin\frac{7\pi t}{3} + \cdots\right] \qquad \text{(figure on next page)}$$

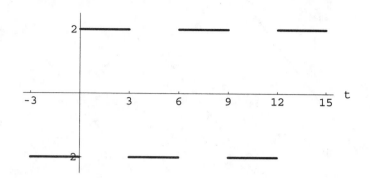

3. $a_0 = \dfrac{1}{2\pi}\displaystyle\int_{-2\pi}^{0}(2)\,dt + \dfrac{1}{2\pi}\int_{0}^{2\pi}(-1)\,dt = 1$

$a_n = \dfrac{1}{2\pi}\displaystyle\int_{-2\pi}^{0}(2)\cos\dfrac{nt}{2}\,dt + \dfrac{1}{2\pi}\int_{0}^{2\pi}(-1)\cos\dfrac{nt}{2}\,dt = \dfrac{\sin n\pi}{n\pi} = 0$

$b_n = \dfrac{1}{2\pi}\displaystyle\int_{-2\pi}^{0}(2)\sin\dfrac{nt}{2}\,dt + \dfrac{1}{2\pi}\int_{0}^{2\pi}(-1)\sin\dfrac{nt}{2}\,dt = \dfrac{3(\cos n\pi - 1)}{n\pi} = \dfrac{3}{n\pi}\left[(-1)^n - 1\right]$

$f(t) = \dfrac{1}{2} - \dfrac{6}{\pi}\left[\sin\dfrac{t}{2} + \dfrac{1}{3}\sin\dfrac{3t}{2} + \dfrac{1}{5}\sin\dfrac{5t}{2} + \dfrac{1}{7}\sin\dfrac{7t}{2} + \cdots\right]$ (figure below)

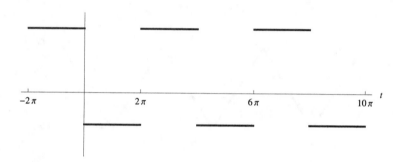

5. $a_0 = \dfrac{1}{2\pi}\displaystyle\int_{-2\pi}^{2\pi}t\,dt = 0, \qquad\qquad a_n = \dfrac{1}{2\pi}\displaystyle\int_{-2\pi}^{2\pi}t\cos\dfrac{nt}{2}\,dt = 0$

$b_n = \dfrac{1}{2\pi}\displaystyle\int_{-2\pi}^{2\pi}t\sin\dfrac{nt}{2}\,dt = \dfrac{4(\sin n\pi - n\pi\cos n\pi)}{n^2\pi} = \dfrac{4(-1)^{n+1}}{n}$

$f(t) = 4\left[\sin\dfrac{t}{2} - \dfrac{1}{2}\sin\dfrac{2t}{2} + \dfrac{1}{3}\sin\dfrac{3t}{2} - \dfrac{1}{4}\sin\dfrac{4t}{2} + \cdots\right]$ (figure on next page)

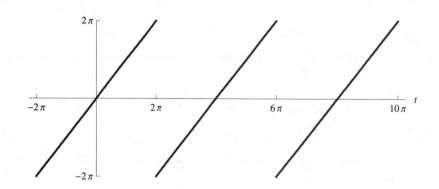

7.

$$a_0 = \int_{-1}^{0} (-t)\, dt + \int_{0}^{1} (t)\, dt = 1$$

$$a_n = \int_{-1}^{0} (-t)\cos n\pi t\, dt + \int_{0}^{1} t\cos n\pi t\, dt = \frac{2\left[\cos n\pi + n\pi \sin n\pi - 1\right]}{n^2\pi^2} = \frac{2}{n^2\pi^2}\left[(-1)^n - 1\right]$$

$$b_n = \int_{-1}^{0} (-t)\sin n\pi t\, dt + \int_{0}^{1} t\sin n\pi t\, dt = 0$$

$$f(t) = \frac{1}{2} - \frac{4}{\pi^2}\left[\cos \pi t + \frac{1}{9}\cos 3\pi t + \frac{1}{25}\cos 5\pi t + \frac{1}{49}\cos 7\pi t + \cdots\right] \qquad \text{(figure below)}$$

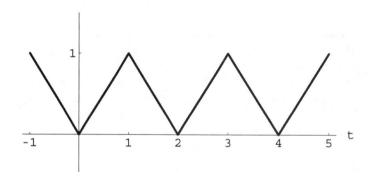

9.

$$a_0 = \int_{-1}^{1} t^2\, dt = \frac{2}{3}$$

$$a_n = \int_{-1}^{1} t^2 \cos n\pi t\, dt = \frac{4n\pi \cos n\pi + 2(n^2\pi^2 - 2)\sin n\pi}{n^3\pi^3} = \frac{4(-1)^n}{n^2\pi^2}$$

$$b_n = \int_{-1}^{1} t^2 \sin n\pi t\, dt = 0$$

$$f(t) = \frac{1}{3} - \frac{4}{\pi^2}\left[\cos \pi t - \frac{1}{4}\cos 2\pi t + \frac{1}{9}\cos 3\pi t - \frac{1}{16}\cos 4\pi t + \cdots\right] \qquad \text{(figure on next page)}$$

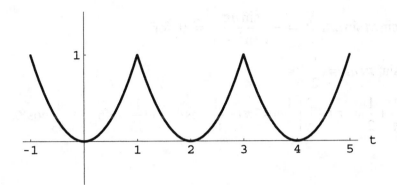

To calculate the Fourier coefficients in Problems 11–14 we use the trigonometric identities for $\sin A \cos B$ and $\sin A \sin B$ that are listed above in Section 9.1 (prior to Problems 24–26 there).

11. $a_0 = \displaystyle\int_{-1}^{1} \cos\frac{\pi t}{2}\, dt = \frac{4}{\pi}$

   $a_n = \displaystyle\int_{-1}^{1} \cos\frac{\pi t}{2}\cos n\pi t\, dt = -\frac{4\cos n\pi}{\pi(4n^2-1)} = \frac{4(-1)^{n+1}}{\pi(4n^2-1)}$

   $b_n = \displaystyle\int_{-1}^{1} \cos\frac{\pi t}{2}\sin n\pi t\, dt = 0$

   $f(t) = \dfrac{2}{\pi} + \dfrac{4}{\pi}\left[\dfrac{1}{3}\cos\pi t - \dfrac{1}{15}\cos 2\pi t + \dfrac{1}{35}\cos 3\pi t - \dfrac{1}{63}\cos 4\pi t + \cdots\right]$ (figure below)

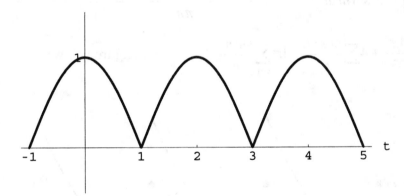

13. $a_0 = \displaystyle\int_{0}^{1} \sin\pi t\, dt = \frac{2}{\pi}$

   $a_n = \displaystyle\int_{0}^{1} \sin\pi t \cos n\pi t\, dt = -\frac{1+\cos n\pi}{\pi(n^2-1)} = -\frac{1+(-1)^n}{\pi(n^2-1)}$ for $n>1$

   $a_1 = \displaystyle\int_{0}^{1} \sin\pi t \cos\pi t\, dt = 0$

$$b_n = \int_0^1 \sin \pi t \sin n\pi t \, dt = -\frac{\sin n\pi}{\pi(n^2-1)} = 0 \text{ for } n > 1$$

$$b_1 = \int_0^1 \sin^2 \pi t \, dt = \frac{1}{2}$$

$$f(t) = \frac{1}{\pi} + \frac{1}{2}\sin \pi t - \frac{2}{\pi}\left[\frac{1}{3}\cos 2\pi t + \frac{1}{15}\cos 4\pi t + \frac{1}{35}\cos 6\pi t + \frac{1}{63}\cos 8\pi t + \cdots\right]$$

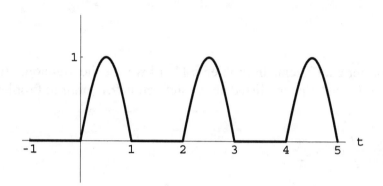

15.   **(a)** $a_0 = \dfrac{1}{\pi}\displaystyle\int_0^{2\pi} t^2 \, dt = \dfrac{8\pi^2}{3}$

$$a_n = \frac{1}{\pi}\int_0^{2\pi} t^2 \cos nt \, dt = \frac{4n\pi \cos 2n\pi + 2(2n^2\pi^2 - 1)\sin 2n\pi}{\pi n^3} = \frac{4}{n^2}$$

$$b_n = \frac{1}{\pi}\int_0^{2\pi} t^2 \sin nt \, dt = \frac{(2 - 4n^2\pi^2)\cos 2n\pi + 4n\pi \sin 2n\pi - 2}{\pi n^3} = -\frac{4\pi}{n}$$

$$f(t) = \frac{4\pi^2}{3} + 4\sum_{n=1}^{\infty}\frac{\cos nt}{n^2} - 4\pi\sum_{n=1}^{\infty}\frac{\sin nt}{n} \qquad \text{(figure below)}$$

(b)   If we substitute $t = 0$ in the Fourier series of part (a) and note that
$f(0) = \frac{1}{2}\left[f(0-) + f(0+)\right] = \frac{1}{2}\left[(2\pi)^2 + (0)^2\right] = 2\pi^2$, we get

$$2\pi^2 = \frac{4\pi^2}{3} + 4\sum_{n=1}^{\infty}\frac{1}{n^2}, \quad \text{so} \quad \sum_{n=1}^{\infty}\frac{1}{n^2} = \frac{\pi^2}{6}.$$

When we substitute $t = \pi$ and $f(\pi) = \pi^2$ in the series of part (a) we get

$$\pi^2 = \frac{4\pi^2}{3} + 4\sum_{n=1}^{\infty}\frac{(-1)^n}{n^2}, \quad \text{so} \quad \sum_{n=1}^{\infty}\frac{(-1)^{n+1}}{n^2} = \frac{\pi^2}{12}.$$

17. **(a)** $a_0 = \displaystyle\int_0^2 t\,dt = 2$

$$a_n = \int_0^2 t\cos n\pi t\,dt = \frac{\cos 2n\pi + 2n\pi\sin 2n\pi - 1}{n^2\pi^2} = 0$$

$$b_n = \int_0^2 t\sin n\pi t\,dt = \frac{\sin 2n\pi - 2n\pi\cos 2n\pi}{n^2\pi^2} = -\frac{2}{n\pi}$$

$$f(t) = 1 - \frac{2}{\pi}\sum_{n=1}^{\infty}\frac{\sin n\pi t}{n}$$

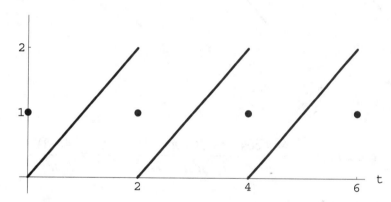

(b) Substitution of $t = 1/2$, $f(t) = 1/2$ in this series gives

$$\frac{1}{2} = 1 - \frac{2}{\pi}\left(1 - \frac{1}{3} + \frac{1}{5} - \frac{1}{7} + \cdots\right), \quad \text{so} \quad 1 - \frac{1}{3} + \frac{1}{5} - \frac{1}{7} + \cdots = \frac{\pi}{4}.$$

The expansions in Problems 19 and 21 are valid on the interval $-\pi < t < \pi$ rather than the interval $0 < t < 2\pi$. When we calculate the Fourier series of the functions $f(t)$ and $g(t)$ of period 2π defined for $-\pi < t < \pi$ by $f(t) = t$ and $g(t) = t^2$, we find that

$$t = 2\left[\sin t - \frac{\sin 2t}{2} + \frac{\sin 3t}{3} - \frac{\sin 4t}{4} + \cdots\right],$$

$$t^2 = \frac{\pi^2}{3} - 4\left[\cos t - \frac{\cos 2t}{4} + \frac{\cos 3t}{9} - \frac{\cos 4t}{16} + \cdots\right]$$

if $-\pi < t < \pi$.

19. $a_0 = \dfrac{1}{\pi}\displaystyle\int_{-\pi}^{\pi} \dfrac{t}{2}\,dt = 0,$ $a_n = \dfrac{1}{\pi}\displaystyle\int_{-\pi}^{\pi} \dfrac{t}{2}\cos nt\,dt = 0$

$b_n = \dfrac{1}{\pi}\displaystyle\int_{-\pi}^{\pi} \dfrac{t}{2}\sin nt\,dt = \dfrac{\sin n\pi - n\pi\cos n\pi}{n^2\pi} = \dfrac{(-1)^{n+1}}{n}$

$\dfrac{t}{2} = \displaystyle\sum_{n=1}^{\infty} \dfrac{(-1)^{n+1}\sin nt}{n}$ $(-\pi < t < \pi)$ (figure below)

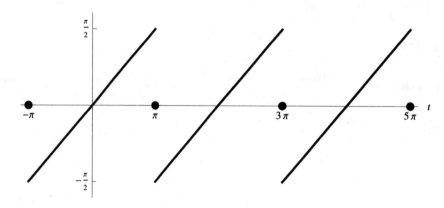

21. $a_0 = \dfrac{1}{\pi}\displaystyle\int_{-\pi}^{\pi} t^2\,dt = \dfrac{\pi^2}{3},$

$a_n = \dfrac{1}{\pi}\displaystyle\int_{-\pi}^{\pi} t^2\cos nt\,dt = \dfrac{4n\pi\cos n\pi + 2(n^2\pi^2 - 1)\sin n\pi}{n^3\pi} = \dfrac{4(-1)^n}{n^2}$

$b_n = \dfrac{1}{\pi}\displaystyle\int_{-\pi}^{\pi} t^2\sin nt\,dt = 0$

$t^2 = \dfrac{\pi^2}{3} + 4\displaystyle\sum_{n=1}^{\infty} \dfrac{(-1)^n\cos nt}{n^2}$ $(-\pi < t < \pi)$

$\dfrac{\pi^2 - 3t^2}{12} = \dfrac{\pi^2}{12} - \dfrac{1}{4}\left(\dfrac{\pi^2}{3} + 4\displaystyle\sum_{n=1}^{\infty} \dfrac{(-1)^n\cos nt}{n^2}\right) = \displaystyle\sum_{n=1}^{\infty} \dfrac{(-1)^{n+1}\cos nt}{n^2}$

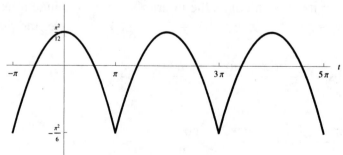

25.   Now we want to sum the alternating series

$$1 - \frac{1}{3^3} + \frac{1}{5^3} - \frac{1}{7^3} + \frac{1}{9^3} + \cdots$$

of reciprocals of odd cubes. Having used a Fourier series of t^4 in Problem 24 to evaluate $\Sigma(1/n^4)$, it is natural to look at a Fourier series of t^3. Let $f(t)$ be the period 2π function with $f(t) = t^3$ if $-\pi < t < \pi$. We calculate the Fourier coefficients of $f(t)$, and get

$$a_0 = \frac{1}{\pi}\int_{-\pi}^{\pi} t^3\, dt = 0, \qquad a_n = \frac{1}{\pi}\int_{-\pi}^{\pi} t^3 \cos nt\, dt = 0$$

$$b_n = \frac{1}{\pi}\int_{-\pi}^{\pi} t^3 \sin nt\, dt = -\frac{2n\pi\left(n^2\pi^2 - 6\right)\cos n\pi - 6(n^2\pi^2 - 2)\sin n\pi}{n^4\pi} = 2\left(\frac{6}{n^3} - \frac{\pi^2}{n}\right)$$

$$t^3 = 2\pi^2 \sum_{n=1}^{\infty}(-1)^{n+1}\frac{\sin nt}{n} - 12\sum_{n=1}^{\infty}(-1)^{n+1}\frac{\sin nt}{n^3}. \qquad \text{(figure below)}$$

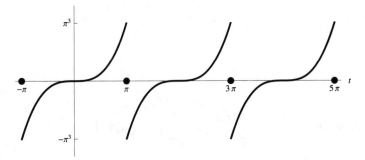

If we substitute $t = \pi/2$ and use Leibniz's series $\Sigma(-1)^{n+1}/n = \pi/4$ of Problem 17 we find that

$$1 - \frac{1}{3^3} + \frac{1}{5^3} - \frac{1}{7^3} + \frac{1}{9^3} + \cdots = \frac{\pi^3}{32}.$$

There is *no* value of t whose substitution in the Fourier series of $f(t) = t^3$ yields the series $\Sigma(1/n^3)$ containing the reciprocal cubes of *both* the odd and even integers. Indeed, the summation in "closed form" of the series

$$\sum_{n=1}^{\infty}\frac{1}{n^3} = 1 + \frac{1}{2^3} + \frac{1}{3^3} + \frac{1}{4^3} + \frac{1}{5^3} + \cdots$$

is a problem that has challenged many fine mathematicians since the time of Euler. Only in modern times (by R. Apery in 1978) has it been shown that this sum is an irrational number. For a delightful account of this work, see the article "A Proof that Euler Missed . . . An Informal Report" by Alfred van der Poorten in the *The Mathematical Intelligencer*, Volume 1 (1979), pages 195–203.

SECTION 9.3

FOURIER SINE AND COSINE SERIES

1. $a_0 = \dfrac{2}{\pi} \displaystyle\int_0^\pi 1 \, dt = 2, \qquad a_n = \dfrac{2}{\pi} \displaystyle\int_0^\pi \cos nt \, dt = \dfrac{2 \sin n\pi}{n\pi} = 0$

Cosine series: $f(t) = 1$

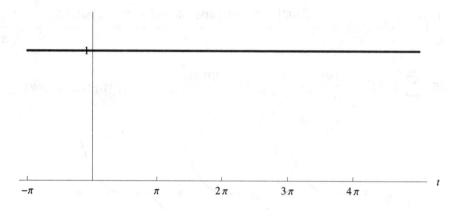

$b_n = \dfrac{2}{\pi} \displaystyle\int_0^\pi \sin nt \, dt = \dfrac{2(1 - \cos n\pi)}{n\pi} = \dfrac{2}{n\pi}\left[1 - (-1)^n\right]$

Sine series: $f(t) = \dfrac{4}{\pi}\left(\sin t + \dfrac{1}{3}\sin 3t + \dfrac{1}{5}\sin 5t + \dfrac{1}{7}\sin 7t + \cdots\right)$

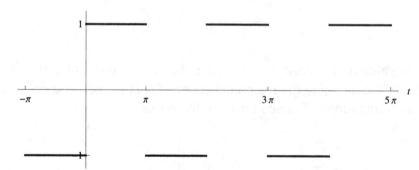

3. $a_0 = \displaystyle\int_0^2 (1 - t)\, dt = 0$

$a_n = \displaystyle\int_0^2 (1 - t)\cos\dfrac{n\pi t}{2}\, dt = \dfrac{4 - 4\cos n\pi - 2n\pi \sin n\pi}{n^2 \pi^2} = \dfrac{4}{n^2 \pi^2}\left[1 - (-1)^2\right]$

Cosine series: $f(t) = \dfrac{8}{\pi^2}\left(\cos\dfrac{\pi t}{2} + \dfrac{1}{3^2}\cos\dfrac{3\pi t}{2} + \dfrac{1}{5^2}\cos\dfrac{5\pi t}{2} + \dfrac{1}{7^2}\cos\dfrac{7\pi t}{2} + \cdots\right)$

See figure at top of next page.

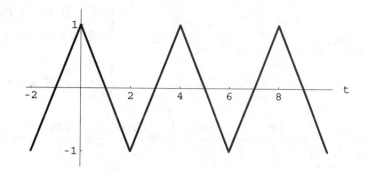

$$b_n = \int_0^2 (1-t)\sin\frac{n\pi t}{2}\, dt = \frac{2n\pi(1+\cos n\pi) - 2\sin n\pi}{n^2\pi^2} = \frac{2}{n\pi}\left[1+(-1)^n\right]$$

Sine series: $f(t) = \dfrac{4}{\pi}\left(\dfrac{\sin\pi t}{2} + \dfrac{\sin 2\pi t}{4} + \dfrac{\sin 3\pi t}{6} + \dfrac{\sin 4\pi t}{8} + \cdots\right)$

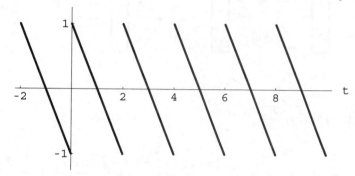

5. $a_0 = \dfrac{2}{3}\displaystyle\int_1^2 1\, dt = \dfrac{2}{3}$

$$a_n = \frac{2}{3}\int_1^2 \cos\frac{n\pi t}{3}\, dt = \frac{2}{n\pi}\left(\sin\frac{2n\pi}{3} - \sin\frac{n\pi}{3}\right) = \begin{cases} -2\sqrt{3}/n\pi & \text{if } n = 2, 8, 14, \cdots \\ +2\sqrt{3}/n\pi & \text{if } n = 4, 10, 16, \cdots \\ \qquad 0 & \text{otherwise} \end{cases}$$

Cosine series: $f(t) = \dfrac{1}{3} - \dfrac{2\sqrt{3}}{\pi}\left[\dfrac{1}{2}\cos\dfrac{2\pi t}{3} - \dfrac{1}{4}\cos\dfrac{4\pi t}{3} + \dfrac{1}{8}\cos\dfrac{8\pi t}{3} - \dfrac{1}{10}\cos\dfrac{10\pi t}{3} + \cdots\right]$

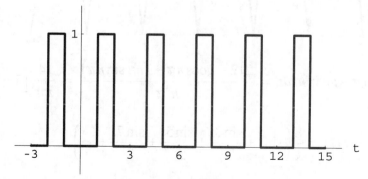

$$b_n = \frac{2}{3}\int_1^2 \sin\frac{n\pi t}{3}\,dt = \frac{2}{n\pi}\left(\cos\frac{n\pi}{3} - \cos\frac{2n\pi}{3}\right) = \begin{cases} 0 \text{ for } n \text{ even} \\ +2/n\pi \text{ if } n = 1, 7, 13, \cdots \\ -4/n\pi \text{ if } n = 3, 9, 15, \cdots \\ +2/n\pi \text{ if } n = 5, 11, 17, \cdots \end{cases}$$

Sine series:

$$f(t) = \frac{2}{\pi}\left[\sin\frac{\pi t}{3} - \frac{2}{3}\sin\frac{3\pi t}{3} + \frac{1}{5}\sin\frac{5\pi t}{3} + \frac{1}{7}\sin\frac{7\pi t}{3} - \frac{2}{9}\sin\frac{9\pi t}{3} + \frac{1}{11}\sin\frac{11\pi t}{3} + \cdots\right]$$

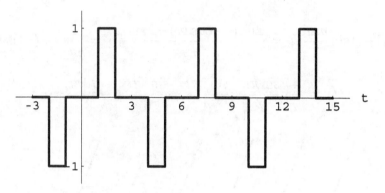

7. $$a_0 = \frac{2}{\pi}\int_0^\pi t(\pi - t)\,dt = \frac{\pi^2}{3},$$

$$a_n = \frac{2}{\pi}\int_0^\pi t(\pi - t)\cos nt\,dt = -\frac{2[n\pi\cos n\pi + n\pi - 2\sin n\pi]}{n^3\pi} = -\frac{2}{n^2}\left[1 + (-1)^n\right]$$

Cosine series: $$f(t) = \frac{\pi^2}{6} - 4\left(\frac{\cos 2t}{2^2} + \frac{\cos 4t}{4^2} + \frac{\cos 6t}{6^2} + \frac{\cos 8t}{8^2} + \cdots\right)$$

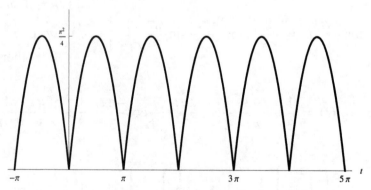

$$b_n = \frac{2}{\pi}\int_0^\pi t(\pi - t)\sin nt\,dt = \frac{2[2 - 2\cos n\pi - 2n\pi\sin n\pi]}{n^3\pi} = \frac{4}{\pi n^3}\left[1 - (-1)^n\right]$$

Sine series: $$f(t) = \frac{8}{\pi}\left(\sin t + \frac{\sin 3t}{3^3} + \frac{\sin 5t}{5^3} + \frac{\sin 7t}{7^3} + \cdots\right)$$

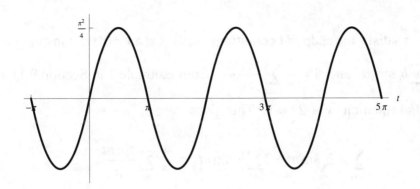

9.
$$a_0 = \frac{2}{\pi} \int_0^\pi \sin t \, dt = \frac{4}{\pi},$$

$$a_n = \frac{2}{\pi} \int_0^\pi \sin t \cos nt \, dt = \frac{2[1 + \cos n\pi]}{\pi(1 - n^2)} = \frac{2[1 + (-1)^n]}{\pi(1 - n^2)} \text{ if } n > 1$$

$$a_1 = \frac{2}{\pi} \int_0^\pi \sin t \cos t \, dt = 0$$

Cosine series: $f(t) = \dfrac{2}{\pi} - \dfrac{4}{\pi^2} \left(\dfrac{\cos 2t}{3} + \dfrac{\cos 4t}{15} + \dfrac{\cos 6t}{35} + \dfrac{\cos 8t}{63} + \cdots \right)$

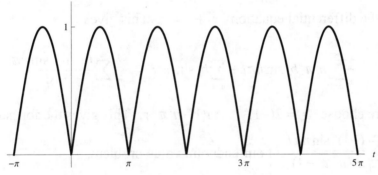

$$b_n = \frac{2}{\pi} \int_0^\pi \sin t \sin nt \, dt = \frac{2 \sin n\pi}{\pi(1 - n^2)} = 0 \text{ if } n > 1$$

$$b_1 = \frac{2}{\pi} \int_0^\pi \sin^2 t \, dt = 1$$

Sine series: $f(t) = \sin t$

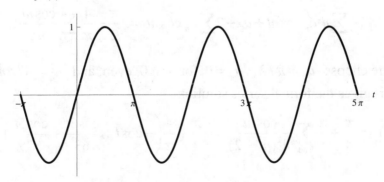

11. In order to satisfy the endpoint conditions $x(0) = x(\pi) = 0$ we substitute the sine series

$$x(t) = \sum_{n=1}^{\infty} b_n \sin nt \quad \text{and} \quad 1 = \frac{4}{\pi} \sum_{n \text{ odd}} \frac{\sin nt}{n} \quad \text{(from Example 1 in Section 9.1) into the}$$

differential equation $x'' + 2x = 1$. This gives

$$-\sum_{n=1}^{\infty} n^2 b_n \sin nt + 2\sum_{n=1}^{\infty} b_n \sin nt = \frac{4}{\pi} \sum_{n \text{ odd}} \frac{\sin nt}{n}.$$

We therefore choose $b_n = 4/\pi n(2 - n^2)$ for n odd, $b_n = 0$ for n even. This gives the formal series solution

$$x(t) = \frac{4}{\pi} \sum_{n \text{ odd}} \frac{\sin nt}{n(2 - n^2)} = \frac{4}{\pi}\left(\sin t - \frac{\sin 3t}{21} - \frac{\sin 5t}{115} - \frac{\sin 7t}{329} - \cdots \right).$$

13. In order to satisfy the endpoint conditions $x(0) = x(1) = 0$ we substitute the sine series

$$x(t) = \sum_{n=1}^{\infty} b_n \sin n\pi t \quad \text{and} \quad t = \frac{2}{\pi} \sum_{n=1}^{\infty} \frac{(-1)^{n+1} \sin n\pi t}{n} \quad \text{(from Example 1 in Section 9.3, with}$$

$L = 1$) into the differential equation $x'' + x = t$. This gives

$$-\sum_{n=1}^{\infty} n^2 \pi^2 b_n \sin n\pi t + \sum_{n=1}^{\infty} b_n \sin n\pi t = \frac{2}{\pi} \sum_{n=1}^{\infty} \frac{(-1)^{n+1} \sin n\pi t}{n}.$$

We therefore choose $b_n = 2(-1)^{n+1}/\pi n(1 - n^2\pi^2)$. This gives the formal series solution

$$x(t) = \frac{2}{\pi} \sum_{n=1}^{\infty} \frac{(-1)^n \sin n\pi t}{n(n^2\pi^2 - 1)} \quad \text{of our endpoint value problem.}$$

15. In order to satisfy the endpoint conditions $x'(0) = x'(2) = 0$ we substitute the cosine

series $x(t) = \frac{a_0}{2} + \sum_{n=1}^{\infty} a_n \cos nt \quad \text{and} \quad t = \frac{\pi}{2} - \frac{4}{\pi} \sum_{n \text{ odd}} \frac{\cos nt}{n^2} \quad \text{(from Example 1 in Section}$

9.3, with $L = \pi$) into the differential equation $x'' + 2x = t$. This gives

$$-\sum_{n=1}^{\infty} n^2 a_n \cos nt + a_0 + 2\sum_{n=1}^{\infty} a_n \cos nt = \frac{\pi}{2} - \frac{4}{\pi} \sum_{n \text{ odd}} \frac{\cos nt}{n^2}.$$

We therefore choose $a_0 = \pi/2$, $a_n = 0$ for $n > 0$ even, and $a_n = 4/\pi n^2(n^2 - 2)$ for n odd. This gives the formal series solution

$$x(t) = \frac{\pi}{4} + \frac{4}{\pi} \sum_{n \text{ odd}} \frac{\cos nt}{n^2(n^2 - 2)} = \frac{\pi}{4} + \frac{4}{\pi}\left(-\cos t + \frac{\cos 3t}{63} + \frac{\cos 5t}{575} + \frac{\cos 7t}{2303} + \cdots \right)$$

of our endpoint value problem.

17.   *Suggestion*: Substitute $u = -t$ in the left-hand integral.

19.   The first termwise integration yields

$$\frac{t^2}{2} = 2\sum_{n=1}^{\infty} \frac{(-1)^n \cos nt}{n^2} + C_1,$$

and substitution of $t = 0$ gives $C_1 = 2\sum_{n=1}^{\infty}(-1)^{n+1}/n^2 = \pi^2/6$, so

$$\frac{t^2}{2} = 2\sum_{n=1}^{\infty} \frac{(-1)^n \cos nt}{n^2} + \frac{\pi^2}{6}.$$

A second termwise integration gives

$$\frac{t^3}{6} = 2\sum_{n=1}^{\infty} \frac{(-1)^n \sin nt}{n^3} + \frac{\pi^2 t}{6} + C_2,$$

and substitution of $t = 0$ gives $C_2 = 0$. The final termwise integration gives

$$\frac{t^4}{24} = -2\sum_{n=1}^{\infty} \frac{(-1)^n \cos nt}{n^4} + \frac{\pi^2 t^2}{12} + C_3,$$

and substitution of $t = 0$ yields $C_3 = 2\sum_{n=1}^{\infty}(-1)^n/n^4$.

21.   We want to calculate the coefficients in the period $4L$ Fourier sine series

$$F(t) = \sum_{n=1}^{\infty} b_n \sin\frac{n\pi t}{2L}$$

which agrees with $f(t)$ if $0 < t < L$. Then

$$b_n = \frac{2}{2L}\int_0^L f(t)\sin\frac{n\pi t}{2L}\,dt + \frac{2}{2L}\int_L^{2L} f(2L-t)\sin\frac{n\pi t}{2L}\,dt.$$

The substitution $u = 2L - t$ yields

$$b_n = \frac{1}{L}\int_0^L f(t)\sin\frac{n\pi t}{2L}\,dt - \frac{1}{L}\int_L^0 f(u)\sin\frac{n\pi(2L-u)}{2L}\,du$$

$$= \frac{1}{L}\int_0^L f(t)\sin\frac{n\pi t}{2L}\,dt - \frac{(-1)^n}{L}\int_0^L f(u)\sin\frac{n\pi u}{2L}\,du.$$

Now it is clear that

$$b_n = \frac{2}{L}\int_0^L f(t)\cos\frac{n\pi t}{2L}\,dt$$

if n is odd, whereas $b_n = 0$ if n is even.

23. $b_n = \dfrac{2}{\pi}\displaystyle\int_0^\pi t\sin\frac{nt}{2}\,dt = \dfrac{4}{\pi n^2}\left(2\sin\frac{n\pi}{2} - n\pi\cos\frac{n\pi}{2}\right) = \dfrac{8(-1)^{(n-1)/2}}{\pi n^2}$ for n odd

$$f(t) = \frac{8}{\pi^2}\left(\sin\frac{t}{2} - \frac{1}{3^2}\sin\frac{3t}{2} + \frac{1}{5^2}\sin\frac{5t}{2} - \frac{1}{7^2}\sin\frac{7t}{2} + \cdots\right)$$

SECTION 9.4

APPLICATIONS OF FOURIER SERIES

1.   We substitute the sine series $x(t) = \displaystyle\sum_{n=1}^\infty b_n\sin nt$ and $F(t) = \dfrac{12}{\pi}\displaystyle\sum_{n\ \text{odd}}\dfrac{\sin nt}{n}$ (from Example 1 in Section 9.1) into the differential equation $x'' + 5x = F(t)$. This gives

$$-\sum_{n=1}^\infty n^2 b_n\sin nt + 5\sum_{n=1}^\infty b_n\sin nt = \frac{12}{\pi}\sum_{n\ \text{odd}}\frac{\sin nt}{n}.$$

We therefore choose $b_n = 0$ for $n > 0$ even, and $b_n = 12/\pi n(5 - n^2)$ for n odd. This gives the formal series solution

$$x_{sp}(t) = \frac{12}{\pi}\sum_{n\ \text{odd}}\frac{\sin nt}{n(5-n^2)} = \frac{12}{\pi}\left(\frac{\sin t}{4} - \frac{\sin 3t}{12} - \frac{\sin 5t}{100} - \frac{\sin 7t}{308} - \cdots\right).$$

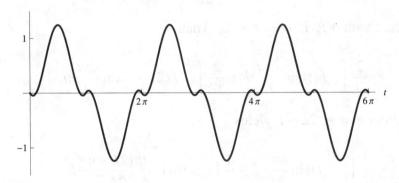

3.   We substitute the sine series $x(t) = \displaystyle\sum_{n=1}^\infty b_n\sin nt$ and $F(t) = 2\displaystyle\sum_{n=1}^\infty \dfrac{(-1)^{n-1}\sin nt}{n}$ (from Example 1 in Section 9.3, with $L = \pi$) into the differential equation $x'' + 3x = F(t)$. This gives

$$-\sum_{n=1}^{\infty} n^2 b_n \sin nt + 3 \sum_{n=1}^{\infty} b_n \sin nt \ = \ 4 \sum_{n=1}^{\infty} \frac{(-1)^{n-1} \sin nt}{n}.$$

We therefore choose $b_n = 4(-1)^{n-1}/n(3-n^2)$. This gives the formal series solution

$$x_{sp}(t) \ = \ 4 \sum_{n=1}^{\infty} \frac{(-1)^{n-1} \sin nt}{n(3-n^2)} = 4\left(\frac{\sin t}{2} + \frac{\sin 2t}{2} - \frac{\sin 3t}{18} + \frac{\sin 4t}{52} - \cdots \right).$$

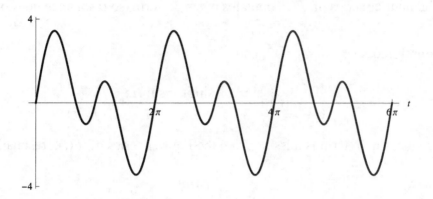

5. We substitute the sine series $x(t) = \sum_{n=1}^{\infty} b_n \sin n\pi t$ and $F(t) = \dfrac{8}{\pi^3} \sum_{n \text{ odd}} \dfrac{\sin n\pi t}{n^3}$ into the

differential equation $x'' + 10x = F(t)$. This gives

$$-\sum_{n=1}^{\infty} n^2 \pi^2 b_n \sin n\pi t + 10 \sum_{n=1}^{\infty} b_n \sin n\pi t \ = \ \frac{8}{\pi^3} \sum_{n \text{ odd}} \frac{\sin n\pi t}{n^3}.$$

We therefore choose $b_n = 8/n^3\pi^3(10-n^2\pi^2)$. This gives the formal series solution

$$x_{sp}(t) \ = \ \frac{8}{\pi^3} \sum_{n \text{ odd}} \frac{\sin n\pi t}{n^3(10-n^2\pi^2)}.$$

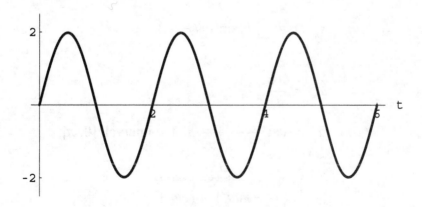

In Problems 7–12 we are dealing with the equation $mx'' + kx \ = \ F(t)$ where $F(t)$ is the external periodic force. The natural frequency is $\omega_0 \ = \ \sqrt{k/m}$. If the Fourier series of $F(t)$ contains a term of the form $\cos(N\pi t/L)$ or $\sin(N\pi t/L)$ with $\omega_0 = N\pi/L$, then pure resonance occurs. Otherwise, it does not.

7.   The natural frequency is $\omega_0 = 3$, and

$$F(t) = \frac{4}{\pi}\left(\sin t + \frac{\sin 3t}{3} + \frac{\sin 5t}{5} + \frac{\sin 7t}{7} + \cdots\right).$$

Thus the Fourier series of $F(t)$ contains a $\sin 3t$ term, so resonance does occur.

9.   The natural frequency is $\omega_0 = 2$, and

$$F(t) = \frac{4}{\pi}\left(\sin t + \frac{\sin 3t}{3} + \frac{\sin 5t}{5} + \frac{\sin 7t}{7} + \cdots\right).$$

Because the $\sin 2t$ term is missing from the Fourier series of $F(t)$, resonance will not occur.

11.   The natural frequency is $\omega_0 = 4$. From Equation (15) in Section 9.3 we see that

$$F(t) = \frac{\pi}{2} - \frac{4}{\pi}\left(\cos t + \frac{\cos 3t}{3^2} + \frac{\cos 5t}{5^2} + \cdots\right).$$

Because the $\cos 4t$ term is missing, we see that resonance will not occur.

Problems 13–18 are based on Equations (14)–(16) in the text, according to which the steady periodic solution of

$$mx'' + cx' + kx = \sum_{n=1}^{\infty} B_n \sin\frac{n\pi t}{L}$$

is given by

$$x_{sp}(t) = \sum_{n=1}^{\infty} b_n \sin(\omega_n t - \alpha_n),$$

where

$$\omega_n = \frac{n\pi}{L},$$

$$\alpha_n = \tan^{-1}\frac{c\omega_n}{k - m\omega_n^2} \quad \text{in the interval } [0, \pi],$$

$$b_n = \frac{B_n}{\sqrt{\left(k - m\omega_n^2\right)^2 + \left(c\omega_n\right)^2}}.$$

This calculation is readily automated. The following MATLAB script was written to calculate the coefficients $\{b_n\}$ for Problem 13. Only the values of m, c, k, L and the calculation of the force function coefficients $\{B_n\}$ need to be changed for Problems 14–18.

```
m = 1;    c = 0.1;    k = 4;
L = pi;
results = ones(0,4);
for n = 1:9
   w = n*pi/L;
   alpha = atan(c*w/(k-m*w^2));
   if k-m*w^2<0
      alpha = pi + alpha;
   end
   B = 12/(pi*n);        % force function coeffs
   if floor(n/2)==n/2    % are nonzero if n is odd,
      B = 0;             % zero if n is even
   end
   b = B/sqrt((k-m*w^2)^2+(c*w)^2);
   results = [results; n, b, w, alpha];
end
results
```

13. $B_n = 12/\pi n$ for n odd, $B_n = 0$ for n even

$x_{sp}(t) \approx 1.2725 \sin(t - 0.0333) + 0.2542 \sin(3t - 3.0817) + 0.0364 \sin(5t - 3.1178) + \cdots$

15. $B_n = 8/n^3\pi^3$ for n odd, $B_n = 0$ for n even

$x_{sp}(t) \approx 0.08150 \sin(\pi t - 1.44692) + 0.00004 \sin(3\pi t - 3.10176) + \cdots$

17. $B_n = 60/n\pi$ for n odd, $B_n = 0$ for n even

$x_{sp}(t) \approx 0.5687 \sin(\pi t - 0.0562) + 0.4271 \sin(3\pi t - 0.3891)$
$\qquad\qquad + 0.1396 \sin(5\pi t - 2.7899) + 0.0318 \sin(7\pi t - 2.9874) + \cdots$

$x_{sp}(5) \approx 0.248$ ft ≈ 2.98 in.

19.   We suppose that $f(t + P) = f(t)$ and $g(t + Q) = g(t)$ for all t. If $P/Q = m/n$ where m and n are integers, let $R = nP = mQ$. Then

$$f(t + R) + g(t + R) = f(t + nP) + g(t + mQ) = f(t) + g(t)$$

for all t, so we see that the sum $f(t) + g(t)$ is periodic with period R.

SECTION 9.5

HEAT CONDUCTION AND SEPARATION OF VARIABLES

1.   From Equation (31) in the text, with $L = \pi$ and $k = 3$, we get

$$u(x,t) = \sum_{n=1}^{\infty} b_n \exp(-3n^2 t) \sin nx \, .$$

With $b_2 = 4$ and $b_n = 0$ otherwise we get the solution

$$u(x, t) = 4e^{-12t} \sin 2x.$$

3.   With $L = 1$ and $k = 2$ in Equation (31), we take $b_1 = 5$, $b_3 = -1/5$, and $b_n = 0$ otherwise. The result is the solution

$$u(x,t) = 5e^{-2\pi^2 t} \sin \pi x - \frac{1}{5} e^{-18\pi^2 t} \sin 3\pi x.$$

5.   From Equation (40) in the text, with $k = 2$ and $L = 3$ we get

$$u(x,t) = \frac{a_0}{2} + \sum_{n=1}^{\infty} a_n \exp\left(-\frac{2n^2 \pi^2 t}{9}\right) \cos \frac{n\pi x}{3}.$$

With $a_0 = 0$, $a_2 = 4$, $a_4 = -2$, and $a_n = 0$ otherwise, and $a_n = 0$ we get the solution

$$u(x,t) = 4\exp\left(-\frac{8\pi^2 t}{9}\right) \cos \frac{2\pi x}{3} - 2\exp\left(-\frac{32\pi^2 t}{9}\right) \cos \frac{4\pi x}{3}.$$

7.   From Equation (40) in the text, with $k = 1/3$ and $L = 2$ we get,

$$u(x,t) = \frac{a_0}{2} + \sum_{n=1}^{\infty} a_n \exp\left(-\frac{n^2 \pi^2 t}{12}\right) \cos \frac{n\pi x}{2}.$$

Because of the identity $\cos^2 2\pi x = (1 + \cos 4\pi x)/2$, we choose $a_0 = 1$, $a_8 = 1/2$, and $a_n = 0$ otherwise. This gives the solution

$$u(x,t) = \frac{1}{2} + \frac{1}{2} \exp\left(-\frac{16\pi^2 t}{3}\right) \cos 4\pi x.$$

9.   Because of the zero endpoint conditions $u(0,t) = u(5,t) = 0$, we use the Fourier sine series expansion

$$u(x,0) = \frac{100}{\pi} \sum_{n \text{ odd}} \frac{1}{n} \sin \frac{n\pi x}{5}$$

of $u(x,0) = 25$ on the interval $0 < x < 5$. When we supply the exponential factors in Eq. (31) with $k = 1/10$ and $L = 5$, we get the solution

$$u(x,t) = \frac{100}{\pi} \sum_{n \text{ odd}} \frac{1}{n} \exp\left(\frac{-n^2\pi^2 t}{250}\right) \sin \frac{n\pi x}{5}$$

11.   Because of the zero-derivative endpoint conditions $u_x(0,t) = u_x(10,t) = 0$, we use the Fourier cosine series expansion

$$u(x,0) = 20 - \frac{160}{\pi^2} \sum_{n \text{ odd}} \frac{1}{n^2} \cos \frac{n\pi x}{10}$$

of $u(x,0) = 4x$ on the interval $0 < x < 10$ (from Eq. (15) in Section 9.3). When we supply the exponential factors in Eq. (40) here with $k = 1/5$ and $L = 10$, we get

$$u(x,t) = 20 - \frac{160}{\pi^2} \sum_{n \text{ odd}} \frac{1}{n^2} \exp\left(\frac{-n^2\pi^2 t}{500}\right) \cos \frac{n\pi x}{10}$$

13.   **(a)**   The boundary value problem is

$$u_t = k u_{xx} \quad (0 < x < 40),$$
$$u_x(0, t) = u_x(40, t) = 0,$$
$$u(x, 0) = 100.$$

By Equation (31) in the text (with $L = 40$) the solution is of the form

$$u(x,t) = \sum_{n=1}^{\infty} b_n \exp\left(-\frac{n^2\pi^2 kt}{1600}\right) \sin \frac{n\pi x}{40}.$$

We use the Fourier sine coefficients $b_n = 400/\pi n$ for n odd, $b_n = 0$ otherwise, of the initial value function $f(x) = 100$ on the interval $0 < x < 100$. This gives

$$u(x,t) = \frac{400}{\pi} \sum_{n \text{ odd}} \frac{1}{n} \exp\left(\frac{-n^2\pi^2 kt}{1600}\right) \sin \frac{n\pi x}{40}.$$

(b)   With $k = 1.15$ for copper we find that

$$u(20,300) \approx 15.1591 - 0.000000204 + \cdots \approx 15.16°C.$$

(c)   With $k = 0.005$ for concrete, the first term of the series gives

$$u(20,t) = \frac{400}{\pi}\exp\left(-\frac{0.00\pi^2 t}{1600}\right) = 15,$$

and we solve for $t \approx 66{,}342$ sec ≈ 19 hr 15 min 42 sec. As a check that the first term suffices for this computation, we find that the next term in the series is then approximately 0.00000019.

15. We need only calculate the coefficients in the usual zero-endpoint series

$$u(x,t) = \sum_{n=1}^{\infty} b_n \exp\left(-\frac{n^2\pi^2 kt}{L^2}\right)\sin\frac{n\pi x}{L}.$$

For the function $f(x) \equiv A$ for $0 < x < L/2$, $f(x) \equiv 0$ for $L/2 < x < L$ we calculate the Fourier sine coefficient

$$b_n = \frac{2}{L}\int_0^{L/2} A\sin\frac{n\pi x}{L}dx = \frac{4A}{n\pi}\sin^2\frac{n\pi}{4} = \frac{4A}{n\pi}\times\begin{cases}1/2 & \text{for } n \text{ odd,}\\ 1 & \text{for } n = 2,6,10,\cdots,\\ 0 & \text{for } n = 4,8,12,\cdots.\end{cases}$$

SECTION 9.6

VIBRATING STRINGS AND THE ONE-DIMENSIONAL WAVE EQUATION

In Problems 1–10 we use the general solution

$$y(x,t) = \sum_{n=1}^{\infty}\left(A_n\cos\frac{n\pi at}{L} + B_n\sin\frac{n\pi at}{L}\right)\sin\frac{n\pi x}{L} \tag{*}$$

of the string equation $y_{tt} = a^2 y_{xx}$ with endpoint conditions $y(0, t) = y(L, t) = 0$. This form of the solution is obtained by superposition of the solutions in Equations (23) and (33) of Problems A and B in this section. It remains only to choose the coefficients $\{A_n\}$ and $\{B_n\}$ so as to satisfy given initial conditions

$$y(x,0) = \sum_{n=1}^{\infty} A_n\sin\frac{n\pi x}{L} = f(x), \quad \text{thus,} \quad A_n = \frac{2}{L}\int_0^L f(x)\sin\frac{n\pi x}{L}dx; \text{ and}$$

$$y_t(x,0) = \sum_{n=1}^{\infty}\frac{n\pi a}{L}B_n\sin\frac{n\pi x}{L} = g(x), \quad \text{thus,} \quad B_n = \frac{2}{n\pi a}\int_0^L g(x)\sin\frac{n\pi x}{L}dx.$$

1. Here $a = 2$ and $L = \pi$. To satisfy the condition $y(x, 0) = (1/10)\sin 2x$ we choose $A_2 = 1/10$ in Eq. (*) above, and $A_n = 0$ otherwise. To satisfy the condition $y_t(x, 0) = 0$ we choose $B_n = 0$ for all n. Thus

$$y(x, t) = \frac{1}{10}\cos 4t \sin 2x.$$

3.   Here $a = 1/2$ and $L = \pi$. Choosing $A_1 = 1/10$ and $A_n = 0$ otherwise, $B_1 = 1/5$ and $B_n = 0$ otherwise, we get

$$y(x, t) = \frac{1}{10}\left(\cos\frac{t}{2} + 2\sin\frac{t}{2}\right)\sin x.$$

5.   Here $a = 5$ and $L = 3$. Choosing $A_3 = 1/4$ and $A_n = 0$ for $n \neq 3$, $B_6 = 1/\pi$ and $B_n = 0$ for $n \neq 6$, we get

$$y(x, t) = \frac{1}{4}\cos 5\pi t \sin \pi x + \frac{1}{\pi}\sin 10\pi t \sin 2\pi x.$$

7.   Here $a = 10$ and $L = 1$. To satisfy the condition $y(x, 0) = 0$ we choose $A_n = 0$ for all n, so

$$y(x, t) = \sum_{n=1}^{\infty} B_n \sin 10n\pi t \sin n\pi x.$$

To satisfy the condition $y_t(x, 0) = x$ we choose

$$B_n = \frac{1}{10n\pi} \cdot \frac{2(-1)^{n+1}}{n\pi} = \frac{(-1)^{n+1}}{5n^2\pi^2}$$

for $n \geq 1$ (see Equation (16) in Section 9.3). This gives

$$y(x, t) = \frac{1}{5\pi^2}\sum_{n=1}^{\infty}\frac{(-1)^{n+1}}{n^2}\sin 10n\pi t \sin n\pi x.$$

9.   Here $a = 2$ and $L = 1$. To satisfy the condition $y(x, 0) = 0$ we choose $A_n = 0$ for all n, so

$$y(x, t) = \sum_{n=1}^{\infty} B_n \sin 2n\pi t \sin n\pi x.$$

To satisfy the condition $y_t(x, 0) = x(1 - x)$ we choose

$$B_n = \frac{1}{n\pi}\int_0^1 x(1-x)\sin n\pi x\, dx = \frac{2 - 2\cos n\pi - n\pi \sin n\pi}{n^4\pi^4}.$$

Hence

$$y(x,t) = \frac{4}{\pi^4}\sum_{n\ \text{odd}}\frac{\sin 2n\pi t \sin n\pi x}{n^4}.$$

11.   Substitution of $L = 2$ ft, $T = 32$ lb, and the *linear* density

$$\rho \;=\; \frac{1/32\,\text{oz}}{2\,\text{ft}} \;=\; \frac{1\,\text{oz}}{64\,\text{ft}}\cdot\frac{1\,\text{lb}}{16\,\text{oz}}\cdot\frac{1\,\text{slug}}{32\,\text{lb}} \;=\; \frac{1}{32^3}\,\frac{\text{slug}}{\text{ft}}$$

in Eqs. (2) and (26) in the text yields the velocity $a = \sqrt{T/\rho} = \sqrt{32^4} = 1024\,\text{ft}/\text{sec}$ with which waves move along the string, and its fundamental frequency

$$v_1 \;=\; \frac{1}{2L}\sqrt{\frac{T}{\rho}} \;=\; \frac{a}{2L} \;=\; 256\,\text{Hz},$$

which is approximately middle C.

13. If $y(x,t) = F(x+at) = F(u)$ with $u = x+at$, then the chain rule gives

$$\frac{\partial y}{\partial x} = \frac{dF}{du}\frac{\partial u}{\partial x} = F'(u)\cdot 1 = F'(x+at);$$

$$\frac{\partial y}{\partial t} = \frac{dF}{du}\frac{\partial u}{\partial t} = F'(u)\cdot a = aF'(x+at) = a\frac{\partial y}{\partial x};$$

$$\frac{\partial^2 y}{\partial x^2} = \frac{dF'}{du}\frac{\partial u}{\partial x} = F''(u)\cdot 1 = F''(x+at);$$

$$\frac{\partial^2 y}{\partial t^2} = a\frac{dF'}{du}\frac{\partial u}{\partial t} = a\cdot F''(u)\cdot a = a^2 F''(x+at) = a^2\frac{\partial^2 y}{\partial x^2}.$$

15. If $y(x,0) = 0$ then the fundamental theorem of calculus gives

$$y(x,t) \;=\; y(x,t) - y(x,0) \;=\; \int_0^t y_t(x,\tau)\,d\tau \;=\; \int_0^t \frac{1}{2}\big[G(x+a\tau) + G(x-a\tau)\big]\,d\tau.$$

19. The general solution of the second-order ordinary differential equation $a^2 y'' = g$ is a second-order polynomial in x with leading coefficient $g/2a^2$. But the polynomial $\phi(x) = gx(x-L)/2a^2$ has this leading coefficient and satisfies the endpoint conditions $y(0) = y(L) = 0$.

23. First, if $0 \le x \le \dfrac{\pi}{4}$, then $\dfrac{\pi}{4} \le x + \dfrac{\pi}{4} \le \dfrac{\pi}{2}$ and $-\dfrac{\pi}{4} \le x - \dfrac{\pi}{4} \le 0$, so that

$$y\left(x, \frac{\pi}{4}\right) = \frac{1}{2}\left[F\left(x + \frac{\pi}{4}\right) + F\left(x - \frac{\pi}{4}\right)\right]$$

$$= \frac{1}{2}\left[1 - \cos 2\left(x + \frac{\pi}{4}\right) + \cos 2\left(x - \frac{\pi}{4}\right) - 1\right]$$

$$= \frac{1}{2}\left[\cancel{1} - \cos\left(2x + \frac{\pi}{2}\right) + \cos\left(2x - \frac{\pi}{2}\right) - \cancel{1}\right]$$

$$= \frac{1}{2}(\sin 2x + \sin 2x)$$

$$= \sin 2x.$$

Next, if $\frac{\pi}{4} \le x \le \frac{3\pi}{4}$, then $\frac{\pi}{2} \le x + \frac{\pi}{4} \le \pi$ and $0 \le x - \frac{\pi}{4} \le \frac{\pi}{2}$, so that

$$y\left(x, \frac{\pi}{4}\right) = \frac{1}{2}\left[F\left(x + \frac{\pi}{4}\right) + F\left(x - \frac{\pi}{4}\right)\right]$$

$$= \frac{1}{2}\left[1 - \cos 2\left(x + \frac{\pi}{4}\right) + 1 - \cos 2\left(x - \frac{\pi}{4}\right)\right]$$

$$= \frac{1}{2}\left[1 - \cos\left(2x + \frac{\pi}{2}\right) + 1 - \cos\left(2x - \frac{\pi}{2}\right)\right]$$

$$= \frac{1}{2}\left(2 + \cancel{\sin 2x} - \cancel{\sin 2x}\right)$$

$$= 1.$$

Finally, if $\frac{3\pi}{4} \le x \le \pi$, then $\pi \le x + \frac{\pi}{4} \le \frac{5\pi}{4}$ and $\frac{\pi}{2} \le x - \frac{\pi}{4} \le \frac{3\pi}{4}$, so that

$$y\left(x, \frac{\pi}{4}\right) = \frac{1}{2}\left[F\left(x + \frac{\pi}{4}\right) + F\left(x - \frac{\pi}{4}\right)\right]$$

$$= \frac{1}{2}\left[\cos 2\left(x + \frac{\pi}{4}\right) - \cancel{1} + \cancel{1} - \cos 2\left(x - \frac{\pi}{4}\right)\right]$$

$$= \frac{1}{2}\left[\cos\left(2x + \frac{\pi}{2}\right) - \cos\left(2x - \frac{\pi}{2}\right)\right]$$

$$= \frac{1}{2}(-\sin 2x - \sin 2x)$$

$$= -\sin 2x.$$

SECTION 9.7

STEADY-STATE TEMPERATURE AND LAPLACE'S EQUATION

1.   Because $Y(0) = Y(b) = 0$ we take our separation of variables in the form

$$X'' - \lambda X = 0 = Y'' + \lambda Y$$

with $\lambda > 0$. Then it follows that

$$Y_n(y) = \sin\frac{n\pi y}{b}, \qquad \lambda_n = \frac{n^2\pi^2}{b^2}$$

and thence that

$$X_n(x) = A_n \cosh\frac{n\pi x}{b} + B_n \sinh\frac{n\pi x}{b}.$$

The condition that $X(0) = 0$ implies that $A_n = 0$ so $X_n(x) = B_n \cosh n\pi x/b$, and hence

$$u(x, y) = \sum_{n=1}^{\infty} C_n \sinh\frac{n\pi x}{b} \sin\frac{n\pi y}{b}.$$

Finally we satisfy the condition $u(a, y) = g(y)$ by choosing $C_n = b_n /(\sinh n\pi a/b)$, where the $\{b_n\}$ are the Fourier sine coefficients of $g(y)$ on $0 \le y \le b$.

3.   Just as in Example 1 of Section 9.7 we have $X_n(x) = \sin n\pi x/a$ and

$$Y_n(y) = A_n \cosh\frac{n\pi y}{a} + B_n \sinh\frac{n\pi y}{a}.$$

The condition $Y(0) = 0$ now yields $A_n = 0$ so $Y_n(y) = B_n \sinh n\pi y/a$, and hence

$$u(x, y) = \sum_{n=1}^{\infty} C_n \sin\frac{n\pi x}{a} \sinh\frac{n\pi y}{a}.$$

Finally we satisfy the condition $u(x, b) = f(x)$ by choosing $C_n = b_n /(\sinh n\pi b/a)$, where the $\{b_n\}$ are the Fourier sine coefficients of $f(x)$ on $0 \le x \le a$.

5.   Now $Y'(0) = Y'(b) = 0$, so we work with the separation of variables

$$X'' - \lambda X = 0 = Y'' + \lambda Y.$$

The eigenvalue problem

$$Y'' + \lambda Y = 0, \qquad\qquad Y'(0) = Y'(b) = 0,$$

has eigenvalues and eigenfunctions $\lambda_0 = 0$, $Y_0(y) = 1$ and

$$\lambda_n = \frac{n^2\pi^2}{b^2}, \qquad Y_n(y) = \cos\frac{n\pi y}{b}$$

for $n = 1, 2, 3, \cdots$. When $n = 0$, $X_0''(x) \equiv 0$ yields $X_0(x) = Ax + B$. Then $X_0(a) = 0$ is satisfied by $X_0(x) = a - x$. For $n > 0$ we have

$$X_n(x) = A_n \cosh\frac{n\pi x}{b} + B_n \sinh\frac{n\pi x}{b},$$

and $X_n(a) = 0$ is satisfied by the particular linear combination

$$X_n(x) = C_n \sinh\frac{n\pi(a-x)}{b},$$

of $\cosh n\pi x/b$ and $\sinh n\pi x/b$. Therefore

$$u(x,y) = C_0(a-x) + \sum_{n=1}^{\infty} C_n \sinh\frac{n\pi(a-x)}{b}\cos\frac{n\pi y}{b}.$$

Finally we satisfy the condition $u(0, y) = g(y)$ by choosing

$$C_0 = \frac{a_0}{2a} \quad \text{and} \quad C_n = \frac{b_n}{\sinh n\pi a/b},$$

where the $\{a_n\}$ are the Fourier cosine coefficients of $g(y)$ on $0 \le y \le b$.

7.   The eigenvalue problem

$$X'' + \lambda X = 0, \qquad X(0) = X(a) = 0$$

yields the eigenvalues and eigenfunctions

$$\lambda_n = \frac{n^2\pi^2}{a^2}, \qquad X_n(x) = \sin\frac{n\pi x}{a}$$

for $n = 1, 2, 3, \cdots$. Then

$$Y_n'' + \lambda_n Y_n = 0$$

yields

$$Y_n(y) = A_n e^{n\pi y/a} + B_n e^{-n\pi y/a}.$$

In order that $Y(y) \to 0$ as $y \to \infty$ we take $A_n = 0$, so

$$u(x,y) = \sum_{n=1}^{\infty} B_n e^{-n\pi y/a} \sin\frac{n\pi x}{a}.$$

Finally we satisfy the condition $u(x,0) = f(x)$ by choosing the constants $\{B_n\}$ as the Fourier sine coefficients of $f(x)$ on $0 \le x \le a$.

9. If in Problem 8 we have $f(x) = 10x$ on $0 < x < 10$, then

$$a_0 = \frac{2}{10}\int_0^{10} 10x\, dx = 100,$$

$$a_n = \frac{2}{10}\int_0^{10} 10x\cos\frac{n\pi x}{10}\, dx = \frac{200(\cos n\pi - 1 + n\pi\sin\pi)}{n^2\pi^2},$$

so

$$u(x,y) = 50 - \frac{400}{\pi^2}\sum_{n\,\text{odd}}\frac{1}{n^2}e^{-n\pi y/10}\cos\frac{n\pi x}{10}.$$

Then

$$u(0,5) \approx 50 - 8.4250 - 0.0405 - 0.0006 - 0.0000 - \cdots \approx 41.53,$$

$$u(5,5) = 50 - 0 - 0 - 0 - 0 - \cdots = 50,$$

$$u(0,5) \approx 50 + 8.4250 + 0.0405 + 0.0006 + 0.0000 + \cdots \approx 58.47.$$

11. Now the boundary value problem is

$$u_{xx} + u_{yy} = 0 \qquad (0 < x < a,\ 0 < y < b)$$

$$u(a,y) = u_y(x,0) = u(x,b) = 0,$$

$$u(0,y) = g(y).$$

The eigenvalue problem

$$Y'' + \lambda Y = 0, \qquad Y'(0) = Y(b) = 0$$

yields (similar to Example 4 in Section 3.8)

$$\lambda_n = \frac{(2n-1)^2\pi^2}{4b^2}, \qquad Y_n(y) = \cos\frac{(2n-1)\pi y}{2b}$$

for $n = 1, 2, 3, \cdots$. Then

$$X_n'' - \lambda_n X_n = 0$$

yields

$$X_n(x) = A_n\cosh\frac{(2n-1)\pi x}{2b} + B_n\sin\frac{(2n-1)\pi x}{2b}.$$

Now $X_n(a) = 0$ is satisfied by the particular linear combination

$$X_n(x) = C_n\sinh\frac{(2n-1)\pi(a-x)}{2b}$$

of $\cosh{(2n-1)\pi x/2b}$ and $\sinh{(2n-1)\pi x/2b}$. Hence

$$u(x,y) = \sum_{n=1}^{\infty} C_n \sinh\frac{(2n-1)\pi(a-x)}{2b}\cos\frac{(2n-1)\pi y}{2b}$$

$$= \sum_{n \text{ odd}} A_n \sinh\frac{n\pi(a-x)}{2b}\cos\frac{n\pi y}{2b}.$$

Finally we satisfy the condition $u(0,y) = g(y)$ by choosing

$$A_n = \frac{a_n}{\sinh n\pi a/2b},$$

where the $\{a_n\}$ are the odd half-multiple cosine coefficients of $g(y)$ on $[0, b]$, as given by Problem 22 in Section 9.3.

13.   We start with the periodic polar-coordinate solution

$$u(r,\theta) = \frac{a_0}{2} + \sum_{n=1}^{\infty} r^n(a_n\cos n\theta + b_n\sin n\theta)$$

and choose $a_n \equiv 0$ in order to satisfy the conditions $u(r,0) = u(r,\pi) = 0$. Then

$$u(r,\theta) = \sum_{n=1}^{\infty} r^n c_n \sin n\theta$$

satisfies the nonhomogeneous boundary condition $u(a,\theta) = f(\theta)$ provided that $a^n c_n$ is is the nth Fourier sine coefficient of $f(\theta)$ on the interval $0 < \theta < \pi$, that is,

$$c_n = \frac{2}{\pi a^n}\int_0^\pi f(\theta)\sin n\theta\, d\theta.$$

15.   As in the textbook discussion of the polar-coordinate Dirichlet problem, the substitution $u(r,\theta) = R(r)\Theta(\theta)$ in Laplace's equation yields the separated ordinary differential equations

$$r^2 R'' + rR' - \lambda R = 0 \qquad\qquad (25)$$

and

$$\Theta'' + \lambda\Theta = 0. \qquad\qquad (26)$$

With $\lambda = \alpha^2$ the general solution of (26) is

$$\Theta(\theta) = A\cos\alpha\theta + B\sin\alpha\theta,$$

and the endpoint condition $\Theta(0) = \Theta'(0) = 0$ yields $A = 0$ and $\theta = (2n-1)/2$, so the nth eigenvalue and eigenfunction are given by

$$\lambda_n = \frac{(2n-1)^2}{4}, \qquad \Theta_n(\theta) = \sin\frac{(2n-1)\theta}{2}.$$

As in the discussion of Eqs. (29) and (30) in the text, the bounded solution of

$$r^2 R_n'' + r R_n' - \frac{(2n-1)^2}{4} R_n = 0$$

is

$$R_n(r) = r^{(2n-1)/2}$$

for $n = 1, 2, 3, \cdots$. We thereby obtain the formal series solution

$$u(r, \theta) = \sum_{n \text{ odd}} c_n r^{n/2} \sin \frac{n\theta}{2}.$$

It remains only to satisfy the nonhomogeneous boundary condition $u(a, \theta) = f(\theta)$ by choosing

$$c_n = \frac{2}{\pi a^{n/2}} \int_0^\pi f(\theta) \sin \frac{n\theta}{2} d\theta,$$

so that (for n odd) $c_n a^{n/2}$ equals the nth odd half-multiple sine coefficient of $f(\theta)$.

17. The substitution $u(r, \theta) = R(r)\Theta(\theta)$ in Laplace's equation yields the same separated solution functions

$$\Theta_0(\theta) = 1, \qquad R_0(r) = C_0 + D_0 \ln r$$

and

$$\Theta_n(\theta) = A_n \cos n\theta + B_n \sin n\theta, \qquad R_n(r) = C_n r^n + \frac{D_n}{r^n}$$

as in Eqs. (28)-(30) in the text. We choose $B_n \equiv 0$ to satisfy the boundary condition $u(r, \theta) = u(r, -\theta)$, and $n = 1$ with $C_1 = U_0$ to satisfy the given limit condition as $r \to \infty$. Then the condition that $u_r(a, \theta) = 0$ requires that $D_1 = U_0 a^2$, so

$$u(r, \theta) = \frac{U_0}{r}\left(r^2 + a^2\right)\cos\theta.$$

21. **(a)** Since we cannot simply substitute $r = 0$, we apply continuity of $u(r, t)$ at $r = 0$ and calculate

$$u(0, t) = \lim_{r \to 0} u(r, t)$$

noting that

$$\lim_{r \to 0} \frac{\sin n\pi r / a}{r} = \frac{n\pi}{a} \lim_{r \to 0} \frac{\sin n\pi r / a}{n\pi r / a} = \frac{n\pi}{a} \lim_{r \to 0} \frac{\sin \theta}{\theta} = \frac{n\pi}{a}$$

by the elementary fact that $(\sin\theta)/\theta \to 1$ as $\theta \to 0$.

(b) With $a = 30$ and $T_0 = 100$ we have

$$u(0,t) = 200\sum_{n=1}^{\infty}(-1)^{n+1}\exp\left(-\frac{n^2\pi^2 kt}{900}\right).$$

If $k = 0.15$ for iron then after 15 minutes = 900 seconds the center temperature is

$$u(0,900) \approx 45.5075 - 0.5361 + 0.0003 - 0.0000 + \cdots \approx 44.97.$$

If $k = 0.005$ for iron then after 15 minutes the center temperature is

$$\begin{aligned}u(0,900) \approx{}& 190.37 - 164.174 + 128.276 - 90.8081 + 58.2426 \\ & -33.8449 + 17.8190 - 8.4998 + 3.6734 - 1.4384 \\ & +0.5103 - 0.1640 + 0.0478 - 0.0126 + 0.0030 \\ & -0.0007 + 0.0001 - 0.00002 + 0.00000 - \cdots \end{aligned}$$

$$u(0,900) \approx 100.00$$

Thus the center of the ball has not yet begun to cool. For the center of this concrete ball to reach $45°$ (as with the iron ball after 15 minutes) would require $(0.15 / 0.005) \times 15 = 450$ minutes, that is, seven and a half hours!

CHAPTER 10

EIGENVALUE METHODS AND
BOUNDARY VALUE PROBLEMS

SECTION 10.1

STURM-LIOUVILLE PROBLEMS
AND EIGENFUNCTION EXPANSIONS

1. In the notation of Equation (9) in Section 10.1 of the text we have $\alpha_1 = \beta_1 = 0$ and $\alpha_2 = \beta_2 = 1$, so Theorem 1 implies that the eigenvalues are all nonnegative. If $\lambda = 0$, then $y'' = 0$ implies that $y(x) = Ax + B$. Then $y'(x) = A$, so the endpoint conditions yield $A = 0$, but B remains arbitrary. Hence $\lambda_0 = 0$ is an eigenvalue with eigenfunction

$$y_0(x) = 1.$$

If $\lambda = \alpha^2 > 0$, then the equation $y'' + \alpha^2 y = 0$ has general solution

$$y(x) = A \cos \alpha x + B \sin \alpha x,$$

with

$$y'(x) = -A\alpha \sin \alpha x + B\alpha \cos \alpha x.$$

Then $y'(0) = 0$ yields $B = 0$ so $A \neq 0$, and then

$$y'(L) = -A\alpha \sin \alpha L = 0,$$

so αL must be an integral multiple of π. Thus the nth positive eigenvalue is

$$\lambda_n = \alpha_n^2 = \frac{n^2 \pi^2}{L^2},$$

and the associated eigenfunction is

$$y_n(x) = \cos \frac{n\pi x}{L}.$$

3. If $\lambda = 0$ then $y'' = 0$ yields $y(x) = Ax + B$ as usual. But $y'(0) = A = 0$, and then $hy(L) + y'(L) = h(B) + 0 = 0$, so $B = 0$ also. Thus $\lambda = 0$ is not an eigenvalue. If $\lambda = \alpha^2 > 0$ so our equation is $y'' + \alpha^2 y = 0$, then

$$y(x) \;=\; A\cos \alpha x + B\sin \alpha x,$$

so

$$y'(x) \;=\; -A\alpha \sin \alpha x + B\alpha \cos \alpha x.$$

Now $y'(0) = 0$ yields $B = 0$, so we may write

$$y(x) \;=\; \cos \alpha x, \qquad y'(x) \;=\; -\alpha \sin \alpha x.$$

The equation

$$hy(L) + y'(L) \;=\; h\cos \alpha L - \alpha \sin \alpha L \;=\; 0$$

then gives

$$\tan \alpha L \;=\; \frac{h}{\alpha} \;=\; \frac{hL}{\alpha L},$$

so $\beta_n = \alpha_n L$ is the nth positive root of the equation

$$\tan x \;=\; \frac{hL}{x}.$$

Thus

$$\lambda_n \;=\; \alpha_n^2 \;=\; \frac{\beta_n^2}{L^2}, \qquad y_n(x) \;=\; \cos \frac{\beta_n x}{L}.$$

Finally, a sketch of the graphs $y = \tan x$ and $y = hL/x$ indicates that $\beta_n \approx (n-1)\pi$ for n large.

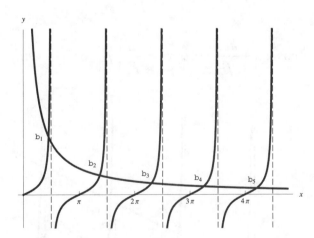

5.   If $\lambda = 0$ then $y'' = 0$ yields $y(x) = Ax + B$ as usual. Then the endpoint condition $hy(0) - y'(0) = 0$ yields $A = hB$. Substitution for A in the other endpoint condition $hy(L) + y'(L) = 0$ then gives $B(h^2 L + 2h) = 0$, so it follows that $A = B = 0$. Thus $\lambda = 0$ is not an eigenvalue.

If $\lambda = \alpha^2 > 0$ so our equation is $y'' + \alpha^2 y = 0$, then

$$y(x) = A\cos\alpha x + B\sin\alpha x,$$

so

$$y'(x) = -A\alpha\sin\alpha x + B\alpha\cos\alpha x.$$

Then the endpoint condition $hy(0) - y'(0) = hA - B\alpha = 0$ gives

$$B = \frac{hA}{\alpha}.$$

Substitution of this value of B in the other endpoint condition $hy(L) + y'(L) = 0$ yields an equation that simplifies to

$$(h^2 - \alpha^2)\sin\alpha L + 2h\alpha\cos\alpha L = 0.$$

Substitution of $\beta = \alpha L$, $\alpha = \beta/L$ in this equation then gives the equation

$$(\beta^2 - h^2 L^2)\sin\beta = 2hL\beta\cos\beta,$$

whence we see that $\beta_n = \alpha_n L$ is the nth positive root of the equation

$$\tan x = \frac{2hLx}{x^2 - h^2 L^2}, \quad \text{or} \quad 2hL\cot x = \frac{x^2 - h^2 L^2}{x}.$$

Thus

$$\lambda_n = \alpha_n^2 = \frac{\beta_n^2}{L^2}$$

and

$$y_n(x) = A\cos\alpha_n x + \frac{hLA}{\beta_n}\sin\alpha_n x = \frac{A}{\beta_n}\left(\beta_n\cos\frac{\beta_n x}{L} + hL\sin\frac{\beta_n x}{L}\right).$$

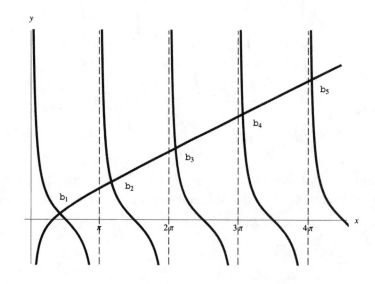

The figure shows the graphs $y = 2hL\cot x$ and $y = (x^2 - h^2 L^2)/x$ (with $hL = 1$ to illustrate the situation). Each intersection point is labeled with its x-coordinate β_n. It is apparent that $\beta_n \approx (n-1)\pi$ for n large.

7. The coefficient c_n in Eq. (23) of this section is given by Formula (25) with $f(x) = r(x) = 1$, $a = 0$, $b = L$, and $y_n(x) = \sin\dfrac{\beta_n x}{L}$. Using the fact that $\tan\beta_n = -\dfrac{\beta_n}{hL}$, so

$$\frac{\sin\beta_n}{\beta_n} = -\frac{\cos\beta_n}{hL},$$ we find that

$$\int_0^L \sin^2\frac{\beta_n x}{L}\,dx = \int_0^L \frac{1}{2}\left(1 - \cos\frac{2\beta_n x}{L}\right)dx = \frac{1}{2}\left[x - \frac{L}{2\beta_n}\sin\frac{2\beta_n x}{L}\right]_0^L$$

$$= \frac{1}{2}\left(L - L\frac{\sin\beta_n}{\beta_n}\cos\beta_n\right) = \frac{1}{2}\left(L + L\frac{\cos\beta_n}{hL}\cos\beta_n\right) = \frac{hL + \cos^2\beta_n}{2h}$$

and $\displaystyle\int_0^L \sin\frac{\beta_n x}{L}\,dx = \frac{L(1 - \cos\beta_n)}{\beta_n}$. Hence the desired eigenfunction expansion is

$$1 = 2hL\sum_{n=1}^{\infty}\frac{1 - \cos\beta_n}{\beta_n\left(hL + \cos^2\beta_n\right)}\sin\frac{\beta_n x}{L}.$$

for $0 < x < L$.

9. The coefficient c_n in (23) is given by Formula (25) with $f(x) = r(x) = 1$, $a = 0$, $b = 1$, and $y_n(x) = \sin\beta_n x$. Using the fact that $\tan\beta_n = -\beta_n/h$, so $h\sin\beta_n = -\beta_n\cos\beta_n$, we find that

$$\int_0^1 \sin^2\beta_n x\,dx = \int_0^1 \frac{1}{2}(1 - \cos 2\beta_n x)\,dx = \frac{1}{2}\left[x - \frac{\sin 2\beta_n x}{2\beta_n}\right]_0^1$$

$$= \frac{1}{2}\left(1 - \frac{\sin\beta_n}{\beta_n}\cos\beta_n\right) = \frac{1}{2}\left(1 + \frac{\cos^2\beta_n}{h}\right) = \frac{h + \cos^2\beta_n}{2h}$$

and

$$\int_0^1 x\sin\beta_n x\,dx = \frac{1}{\beta_n^2}\int_0^1 \beta_n x\sin\beta_n x\cdot\beta_n\,dx = \frac{1}{\beta_n^2}\int_0^{\beta_n} u\sin u\,du$$

$$= \frac{1}{\beta_n^2}\left[\sin u - u\cos u\right]_0^{\beta_n} = \frac{\sin\beta_n - \beta_n\cos\beta_n}{\beta_n^2}$$

$$= \frac{\sin\beta_n - \beta_n\cos\beta_n}{\beta_n^2} = \frac{(1 + h)\sin\beta_n}{\beta_n^2}.$$

It follows that the desired expansion is given by

$$x = 2h(1+h)\sum_{n=1}^{\infty}\frac{\sin\beta_n\,\sin\beta_n x}{\beta_n^2\left(h+\cos^2\beta_n\right)}$$

for $0 < x < 1$.

11. If $\lambda = 0$ then $y'' = 0$ implies that $y(x) = Ax + B$. Then $y(0) = 0$ gives $B = 0$, so $y(x) = Ax$. Hence

$$hy(L) - y'(L) = h(AL) - A = A(hL - 1) = 0$$

if and only if $hL = 1$, in which case $\lambda_0 = 0$ has associated eigenfunction $y_0(x) = x$.

13. If $\lambda = +\alpha^2 > 0$, then the general solution of $y'' + \alpha^2 y = 0$ is

$$y(x) = A\cos\alpha x + B\sin\alpha x.$$

But then $y(0) = A = 0$, so we may take $y(x) = \sin\alpha x$. Now the condition $hy(L) = y'(L)$ yields

$$h\sin\alpha L = \alpha\cos\alpha L.$$

It follows that $\beta = \alpha L$ must be a root of the equation

$$\tan x = \frac{x}{hL}.$$

So if β_n is the nth positive root of this equation, then $\lambda_n = \alpha_n^2 = \beta_n^2/L^2$ and the corresponding eigenfunction is $y_n(x) = \sin\beta_n x/L$.

15. If $\lambda_0 = 0$, then a general solution of $y'' = 0$ is $y(x) = Ax + B$. The conditions

$$y(0) + y'(0) = B + A = 0, \qquad y(1) = A + B = 0$$

both say that $B = -A$, so we may take $y_0(x) = x - 1$ as the eigenfunction associated with $\lambda_0 = 0$. If $\lambda = +\alpha^2 < 0$, then the general solution of $y'' + \alpha^2 y = 0$ is

$$y(x) = A\cos\alpha x + B\sin\alpha x.$$

But $y(0) + y'(0) = A + B\alpha = 0$, so $A = -B\alpha$, and then

$$y(1) = A\cos\alpha + B\sin\alpha = -B(\alpha\cos\alpha - \sin\alpha) = 0.$$

Thus the possible values of α are the positive roots $\{\beta_n\}$ of the equation $\tan x = x$, and the nth eigenfunction is $y_n(x) = \beta_n \cos \beta_n x - \sin \beta_n x$,

17.   The Fourier sine series of the constant function $f(x) \equiv w$ for $0 < x < L$ is

$$w = \frac{4w}{\pi} \sum_{n \text{ odd}} \frac{1}{n} \sin \frac{n\pi x}{L}.$$

If $y = \sum b_n \sin n\pi x / L$, then

$$EI\, y^{(4)} = EI \sum_{n=1}^{\infty} \frac{n^4 \pi^4 b_n}{L^4} \sin \frac{n\pi x}{L}.$$

Upon equating coefficients in these two series and solving for b_n, we see that

$$y(x) = \frac{4wL^4}{EI\pi^5} \sum_{n \text{ odd}} \frac{1}{n^5} \sin \frac{n\pi x}{L}.$$

19.   With $\lambda = \alpha^4$, the general solution of $y^{(4)} - \alpha^4 y = 0$ is

$$y(x) = A \cosh \alpha x + B \sinh \alpha x + C \cos \alpha x + D \sin \alpha x,$$

and then

$$y'(x) = \alpha(A \sinh \alpha x + B \cosh \alpha x - C \sin \alpha x + D \cos \alpha x).$$

The conditions $y(0) = 0$ and $y'(0) = 0$ yield $C = -A$ and $D = -B$, so now

$$y(x) = A(\cosh \alpha x - \cos \alpha x) + B(\sinh \alpha x - \sin \alpha x).$$

The conditions $y(L) = 0$ and $y'(L) = 0$ yield the two linear equations

$$A(\cosh \alpha L - \cos \alpha L) + B(\sinh \alpha L - \sin \alpha L) = 0,$$

$$A(\sinh \alpha L + \sin \alpha L) + B(\cosh \alpha L - \cos \alpha L) = 0.$$

This linear system can have a non-trivial solution for A and B only if its coefficient determinant vanishes,

$$(\cosh \alpha L - \cos \alpha L)^2 - (\sinh^2 \alpha L - \sin^2 \alpha L) = 0.$$

Using the facts that $\cosh^2 A - \sinh^2 A = 1$ and $\cos^2 A + \sin^2 A = 1$, this equation simplifies to

$$\cosh \alpha L \cos \alpha L - 1 = 0,$$

so $\beta = \alpha L = x$ satisfies the equation

$$\cosh x \cos x = 1.$$

The eigenvalue corresponding to the nth positive root β_n is

$$\lambda_n = \alpha_n^4 = \left(\frac{\beta_n}{L}\right)^4.$$

Finally the first equation in the pair above yields

$$B = -\frac{\cosh \alpha L - \cos \alpha L}{\sinh \alpha L - \sin \alpha L},$$

so we may take

$$y_n(x) = \left(\sinh \beta_n - \sin \beta_n\right)\left(\cosh \frac{\beta_n x}{L} - \cos \frac{\beta_n x}{L}\right)$$
$$- \left(\cosh \beta_n - \cos \beta_n\right)\left(\sinh \frac{\beta_n x}{L} - \sin \frac{\beta_n x}{L}\right)$$

as the eigenfunction associated with the eigenvalue λ_n.

$$- \left(\cosh \beta_n + \cos \beta_n\right)\left(\sinh \frac{\beta_n x}{L} - \sin \frac{\beta_n x}{L}\right)$$

as the eigenfunction associated with the eigenvalue λ_n.

21.   As in Problem 19, the solution of $y^{(4)} - \alpha^4 y = 0$ satisfying the left-endpoint conditions $y(0) = 0$ and $y'(0) = 0$ is given by

$$y(x) = A(\cosh \alpha x - \cos \alpha x) + B(\sinh \alpha x - \sin \alpha x).$$

The right-endpoint conditions $y(L) = 0$ and $y''(L) = 0$ yield the two linear equations

$$A(\cosh \alpha L - \cos \alpha L) + B(\sinh \alpha L - \sin \alpha L) = 0,$$
$$A(\cosh \alpha L + \cos \alpha L) + B(\sinh \alpha L + \sin \alpha L) = 0.$$

This linear system can have a non-trivial solution for A and B only if its coefficient determinant vanishes,

$$(\cosh \alpha L - \cos \alpha L)(\sinh \alpha L + \sin \alpha L)$$
$$- (\cosh \alpha L + \cos \alpha L)(\sinh \alpha L - \sin \alpha L) = 0.$$

This equation simplifies to $2\cosh\alpha L\sin\alpha L - 2\cos\alpha L\sinh\alpha L = 0$, which is equivalent to $\tanh\alpha L = \tan\alpha L$. Hence $\beta = \alpha L = x$ satisfies the equation $\tanh x = \tan x$, and the eigenvalue corresponding to the nth positive root β_n is $\lambda_n = \alpha_n^4 = (\beta_n/L)^4$.

SECTION 10.2

APPLICATIONS OF EIGENFUNCTION SERIES

1.   The substitution $u(x,t) = X(x)T(t)$ yields the separated equations

$$X'' + \alpha^2 X = 0 \quad\text{and}\quad T' = -k\lambda T$$

with separation constant $\lambda = \alpha^2$. In Problem 3 of Section 10.1 we saw that the Sturm-Liouville problem

$$X'' + \alpha^2 X = 0, \qquad X'(0) = hX(L) + X'(L) = 0$$

has eigenvalues $\lambda_n = \alpha_n^2 = \beta_n^2/L^2$ and eigenfunctions

$$X_n(x) = \cos\frac{\beta_n x}{L}$$

for $n = 1, 2, 3, \cdots$, with $\{\beta_n\}$ being the positive roots of the equation $\tan x = hL/x$. The solution of $T_n' = -k\lambda_n T_n$ is then

$$T_n(t) = \exp\left(-\frac{\beta_n^2 kt}{L^2}\right),$$

so the resulting formal series solution is

$$u(x,t) = \sum_{n=1}^{\infty} c_n \exp\left(-\frac{\beta_n^2 kt}{L^2}\right)\cos\frac{\beta_n x}{L}.$$

The coefficients in the eigenfunction expansion are given by

$$c_n = \frac{\displaystyle\int_0^L f(x)\cos\frac{\beta_n x}{L}\,dx}{\displaystyle\int_0^L \cos^2\frac{\beta_n x}{L}\,dx} = \frac{2h}{hL + \sin^2\beta_n}\int_0^L f(x)\cos\frac{\beta_n x}{L}\,dx,$$

because

$$\int_0^L \cos^2\frac{\beta_n x}{L}\,dx = \int_0^L \frac{1}{2}\left(1 + \cos\frac{2\beta_n x}{L}\right)dx = \left[\frac{1}{2}\left(x + \frac{L}{2\beta_n}\sin\frac{2\beta_n x}{L}\right)\right]_0^L$$

$$= \frac{1}{2}\left(L + \frac{L}{2\beta_n}\sin 2\beta_n\right) = \frac{1}{2h}\left(hL + \sin\beta_n \cdot \frac{hL\cos\beta_n}{\beta_n}\right)$$

$$= \frac{hL + \sin^2\beta_n}{2h}.$$

In the final step here we use the fact that $(hL\cos\beta_n)/\beta_n = \sin\beta_n$ because $\tan\beta_n = hL/\beta_n$.

3. The substitution $u(x,y) = X(x)Y(y)$ yields the separated equations

$$X'' - \alpha^2 X = 0 \quad \text{and} \quad Y'' + \alpha^2 Y = 0$$

with separation constant $\lambda = \alpha^2$. In Problem 3 of Section 10.1 we saw that the Sturm-Liouville problem

$$Y'' + \alpha^2 Y = 0, \qquad Y'(0) = hY(L) + Y'(L) = 0$$

has eigenvalues $\lambda_n = \alpha_n^2 = \beta_n^2/L^2$ and eigenfunctions

$$Y_n(y) = \cos\frac{\beta_n y}{L}$$

for $n = 1, 2, 3, \cdots$, with $\{\beta_n\}$ being the positive roots of the equation $\tan x = hL/x$. The solution of

$$X_n'' - \frac{\beta_n^2}{L^2}X_n = 0, \qquad X(L) = 0$$

is

$$X_n(x) = \sinh\frac{\beta_n(L-x)}{L},$$

so the resulting formal series solution is

$$u(x,y) = \sum_{n=1}^{\infty} c_n \sinh\frac{\beta_n(L-x)}{L}\cos\frac{\beta_n y}{L}.$$

The coefficients in the eigenfunction expansion are given by

$$c_n = \frac{\displaystyle\int_0^L g(y)\cos\frac{\beta_n y}{L}\,dy}{(\sinh\beta_n)\displaystyle\int_0^L \cos^2\frac{\beta_n y}{L}\,dy} = \frac{2h}{(\sinh\beta_n)\left(hL + \sin^2\beta_n\right)}\int_0^L g(y)\cos\frac{\beta_n y}{L}\,dy,$$

because

$$\int_0^L \cos^2 \frac{\beta_n y}{L} \, dy = \int_0^L \frac{1}{2}\left(1 + \cos\frac{2\beta_n y}{L}\right) dy = \left[\frac{1}{2}\left(y + \frac{L}{2\beta_n}\sin\frac{2\beta_n y}{L}\right)\right]_0^L$$

$$= \frac{1}{2}\left(L + \frac{L}{2\beta_n}\sin 2\beta_n\right) = \frac{hL + \sin^2\beta_n}{2h}.$$

The final step here is the same as in Problem 1, using the fact that $(hL\cos\beta_n)/\beta_n = \sin\beta_n$ because $\tan\beta_n = hL/\beta_n$.

5.   The substitution $u(x,t) = X(x)T(t)$ yields the separated equations

$$X'' + \alpha^2 X = 0 \quad \text{and} \quad T' = -k\lambda T$$

with separation constant $\lambda = \alpha^2$. In Problem 4 of Section 10.1 we saw that the Sturm-Liouville problem

$$X'' + \alpha^2 X = 0, \qquad hX(0) - X'(0) = X(L) = 0$$

has eigenvalues $\lambda_n = \alpha_n^2 = \beta_n^2/L^2$ and eigenfunctions

$$X_n(x) = \beta_n \cos\frac{\beta_n x}{L} + hL\sin\frac{\beta_n x}{L}$$

for $n = 1, 2, 3, \cdots$, with $\{\beta_n\}$ being the positive roots of the equation $\tan x = -x/hL$. The solution of $T_n' = -k\lambda_n T_n$ is then

$$T_n(t) = \exp\left(-\frac{\beta_n^2 kt}{L^2}\right),$$

so the resulting formal series solution is

$$u(x,t) = \sum_{n=1}^{\infty} c_n \exp\left(-\frac{\beta_n^2 kt}{L^2}\right)\left(\beta_n \cos\frac{\beta_n x}{L} + hL\sin\frac{\beta_n x}{L}\right).$$

The coefficients in the eigenfunction expansion are given by

$$c_n = \frac{\displaystyle\int_0^L f(x)\left(\beta_n \cos\frac{\beta_n x}{L} + hL\sin\frac{\beta_n x}{L}\right) dx}{\displaystyle\int_0^L \left(\beta_n \cos\frac{\beta_n x}{L} + hL\sin\frac{\beta_n x}{L}\right)^2 dx}.$$

The evaluation of the denominator integral here is elementary, but there seems little point in carrying it out explicitly.

7. The boundary value problem here is

$$u_{xx} + u_{yy} = 0 \quad (0 < x < 1, \quad y > 0)$$
$$u_x(0, y) = u(1, y) + u_x(1, t) = 0,$$
$$u(x, 0) = 100.$$

The substitution $u(x, y) = X(x)Y(y)$ yields the separated equations

$$X'' + \alpha^2 X = 0 \quad \text{and} \quad Y'' - \alpha^2 Y = 0$$

with separation constant $\lambda = \alpha^2$. In Problem 3 of Section 10.1 we saw (taking $h = L = 1$) that the Sturm-Liouville problem

$$X'' + \alpha^2 X = 0, \qquad X'(0) = X(1) + X'(1) = 0$$

has eigenvalues $\lambda_n = \alpha_n^2$ and eigenfunctions

$$X_n(x) = \cos \alpha_n x$$

for $n = 1, 2, 3, \cdots$, with $\{\alpha_n\}$ being the positive roots of the equation $\tan x = 1/x$. The bounded solution of $Y_n'' - \alpha_n^2 Y_n = 0$ is then

$$Y_n(y) = \exp(-\alpha_n y),$$

so the resulting formal series solution is

$$u(x, y) = \sum_{n=1}^{\infty} c_n \cos \alpha_n x \exp(-\alpha_n y).$$

The coefficients in the eigenfunction expansion are given by

$$c_n = \frac{\int_0^L 100 \cos \alpha_n x \, dx}{\int_0^L \cos^2 \alpha_n x \, dx} = \frac{\left[\dfrac{100}{\alpha_n} \sin \alpha_n x\right]_0^1}{\left[\dfrac{1}{2}\left(x + \dfrac{1}{2\alpha_n} \sin 2\alpha_n x\right)\right]_0^1} = \frac{200 \sin \alpha_n}{\alpha_n + \sin \alpha_n \cos \alpha_n},$$

so

$$u(x, y) = 200 \sum_{n=1}^{\infty} \frac{\sin \alpha_n \cos \alpha_n x \exp(-\alpha_n y)}{\alpha_n + \sin \alpha_n \cos \alpha_n}.$$

The first five positive solutions of $\tan x = 1/x$ are $0.8603, 3.4256, 7.4373, 9.5293,$ and $12.6453,$ and we find that

$$u(1,1) \approx 30.8755 + 0.4737 + 0.0074 + 0.0002 + 0.0000 + \cdots \approx 31.4°C.$$

9. **(a)** With $\lambda = 0,$ the endpoint-value problem in (19) is $X'' = 0,$ $X(0) = X'(0) = 0,$ which has only the trivial solution $X(x) \equiv 0.$ Thus $\lambda = 0$ is not an eigenvalue.

(b) With $\lambda = -\alpha^2 < 0,$ the endpoint-value problem in (19) is

$$X'' - \alpha^2 X = 0, \qquad X(0) = 0, \qquad -m\alpha^2 X(L) = A\delta X'(L).$$

The differential equation and the left-endpoint condition here give $X(x) = \sinh \alpha x,$ and substitution in the right-endpoint condition gives

$$-m\alpha^2 \sinh \alpha L = A\delta \alpha \cosh \alpha L, \text{ that is, } \tanh \alpha L = -\frac{k}{\alpha L}$$

with $k = A\delta L/m > 0.$ But the graph $y = \tanh x$ lies (aside from the origin) in the first and third quadrants, while the graph $y = -k/x$ lies interior to the second and fourth quadrants. Hence the two cannot intersect, and it follows that there cannot be an eigenvalue of the assumed form $\lambda = -\alpha^2 < 0,.$

11. **(a)** $a = \sqrt{\dfrac{K}{\delta}} = \sqrt{\dfrac{\lambda p}{m/V}} = \sqrt{\dfrac{\gamma p V}{m}} = \sqrt{\dfrac{\gamma n R T_K}{n m_0}} = \sqrt{\dfrac{\gamma R T_K}{m_0}}$

(b) $a = \sqrt{\dfrac{\gamma R T_K}{m_0}} = \sqrt{\dfrac{1.4 \times 8314(273 + T_C)}{29}} = \sqrt{\dfrac{1.4 \times 8314 \times 273}{29}\left(1 + \dfrac{T_C}{273}\right)}$

$$\approx 331.02\sqrt{1 + \dfrac{T_C}{273}} \ \dfrac{m}{sec} \approx 740.47\sqrt{1 + \dfrac{T_C}{273}} \ \dfrac{miles}{hour}$$

$$\approx 740.47\left[1 + \dfrac{1}{2}\left(\dfrac{T_C}{273}\right) + \cdots\right] \approx 740.47 + 1.356 T_C$$

15. $\displaystyle\int_0^L \sin\dfrac{\beta_m x}{L} \sin\dfrac{\beta_n x}{L} \, dx = \dfrac{L}{2}\left[\dfrac{\sin(\beta_m - \beta_n)}{\beta_m - \beta_n} - \dfrac{\sin(\beta_m + \beta_n)}{\beta_m + \beta_n}\right]$

$$= \dfrac{L}{2(\beta_m^2 - \beta_n^2)}\left[\begin{array}{c}(\beta_m + \beta_n)(\sin \beta_m \cos \beta_n - \sin \beta_n \cos \beta_m) \\ -(\beta_m - \beta_n)(\sin \beta_m \cos \beta_n + \sin \beta_n \cos \beta_m)\end{array}\right]$$

$$= \frac{L}{\beta_m^2 - \beta_n^2}\left[\beta_n \sin\beta_m \cos\beta_n - \beta_m \sin\beta_n \cos\beta_m\right]$$

$$= \frac{L}{\beta_m^2 - \beta_n^2}\left[\beta_n \cdot \frac{M\cos\beta_m}{m\beta_m}\cdot\cos\beta_n - \beta_m \cdot \frac{M\cos\beta_n}{m\beta_n}\cdot\cos\beta_m\right]$$

$$= \frac{LM}{m\left(\beta_m^2 - \beta_n^2\right)}\cos\beta_m \cos\beta_n\left(\frac{\beta_n}{\beta_m} - \frac{\beta_m}{\beta_n}\right) = -\frac{LM\cos\beta_m\cos\beta_n}{m\beta_m\beta_n} \neq 0$$

19. With the given initial velocity function $g(x)$ with constant value $P/2\rho\varepsilon$ concentrated in the interval $L/2 - \varepsilon < x < L/2 + \varepsilon$, the coefficient formula of Problem 18 gives

$$c_n = \frac{2L}{n^2\pi^2 a^2}\int_{L/2-\varepsilon}^{L/2+\varepsilon}\frac{P}{2\rho\varepsilon}\sin\frac{n\pi x}{L}\,dx$$

$$= \frac{L^2 P}{n^3\pi^3 a^2\rho\varepsilon}\left[\cos\left(\frac{n\pi}{2} - \frac{n\pi\varepsilon}{L}\right) - \cos\left(\frac{n\pi}{2} + \frac{n\pi\varepsilon}{L}\right)\right] = \frac{2L^2 P}{n^3\pi^3 a^2\rho\varepsilon}\sin\frac{n\pi}{2}\sin\frac{n\pi\varepsilon}{L}.$$

This gives the ε-dependent solution

$$y(x,t,\varepsilon) = \frac{2L^2 P}{\pi^3 a^2\rho\varepsilon}\sum_{n=1}^{\infty}\frac{1}{n^3}\sin\frac{n\pi}{2}\sin\frac{n\pi\varepsilon}{L}\sin\frac{n^2\pi^2 a^2 t}{L^2}\sin\frac{n\pi x}{L}.$$

Because

$$\frac{L}{n\pi\varepsilon}\sin\frac{n\pi\varepsilon}{L} = \frac{\sin\left(n\pi\varepsilon/L\right)}{n\pi\varepsilon/L} \to 1 \quad \text{as} \quad \varepsilon \to 0,$$

the limit $y(x,t) = \lim_{\varepsilon\to 0} y(x,t,\varepsilon)$ has the expansion

$$y(x,t) = \frac{2LP}{\pi^2 a^2\rho}\sum_{n=1}^{\infty}\frac{1}{n^2}\sin\frac{n\pi}{2}\sin\frac{n^2\pi^2 a^2 t}{L^2}\sin\frac{n\pi x}{L}.$$

SECTION 10.3

STEADY PERIODIC SOLUTIONS AND NATURAL FREQUENCIES

In Problems 1–6 we substitute $u(x,t) = X(x)\cos\omega t$ in

$$u_{tt} = a^2 u_{xx} \qquad (a^2 = E/\delta)$$

and then cancel the factor $\cos \omega t$ to obtain the ordinary differential equation

$$a^2 X'' + \omega^2 X = 0$$

with general solution

$$X(x) = A \cos \frac{\omega x}{a} + B \sin \frac{\omega x}{a}. \qquad (*)$$

It then remains only to apply the given endpoint conditions to determine the natural (circular) frequencies — the values of ω for which a non-trivial solution exists.

1.   Endpoint conditions: $X(0) = X(L) = 0$

The condition $X(0) = 0$ in $(*)$ implies that $A = 0$, so $X(x) = \sin(\omega x / a)$.   Then $X(L) = \sin(\omega L / a) = 0$ implies that $\omega L / a = n\pi$, an integral multiple of π. Hence the nth natural frequency is $\omega_n = \dfrac{n\pi a}{L} = \dfrac{n\pi}{L}\sqrt{\dfrac{E}{\delta}}.$

3.   Endpoint conditions: $X(0) = X'(L) = 0$

The condition $X(0) = 0$ gives $A = 0$ in $(*)$, so we have

$$X(x) = \sin \frac{\omega x}{a}, \qquad \text{so} \qquad X'(x) = \frac{\omega}{a} \cos \frac{\omega x}{a}.$$

Hence the condition $X'(L) = 0$ implies that $\omega L / a$ is an *odd* integral multiple of $\pi/2$. Thus the nth natural frequency is $\omega_n = \dfrac{(2n-1)\pi a}{2L} = \dfrac{(2n-1)\pi}{2L}\sqrt{\dfrac{E}{\delta}}.$

5.   Endpoint conditions: $u_x(0, t) = ku(L, t) + AEu_x(L, t) = 0$

The condition $X'(0) = 0$ gives $B = 0$ in $(*)$, so we have

$$X(x) = \cos \frac{\omega x}{a}, \qquad \text{so} \qquad u(x,t) = \cos \frac{\omega x}{a} \cos \omega t.$$

Then

$$u_x(x,t) = -\frac{\omega}{a} \sin \frac{\omega x}{a} \cos \omega t,$$

so the other endpoint condition is

$$k \cos \frac{\omega L}{a} \cos \omega t - AE \frac{\omega}{a} \sin \frac{\omega L}{a} \cos \omega t = 0.$$

Upon canceling the $\cos \omega t$ factor, we find that

$$AE\frac{\omega L}{a}\tan\frac{\omega L}{a} = kL.$$

Thus $\beta = \omega L / a$ is a positive root of the equation $AEx \tan x = kL,$ and the nth natural frequency is given by

$$\omega_n = \frac{\beta_n a}{L} = \frac{\beta_n}{L}\sqrt{\frac{E}{\delta}}$$

where β_n is the nth positive root of this equation.

7. Endpoint conditions:

$$u(0, t) = mu_{tt}(L, t) + AEu_x(L, t) + ku(L, t) = 0$$

The condition $u(0, t) = 0$ implies that

$$X(x) = \sin\frac{\omega x}{a}, \quad \text{so} \quad X'(x) = \frac{\omega}{a}\cos\frac{\omega x}{a}.$$

When we substitute $u(x, t) = X(x)\cos \omega t$ in the endpoint condition at $x = L$ and cancel the $\cos \omega t$ factor we get

$$-m\omega^2 X(L) + AEX'(L) + kX(L) = 0.$$

Next we substitute

$$z = \omega L / a, \qquad \omega = az/L, \qquad a^2 = E/\delta,$$

$$X(L) = \sin z, \qquad X'(L) = (z/L)\cos z.$$

The result simplifies readily to the frequency equation

$$(mEz^2 - k\delta L^2)\sin z = MEz \cos z.$$

If β_n is the nth positive root, then the nth natural frequency is $\omega_n = \dfrac{\beta_n a}{L} = \dfrac{\beta_n}{L}\sqrt{\dfrac{E}{\delta}}.$

In Problems 8–14 we substitute $y(x, t) = X(x)\cos \omega t$ in

$$y_{tt} + a^4 y_{xxxx} = 0 \qquad (a^4 = EI/\rho)$$

and then cancel the factor $\cos \omega t$ to obtain the ordinary differential equation

$$a^4 X^{(4)} - \omega^2 X = 0$$

with general solution

$$X(x) \;=\; A\cosh\frac{\theta x}{a}+B\sinh\frac{\theta x}{a}+C\cos\frac{\theta x}{a}+D\sin\frac{\theta x}{a} \qquad\qquad (**)$$

where $\theta = \sqrt{\omega}$. We then get the natural frequencies of vibration by applying the given endpoint conditions.

9.   Endpoint conditions: $y(0, t) = y_x(0, t) = 0, \quad y(L, t) = y_{xx}(L, t) = 0$

Just as in Problem 21 of Section 10.1, the endpoint conditions $X(0) = X'(0) = 0$ and $X(L) = X''(L) = 0$ imply that

$$\lambda_n \;=\; \frac{\omega_n^2}{a^4} \;=\; \left(\frac{\beta_n}{L}\right)^4$$

where β_n is the nth positive zero of the frequency equation

$$\tanh x \;=\; \tan x.$$

Therefore the nth natural frequency ω_n is given by

$$\omega_n \;=\; \left(\frac{\beta_n}{L}\right)^2 a^2 \;=\; \frac{\beta_n^2}{L^2}\sqrt{\frac{EI}{\rho}}.$$

11.   Endpoint conditions: $y(0, t) = y_x(0, t) = 0, \quad y_x(L, t) = y_{xxx}(L, t) = 0$

Here we have the equation

$$X^{(4)} - \lambda X \;=\; 0$$

with $\lambda = \omega^2/a^4 = \theta^4/a^4 = \alpha^4$ and endpoint conditions

$$X(0) \;=\; X'(0) \;=\; X'(L) \;=\; X^{(3)}(L) \;=\; 0.$$

The left-endpoint conditions readily give $C = -A$ and $D = -B$ in $(**)$, so

$$X(x) \;=\; A\cosh\alpha x + B\sinh\alpha x - A\cos\alpha x - B\sin\alpha x.$$
$$=\; A\big(\cosh\alpha x - \cos\alpha x\big) + B\big(\sinh\alpha x - \sin\alpha x\big).$$

Then the right-endpoint conditions give

$$A\big(\sinh\alpha L + \sin\alpha L\big) + B\big(\cosh\alpha L - \cos\alpha L\big) \;=\; 0,$$
$$A\big(\sinh\alpha L - \sin\alpha L\big) + B\big(\cosh\alpha L + \cos\alpha L\big) \;=\; 0.$$

The determinant of coefficients of A and B must vanish if there is to be a nontrivial solution, so

$$\big(\sinh \alpha L + \sin \alpha L\big)\big(\cosh \alpha L + \cos \alpha L\big)$$
$$-\big(\sinh \alpha L - \sin \alpha L\big)\big(\cosh \alpha L - \cos \alpha L\big) \;=\; 0.$$

This equation simplifies to $2\sinh \alpha L \cos \alpha L + 2\cosh \alpha L \sin \alpha L = 0,$ which upon division by $\cosh \alpha L \cos \alpha L$ gives the frequency equation

$$\tanh x + \tan x \;=\; 0$$

for $\beta = \alpha L.$ Then the nth frequency is given as usual by

$$\omega_n \;=\; \alpha_n^2 a^2 \;=\; \frac{\beta_n^2}{L^2}\sqrt{\frac{EI}{\rho}}.$$

13.   This problem is the special case $m = 0$ of Problem 14 below.

14.   Endpoint conditions:

$$y(0, t) \;=\; y_x(0, t) \;=\; y_{xxx}(L, t) \;=\; 0$$
$$m y_{tt}(L, t) \;=\; EI y_{xxx}(L, t) - k y(L, t)$$

With $p = \theta/a,\ \theta = \sqrt{\omega}$ we may write

$$X(x) \;=\; A \cosh px + B \sinh px + C \cos px + D \sin px.$$

The conditions $X(0) = X'(0) = 0$ readily imply that $C = -A$ and $D = -B,$ so

$$
\begin{aligned}
X &= A(\cosh px - \cos px) + B(\sinh px - \sin px),\\
X' &= pA(\sinh px + \sin px) + pB(\cosh px - \cos px),\\
X'' &= p^2 A(\cosh px + \cos px) + p^2 B(\sinh px + \sin px),\\
X^{(3)} &= p^3 A(\sinh px - \sin px) + p^3 B(\cosh px + \cos px).
\end{aligned}
$$

The endpoint conditions at $x = L$ are

$$X''(L) \;=\; 0,$$
$$(k - m\omega^2)X(L) - EI X^{(3)}(L) \;=\; 0.$$

When we substitute the derivatives above and write $z = pL$ we get

$$A\,(\cosh z + \cos z) + B\,(\sinh z + \sin z) \;=\; 0,$$

$$A[(k - m\omega^2)(\cosh z - \cos z) - EIp^3(\sinh z - \sin z)]$$
$$+ B[(k - m\omega^2)(\sinh z - \sin z) - EIp^3(\cosh z + \cos z)] = 0.$$

If Δ denotes the coefficient determinant of these two linear equations in A and B, then the necessary condition $\Delta = 0$ for a non-trivial solution reduces eventually to the equation

$$EIp^3(1 + \cosh z \cos z) - (k - m\omega^2)(\sinh z \cos z - \cosh z \sin z) = 0.$$

Finally we substitute $p = z/L$, $M = \rho L$, and

$$\omega^2 = p^4 a^4 = (z^4/L^4)(EI/\rho)$$

to get the frequency equation

$$MEIz^3(1 + \cosh z \cos z) = (kML^3 - mEIz^4)(\sinh z \cos z - \cosh z \sin z).$$

We may divide by $\cosh z \cos z$ to write this equation in the form

$$MEIz^3(1 + \operatorname{sech} z \sec z) = (kML^3 - mEIz^4)(\tanh z - \tan z).$$

If β_n denotes the nth positive root of this equation, then as usual the nth natural frequency is

$$\omega_n = \frac{\beta_n^2}{L^2}\sqrt{\frac{EI}{\rho}}.$$

15. We want to calculate the fundamental frequency of transverse vibration of a cantilever with the numerical parameters

$$L = 400 \text{ cm}$$
$$E = 2 \cdot 10^{12} \text{ gm/cm-sec}^2$$
$$I = (1/12)(30 \text{ cm})(2 \text{ cm})^3 = 20 \text{ cm}^4$$
$$\rho = (7.75 \text{ gm/cm}^3)(60 \text{ cm}^2) = 465 \text{ gm/cm}.$$

When we substitute these values and $\beta_1 = 1.8751$ in the frequency formula

$$\omega_1 = \frac{\beta_1^2}{L^2}\sqrt{\frac{EI}{\rho}},$$

we find that $\omega_1 \approx 6.45$ rad/sec, so the fundamental frequency is $\omega_1/2\pi \approx 1.03$ cycles/sec. Thus the diver should bounce up and down on the end of the diving board about once every second.

17. When we substitute $y(x, t) = X(x) \cos \omega t$ in the given partial differential equation

$$\rho \frac{\partial^2 y}{\partial t^2} - \frac{I}{A} \frac{\partial^4 y}{\partial x^2 \partial t^2} + EI \frac{\partial^4 y}{\partial x^4} = 0$$

and cancel the factor $\cos \omega t$, we get the ordinary differential equation

$$EIX^{(4)} + PX'' - \lambda X = 0$$

where $P = \lambda I / \rho A$ and $\lambda = \rho \omega^2$. By solving the characteristic equation

$$EI \, r^4 + Pr^2 - \lambda = EI(r^2 - \alpha^2)(r^2 + \beta^2) = 0$$

we find the general solution

$$X(x) = A \cosh \alpha x + B \sinh \alpha x + C \cos \beta x + D \sin \beta x$$

where

$$\alpha^2 = \frac{-P + \sqrt{P^2 + 4\lambda EI}}{2EI}, \qquad \beta^2 = -\frac{-P - \sqrt{P^2 + 4\lambda EI}}{2EI}.$$

The endpoint conditions $X(0) = X''(0) = 0$ imply that $A = C = 0$, so

$$X(x) = B \sinh \alpha x + D \sin \beta x.$$

Then the conditions $X(L) = X''(L) = 0$ yield the equations

$$B \sinh \alpha L + D \sin \beta L = 0,$$

$$\alpha^2 B \sinh \alpha L - \beta^2 D \sin \beta L = 0.$$

The determinant of these two linear equations in B and D must vanish in order that a nontrivial solution exist, so

$$(\alpha^2 + \beta^2)\sinh \alpha L \sin \beta L = 0.$$

It follows that $\sin \beta L = 0$, so βL must be an integral multiple of π. The definitions of α^2 and β^2 imply that

$$\beta^2 - \alpha^2 = \frac{P}{EI} = \frac{\lambda}{\rho AE}, \qquad \alpha^2 \beta^2 = \frac{\lambda}{EI}.$$

Hence if $\beta_n = n\pi/L$, the corresponding value of α_n is given by

$$\alpha_n^2 = \frac{n^2 \pi^2}{L^2} - \frac{\lambda_n}{\rho AE}.$$

Then $\alpha_n^2 \beta_n^2 = \lambda_n / EI$ gives the equation

$$\left(\frac{n^2 \pi^2}{L^2} - \frac{\lambda_n}{\rho AE} \right) \frac{n^2 \pi^2}{L^2} = \frac{\lambda_n}{EI}$$

that we readily solve for λ_n. The resulting value of the nth natural frequency is

$$\omega_n = \sqrt{\frac{\lambda_n}{\rho}} = \frac{n^2 \pi^2}{L^2} \left(1 + \frac{n^2 \pi^2 I}{\rho A L^2} \right)^{-1/2} \sqrt{\frac{EI}{\rho}}.$$

19.   Substitution of $y(x,t) = X(x) \sin \omega t$ in the transverse bar problem

$$\frac{\partial^2 y}{\partial t^2} + a^4 \frac{\partial^4 y}{\partial x^4} = 0 \qquad \left(a^4 = \frac{EI}{\rho} \right)$$

$$y(0,t) = y_x(0,t) = 0,$$
$$y_{xx}(L,t) = EI\, y_{xxx}(L,t) + F_0 \sin \omega t = 0$$

yields the endpoint problem

$$X^{(4)} - p^4 X = 0 \qquad \text{(where } p^2 = \omega / a^2),$$
$$X(0) = X'(0) = 0,$$
$$X''(L) = EI\, X'''(L) + F_0 = 0.$$

When we impose the fixed-end conditions $X(0) = X'(0) = 0$ on the general solution

$$X(x) = A \cosh px + B \sinh px + C \cos px + D \sin px$$

we find readily that $C = -A$ and $D = -B$, so

$$X(x) = A(\cosh px - \cos px) + B(\sinh px - \sin px).$$

It remains only to find A and B. But the free-end conditions yield the linear equations

$$A(\cosh pL + \cos pL) + B(\sinh pL + \sin pL) = 0$$
$$A(\sinh pL - \sin pL) + B(\cosh pL + \cos pL) = -F_0 / p^3 EI$$

that can be solved for

$$A = K(\sinh pL + \sin pL), \quad B = -K(\cosh pL + \cos pL)$$

where

$$K = \frac{F_0}{2EIp^3(1 + \cosh pL \cos pL)}.$$

SECTION 10.4

CYLINDRICAL COORDINATE PROBLEMS

1. Substitution of $u(r,t) = R(r)T(t)$ in the wave equation

$$\frac{\partial^2 u}{\partial t^2} = a^2\left(\frac{\partial^2 u}{\partial r^2} + \frac{1}{r}\frac{\partial u}{\partial r}\right)$$

yields the separation

$$\frac{T''}{a^2 T} = \frac{R'' + \frac{1}{r}R'}{R} = \lambda = -\alpha^2.$$

The t-equation has general solution

$$T(t) = A\cos\alpha at + B\sin\alpha at,$$

and we choose $B = 0$, so that $T'(0) = 0$ (because the membrane is initially at rest). The r-equation can be written in the form

$$r^2 R'' + rR' + \alpha^2 r^2 R = 0,$$

which is the parametric Bessel equation of order zero, with continuous solution $R(r) = J_0(\alpha r)$. In order that the fixed boundary condition $R(c) = 0$ be satisfied, we choose $\alpha = \gamma_n/c$, where γ_n is the nth positive solution of $J_0(x) = 0$. At this point we have product functions of the form $J_0(\gamma_n r/c)\cos(\gamma_n at/c)$ that satisfy the wave equation and the homogeneous boundary conditions, so we form the formal series solution

$$u(r,t) = \sum_{n=1}^{\infty} c_n J_0\left(\frac{\gamma_n r}{c}\right)\cos\frac{\gamma_n at}{c}.$$

In order to satisfy the initial position condition $u(r,0) = f(x)$ is suffices that the $\{c_n\}$ be the Fourier-Bessel coefficients of the function $f(x)$ given by

$$c_n = \frac{2}{c^2 [J_1(\gamma_n)]^2} \int_0^c r f(r) J_0\left(\frac{\gamma_n r}{c}\right) dr.$$

2.   This is the same as Problem 1, except that the membrane has initial position $u(r,0) = 0$, so in the t-factor $T(t) = A\cos\alpha a t + B\sin\alpha a t$ we choose $A = 0$ so that $T(0) = 0$. We then get product functions of the form $J_0(\gamma_n r/c)\sin(\gamma_n a t/c)$ that satisfy the wave equation and the homogeneous boundary conditions, so we form the formal series solution

$$u(r,t) = \sum_{n=1}^{\infty} c_n J_0\left(\frac{\gamma_n r}{c}\right)\sin\frac{\gamma_n a t}{c}.$$

In order that the initial velocity condition $u_t(r,0) \equiv v_0$ we satisfied, we want

$$v_0 = \sum_{n=1}^{\infty} \frac{\gamma_n a}{c} \cdot c_n J_0\left(\frac{\gamma_n r}{c}\right),$$

and hence

$$c_n = \frac{c}{\gamma_n a} \cdot \frac{2}{c^2 J_1(\gamma_n)^2} \int_0^c r v_0 J_0\left(\frac{\gamma_n r}{c}\right) dr$$

$$= \frac{2cv_0}{a\gamma_n^3 J_1(\gamma_n)^2} \int_0^{\gamma_n} x J_0(x)\, dx \qquad \left(\text{with } x = \gamma_n r/c\right)$$

$$= \frac{2cv_0}{a\gamma_n^3 J_1(\gamma_n)^2}\left[x J_1(x)\right]_0^{\gamma_n} = \frac{2cv_0}{a\gamma_n^2 J_1(\gamma_n)}.$$

This gives the desired solution

$$u(r,t) = \frac{2cv_0}{a}\sum_{n=1}^{\infty}\frac{J_0(\gamma_n r/c)\sin(\gamma_n a t/c)}{\gamma_n^2 J_1(\gamma_n)}.$$

3.   **(a)**   As in Problem 2,

$$u(r,t) = \sum_{n=1}^{\infty} c_n J_0\left(\frac{\gamma_n r}{c}\right)\sin\frac{\gamma_n a t}{c}.$$

In order to satisfy the given initial condition we must choose

$$c_n = \frac{c}{\gamma_n a} \cdot \frac{2}{c^2 J_1(\gamma_n)^2} \int_0^\varepsilon \left(\frac{P_0}{\rho\pi\varepsilon^2}\right) r J_0\left(\frac{\gamma_n r}{c}\right) dr$$

$$= \frac{2P_0 c}{\rho\pi\varepsilon^2 \gamma_n^3 a J_1(\gamma_n)^2} \int_0^{\gamma_n \varepsilon/c} x J_0(x)\, dx \qquad \left(\text{with } x = \gamma_n r/c\right)$$

$$= \frac{2P_0 c}{\rho \pi \varepsilon^2 \gamma_n^3 a\, J_1(\gamma_n)^2} \cdot \frac{\gamma_n \varepsilon}{c}\, J_1\!\left(\frac{\gamma_n \varepsilon}{c}\right).$$

$$c_n = \frac{2aP_0}{\pi c \rho a^2 \gamma_n J_1(\gamma_n)^2} \cdot \frac{J_1(\gamma_n \varepsilon / c)}{\gamma_n \varepsilon / c}.$$

(b)   The final formula given in the text for $u(r, t)$ now follows because $\rho a^2 = T$ and $J_1(x)/x \to 1/2$ as $x \to 0$.

5.   **(a)**   We start with the steady-state boundary value problem

$$\frac{\partial^2 u}{\partial r^2} + \frac{1}{r}\frac{\partial u}{\partial r} + \frac{\partial^2 u}{\partial z^2} = 0 \qquad (r < c, \quad 0 < z < L)$$

$$u(c, z) = 0$$

$$u(r, 0) = 0,$$

$$u(r, L) = u_0.$$

The substitution $u(r, z) = R(r)Z(z)$ yields the equations

$$rR'' + R' + \alpha^2 rR = 0, \qquad Z'' - \alpha^2 Z = 0$$

with separation constant $\lambda = \alpha^2$. The homogeneous endpoint conditions are

$$R(c) = Z(0) = 0.$$

If $\lambda = \alpha^2 = 0$ then $rR'' + R = 0$ implies

$$R(r) = A + B \ln r.$$

We choose $B = 0$ for continuity at $r = 0$, so $R(r) = A$. Then $R(c) = 0$, so $A = 0$ also, and hence 0 is not an eigenvalue.

If $\lambda = \alpha^2 > 0$ then we have the parametric Bessel equation with general solution

$$R(r) = AJ_0(\alpha r) + BY(\alpha r).$$

In order that $R(r)$ be continuous at $r = 0$ we choose $B = 0$, so $R(r) = AJ_0(\alpha r)$. Then

$$R(c) = \alpha A J_0(\alpha c) = 0$$

requires that $\gamma = \alpha c$ be a root of the equation

$$J_0(x) = 0.$$

If $\alpha_n = \gamma_n/c$ where γ_n is the nth positive root of this equation, then

$$R_n(r) = J_0\left(\frac{\gamma_n r}{c}\right).$$

The corresponding function $Z(z)$ of z is

$$Z_n(z) = A_n \cosh\frac{\gamma_n z}{c} + B_n \sinh\frac{\gamma_n z}{c},$$

and we choose $A_n = 0$ because $Z(0) = 0$. Thus we get the formal series solution

$$u(r,z) = \sum_{n=1}^{\infty} c_n J_0\left(\frac{\gamma_n r}{c}\right)\sinh\frac{\gamma_n z}{c}$$

where $J_0(\gamma_n) = 0$. To satisfy the condition $u(r, L) = u_0$, we need (by Eq. (22) in the text)

$$c_n = \frac{1}{\sinh(\gamma_n L / c)}\cdot\frac{2}{c^2 J_1(\gamma_n)^2}\int_0^c r u_0 J_0\left(\frac{\gamma_n r}{c}\right)dr$$

$$= \frac{2u_0}{\gamma_n^2 J_1(\gamma_n)^2 \sinh(\gamma_n L / c)}\int_0^{\gamma_n} x J_0(x)\,dx \qquad \left(\text{with } x = \gamma_n r / c\right)$$

$$= \frac{2u_0}{\gamma_n^2 J_1(\gamma_n)^2 \sinh(\gamma_n L / c)}\left[x J_1(x)\right]_0^{\gamma_n} = \frac{2u_0}{\gamma_n J_1(\gamma_n)}.$$

This gives the desired solution

$$u(r,t) = 2u_0\sum_{n=1}^{\infty}\frac{J_0\left(\gamma_n r / c\right)\sinh\left(\gamma_n z / c\right)}{\gamma_n J_1(\gamma_n)\sinh\left(\gamma_n L / c\right)}.$$

7.   We want to solve the boundary value problem

$$\frac{\partial^2 u}{\partial r^2} + \frac{1}{r}\frac{\partial u}{\partial r} + \frac{\partial^2 u}{\partial z^2} = 0 \qquad (r < 1,\ z > 0)$$

$$hu(1, z) + u_r(1, z) = 0$$

$$u(r, z)\ \text{bounded as } z \to \infty$$

$$u(r, 0) = u_0.$$

We start with the separation of variables in Problem 5,

$$rR'' + R' + \alpha^2 rR = 0, \qquad Z'' - \alpha^2 Z = 0$$

and readily see that $\alpha = 0$ is not an eigenvalue. When we impose the condition

$$hR(1) + R'(1) = 0$$

on $R(r) = J_0(\alpha r)$, we find that α must satisfy the equation

$$hJ_0(x) + xJ_0'(x) = 0$$

that corresponds to Case 2 with $n = 0$ in Figure 10.4.2 of the text. If $\{\gamma_n\}$ are the positive roots of this equation then

$$R_n(r) = J_0(\gamma_n r).$$

The general solution of $Z'' = \gamma_n^2 Z$ is

$$Z_n(z) = A_n \exp(-\gamma_n z) + B_n \exp(\gamma_n z),$$

and we choose $B_n = 0$ so that $Z_n(z)$ will be bounded as $z \to \infty$. Thus we obtain a solution of the form

$$u(r,z) = \sum_{n=1}^{\infty} c_n \exp(-\gamma_n z) J_0(\gamma_n r)$$

where

$$hJ_0(\gamma_n) + \gamma_n J_0'(\gamma_n) = 0,$$

so $\gamma_n J_1(\gamma_n) = hJ_0(\gamma_n)$ because $J_0' = -J_1$. Finally, Eq. (23) in the text gives

$$c_n = \frac{2\gamma_n^2}{c^2 \left(\gamma_n^2 + h^2 \right) J_0(\gamma_n)^2} \int_0^c r u_0 J_0\left(\frac{\gamma_n r}{c} \right) dr$$

$$= \frac{2u_0}{\left(\gamma_n^2 + h^2 \right) J_0(\gamma_n)^2} \int_0^{\gamma_n} x J_0(x) \, dx \qquad (\text{with } x = \gamma_n r / c)$$

$$= \frac{2u_0}{\left(\gamma_n^2 + h^2 \right) J_0(\gamma_n)^2} \left[x J_1(x) \right]_0^{\gamma_n}$$

$$= \frac{2u_0 \gamma_n J_1(\gamma_n)}{\left(\gamma_n^2 + h^2 \right) J_0(\gamma_n)^2} = \frac{2hu_0}{\left(\gamma_n^2 + h^2 \right) J_0(\gamma_n)},$$

so

$$u(r,z) = 2hu_0 \sum_{n=1}^{\infty} \frac{\exp\left(-\gamma_n z \right) J_0\left(\gamma_n r \right)}{\left(\gamma_n^2 + h^2 \right) J_0(\gamma_n)}.$$

11.   When we substitute $u(r, t) = R(r)\sin \omega t$ in the given partial differential equation and cancel the factor $\sin \omega t$, we get the ordinary differential equation

$$R'' + \frac{1}{r}R' + \left(\frac{\omega}{a}\right)^2 R = -\frac{F_0}{a^2}$$

The associated homogeneous equation is the Bessel equation of order zero with parameter ω/a. Hence it follows readily that the solution that is continuous at $r = 0$ is

$$R(r) = A J_0\left(\frac{\omega r}{a}\right) - \frac{F_0}{\omega^2}.$$

The condition $R(b) = 0$ yields $A = F_0 / \omega^2 J_0(\omega b / a)$, so it follows that the desired steady periodic solution is

$$u(r,t) = \frac{F_0}{\omega^2 J_0(\omega b / a)}\left[J_0\left(\frac{\omega r}{a}\right) - J_0\left(\frac{\omega b}{a}\right)\right]\sin \omega t.$$

13.   With $w(x) = wx$ and $h(x) = h$ (where w and h on the right are constants) the given partial differential equation

$$\frac{w(x)}{g}\frac{\partial^2 y}{\partial t^2} = \frac{\partial}{\partial x}\left(w(x)h(x)\frac{\partial y}{\partial x}\right) \qquad (*)$$

reduces to

$$x\frac{\partial^2 y}{\partial t^2} = gh\left(\frac{\partial y}{\partial x} + \frac{\partial^2 y}{\partial x^2}\right).$$

When we substitute $y(x, t) = X(x)\cos \omega t$ we get the parametric Bessel equation

$$x^2 X'' + xX' + \frac{\omega^2 x^2}{gh}X = 0$$

with bounded solution

$$X(x) = A J_0\left(\frac{\omega x}{\sqrt{gh}}\right).$$

The condition $X(L) = y_0$ implies that $A = y_0 / J_0\left(\omega L / \sqrt{gh}\right)$, so

$$y(x,t) = y_0 \frac{J_0\left(\omega x / \sqrt{gh}\right)}{J_0\left(\omega L / \sqrt{gh}\right)}\cos \omega t.$$

15. With $w(x) = wx$ and $h(x) = hx$ (with w and h being constants on the right) the partial differential equation in (*) above reduces to

$$\frac{\partial^2 y}{\partial t^2} = gh\left(2\frac{\partial y}{\partial x} + x\frac{\partial^2 y}{\partial x^2}\right).$$

When we substitute $y(x, t) = X(x)\cos \omega t$ we get the ordinary differential equation

$$x^2 X'' + 2xX' + \frac{\omega^2 x}{gh}X = 0.$$

This has the form of Equation (3) in Section 8.6 with $A = 2$, $B = 0$, $C = \omega^2/gh$, and $q = 1$, so its (bounded) solution is given by

$$X(x) = \frac{A}{\sqrt{x}}J_1\left(2\omega\sqrt{\frac{x}{gh}}\right).$$

The condition $X(L) = y_0$ now implies that $A = y_0\sqrt{L}\,/\,J_1\left(2\omega\sqrt{L/gh}\right)$, so

$$y(x,t) = y_0\sqrt{\frac{L}{x}}\frac{J_1\left(2\omega\sqrt{x/gh}\right)}{J_1\left(2\omega\sqrt{L/gh}\right)}\cos \omega t.$$

17. Just as in Problem 1 above, substitution of $u(r,t) = R(r)T(t)$ in the wave equation

$$\frac{\partial^2 u}{\partial t^2} = a^2\left(\frac{\partial^2 u}{\partial r^2} + \frac{1}{r}\frac{\partial u}{\partial r}\right)$$

yields the separation

$$\frac{T''}{a^2 T} = \frac{R'' + \dfrac{1}{r}R'}{R} = \lambda = -\alpha^2.$$

The t-equation has general solution

$$T(t) = A\cos \alpha at + B\sin \alpha at,$$

and we choose $B = 0$, so that $T'(0) = 0$ (assuming, for instance, that the membrane is initially at rest). The r-equation can be written in the form

$$r^2 R'' + rR' + \alpha^2 r^2 R = 0,$$

which is the parametric Bessel equation of order zero. By Problem 16, its solutions satisfying $R(a) = R(b) = 0$ are of the form $R_n(x) = Y_0(\gamma_n a)J_0(\gamma_n x) - J_0(\gamma_n a)Y_0(\gamma_n x)$ with $\alpha = \gamma_n$ being one of the positive roots of Equation (#) there. This leads to a formal series solution of the form

$$u(r,t) = \sum_{n=1}^{\infty} R_n(x)\left(A_n \cos \gamma_n at + B_n \sin \gamma_n at\right),$$

where the frequency of the nth term is $\omega_n = \gamma_n a = \gamma_n \sqrt{T/\rho}$.

19. We want to solve the boundary value problem

$$\frac{\partial^2 u}{\partial r^2} + \frac{1}{r}\frac{\partial u}{\partial r} + \frac{\partial^2 u}{\partial z^2} = 0 \qquad (a < r < 1,\ z > 0)$$

$$u(a,z) = u(b,z) = 0$$

$$u(r,z) \text{ bounded as } z \to \infty$$

$$u(r,0) = f(r).$$

Just as in Problem 5, the substitution $u(r,z) = R(r)Z(z)$ yields the equations

$$rR'' + R' + \alpha^2 rR = 0, \qquad Z'' - \alpha^2 Z = 0$$

with separation constant $\lambda = \alpha^2$. When we impose the conditions $R(a) = R(b) = 0$ on the r-equation here, we have the Sturm-Liouville problem of Problem 16, so $R(r)$ must be one of the eigenfunctions $\{R_n(r)\}$ corresponding to the positive roots $\{\gamma_n\}$ of Equation (#) there. The general solution of $Z'' = \gamma_n^2 Z$ is

$$Z_n(z) = A_n \exp(-\gamma_n z) + B_n \exp(\gamma_n z),$$

and we choose $B_n = 0$ so that $Z_n(z)$ will be bounded as $z \to \infty$. Thus we obtain a solution of the form

$$u(r,z) = \sum_{n=1}^{\infty} c_n \exp(-\gamma_n z)R_n(r)$$

with the coefficients $\{c_n\}$ calculated as in Problem 18.

SECTION 10.5

HIGHER-DIMENSIONAL PHENOMENA

This section provides the interested student with an opportunity to study several applications at greater depth than is afforded by the usual textbook exercises. The problem sets outlined in Section 10.5 can serve as the basis for several fairly substantial computational projects. Because these problem sets and projects are rather heavily annotated in the text, further outlines of solutions are not included in this manual. However, additional discussion — particularly regarding computer implementations — may be found in the applications manual that accompanies the text.

APPENDIX

EXISTENCE AND UNIQUENESS OF SOLUTIONS

In Problems 1–12 we apply the iterative formula

$$y_{n+1} = b + \int_a^x f(t, y_n(t))\, dt$$

to compute successive approximations $\{y_n(x)\}$ to the solution of the initial value problem

$$y' = f(x, y), \qquad\qquad y(a) = b.$$

starting with $y_0(x) = b$.

1. $y_0(x) = 3$

  $y_1(x) = 3 + 3x$

  $y_2(x) = 3 + 3x + 3x^2/2$

  $y_3(x) = 3 + 3x + 3x^2/2 + x^3/2$

  $y_4(x) = 3 + 3x + 3x^2/2 + x^3/2 + x^4/8$

  $y(x) = 3 - 3x + 3x^2/2 + x^3/2 + x^4/8 + \cdots = 3e^x$

3. $y_0(x) = 1$

  $y_1(x) = 1 - x^2$

  $y_2(x) = 1 - x^2 + x^4/2$

  $y_3(x) = 1 - x^2 + x^4/2 - x^6/6$

  $y_4(x) = 1 - x^2 + x^4/2 - x^6/6 + x^8/24$

  $y(x) = 1 - x^2 + x^4/2 - x^6/6 + x^8/24 - \cdots = \exp(-x^2)$

5. $y_0(x) = 0$

  $y_1(x) = 2x$

  $y_2(x) = 2x + 2x^2$

  $y_3(x) = 2x + 2x^2 + 4x^3/3$

  $y_4(x) = 2x + 2x^2 + 4x^3/3 + 2x^4/3$

$$y(x) = 2x + 2x^2 + 4x^3/3 + 2x^4/3 + \cdots = e^{2x} - 1$$

7.

$$y_0(x) = 0$$

$$y_1(x) = x^2$$

$$y_2(x) = x^2 + x^4/2$$

$$y_3(x) = x^2 + x^4/2 + x^6/6$$

$$y_4(x) = x^2 + x^4/2 + x^6/6 + x^8/24$$

$$y(x) = x^2 + x^4/2 + x^6/6 + x^8/24 + \cdots = \exp(x^2) - 1$$

9.

$$y_0(x) = 1$$

$$y_1(x) = (1 + x) + x^2/2$$

$$y_2(x) = (1 + x + x^2) + x^3/6$$

$$y_3(x) = (1 + x + x^2 + x^3/3) + x^4/24$$

$$y(x) = 1 + x + x^2 + x^3/3 + x^4/12 + \cdots = 2e^x - 1 - x$$

11.

$$y_0(x) = 1$$

$$y_1(x) = 1 + x$$

$$y_2(x) = (1 + x + x^2) + x^3/3$$

$$y_3(x) = (1 + x + x^2 + x^3) + 2x^4/3 + x^5/3 + x^6/9 + x^7/63$$

$$y(x) = 1 + x + x^2 + x^3 + x^4 + \cdots = 1/(1 - x)$$

13.

$$\begin{bmatrix} x_0(t) \\ y_0(t) \end{bmatrix} = \begin{bmatrix} 1 \\ -1 \end{bmatrix}$$

$$\begin{bmatrix} x_1(t) \\ y_1(t) \end{bmatrix} = \begin{bmatrix} 1 + 3t \\ -1 + 5t \end{bmatrix}$$

$$\begin{bmatrix} x_2(t) \\ y_2(t) \end{bmatrix} = \begin{bmatrix} 1 + 3t + \frac{1}{2}t^2 \\ -1 + 5t - \frac{1}{2}t^2 \end{bmatrix}$$

$$\begin{bmatrix} x_3(t) \\ y_3(t) \end{bmatrix} = \begin{bmatrix} 1 + 3t + \frac{1}{2}t^2 + \frac{1}{3}t^3 \\ -1 + 5t - \frac{1}{2}t^2 + \frac{5}{6}t^3 \end{bmatrix}$$